Dedicated to IUIH for promoting excellence in healthcare:

www.iuih.co.uk

Dr Limci Gupta is a Joint Director of UK Global Healthcare Limited (Indo UK Institute of Health). She is also Head of Diagnostics at Indo UK Institute of Health (IUIH). She is a Fellow of Royal College of Pathologists (UK) and practises as a part time consultant histopathologist and cytopathologist, currently based at St George's Hospital, London. She graduated from BJ Medical College, Pune having completed MBBS and a Diploma in Clinical Pathology from Pravara Medical College (Pune University). She then pursued her further studies in UK and was awarded FRCPath whilst at Royal Free Hospital, London. Her special interest is in gynaecological and breast pathology. She is on the international editorial board for Indian Journal of Pathology & Oncology and is a member of British Association of Gynaecological Pathologists.

Dr Jayson Wang is a Fellow of Royal College of Pathologists (UK), and a consultant histopathologist and honorary senior lecturer at St George's Hospital and University, where he is the Lead in Molecular Pathology within the Cellular Pathology Department. He has both research and clinical interests in urological, gastrointestinal and gynaecological pathology. He graduated with honours from the University of Aberdeen, and has specialist accreditation both in Histopathology and Medical Oncology. He obtained a PhD in Cell Biology from Cancer Research UK, and was an Academic Clinical Lecturer at Imperial College London. He is a member of the National Cancer Research Institute Cellular-Molecular Pathology Initiative working group.

Dr Val Thomas is a Fellow of the Royal College of Pathologists (UK) and is a consultant and honorary senior lecturer for South West London Pathology based at St George's Universities Foundation Trust, London. She is lead breast pathologist and medical laboratory lead for cervical screening. She graduated from St George's Hospital Medical School (now St George's University London) having completed MBBS and a PhD in developmental biology. Interests include translation of research into clinical practice. She was the cellular pathology lead for SWLP during transition and has been Clinical Director for Cancer and a Divisional Chair at St George's Hospital. She is currently an examiner for the conjoint examination board of the Royal College of Pathology and the Institute of Biomedical Science in cervical and diagnostic cytology.

IP Innovative Publication Pvt. Ltd.

Editorial Office:
IP Innovative Publication Pvt. Ltd.
First Floor, RZ – 1/4 – A, Vijay Enclave, Palam-Dabri Marg, New Delhi-110045, India.
Ph: +91-11-25052216 / 25051061. Mob. : +91-8826859373
E-mail: subscription@innovativepublication.com, rakesh.its@gmail.com
Web: www.innovativepublication.com

Practical Applications in Histopathology, Cytopathology and Autopsy: An MCQ/EMQ Resource
First Edition: 2016
ISBN: 978-81-932450-2-6
Price in ₹ 1850   £ 78 $95
Published by- IP Innovative Publication Pvt. Ltd.

Copyright © 2016, IP Innovative Publication Pvt. Ltd.
All right reserved.

No part of this book may be reproduced in any form, by Photostat, microfilm, xerography, or any other means, or incorporated into any information retrieval system, electronic or mechanical, without the written permission of the copyright owner.

Enquiries for bulk sales may be solicited at: subscription@innovativepublication.com

This book has been published in good faith that the contents provided by the contributor's contained herein are original, and is intended for educational purposes only.

Printed at: **New Delhi, India**

FOREWORD

I am delighted to have been asked to provide the foreword for 'Practical applications in histopathology, cytopathology and autopsy: an MCQ/EMQ resource'. Having been asked to provide editorial assistance in the preparation of the book, I am pleased to report that very little editing was needed! The quality of the chapters is extremely high, and the questions for each subject area testing, but not impossible. The breadth of subject matter covered by the chapters is comprehensive and the layout is very easy to use. Recent developments in molecular pathology are also present throughout the appropriate chapters which increase the value of this publication dramatically.

This is a resource that will obviously be popular with and useful to trainees preparing for the part 1 FRCPath in Histopathology, but which will also be helpful to established histopathologists. It is difficult to keep up to date with recent changes in pathology in the context of a busy NHS consultant post, and I am sure that this book will provide useful reading material that can be kept in a briefcase for quiet moments on the train or between meetings, or on the bedside table for a quick evening refresher. My sincere compliments go to the authors who have produced a very useful text that I am proud to be associated with.

David M Bailey FRCPath
April 2016

Consultant Histopathologist
Vice President for Communications
Royal College of Pathologists
UK

PREFACE

'Practical applications in histopathology, cytopathology and autopsy: an MCQ/EMQ resource' is a comprehensive collection of MCQ/EMQs with explanations and correct answers at the end of each chapter. The authors of all the chapters are consultants in pathology practising in UK with the professional qualification of FRCPath.

This book will have a practical use in day to day pathology practice and would be a very helpful tool for pathology exam preparation (FRCPath part 1 and similar exams).

This book touches most of the important entities / aspects of cellular pathology covering histopathology, cytopathology and autopsy. It includes points of practical importance and addresses some aspects of frozen section diagnoses. It includes questions on important points in 'Minimum Datasets for Cancer Reporting' as recommended by Royal College of Pathologists, UK. It incorporates clinical and surgical aspects of tumour grading and staging which are discussed routinely at multi-disciplinary team meetings with surgeons, physicians, oncologists, radiologists and specialist nurses.

This book has tried to cover tried various styles in which questions can be framed for FRCPath part 1 exam. A unique feature of this book is the inclusion of a section addressing clinical management issues in cellular pathology.

All the questions are original and based on routine pathology practice in UK.

Best wishes,

L. Gupta
V. Thomas.

Dr Valerie Thomas & Dr Limci Gupta
Editors-in-chief

Acknowledgements

We would like to acknowledge all our authors for their respective chapters and to all the other contributors to this book.

We would like to thank the Fellows of the 'Royal College of Pathologists' for producing excellent guides to reporting of malignant tumours in the form of 'Minimum Datasets for cancer reporting' which are referred to throughout this book.

We would like to sincerely thank Dr David Bailey, the Vice President of Communications, Royal College of Pathologists, UK for writing a foreword for the book and also reading the entire book.

Thanks also to Dr Martin Young (Clinical Director Pathology Services RFL, Deputy Head School of Pathology, STC Chair London/Eastern/KSS Deaneries), for his input into the section on clinical management.

Thanks to Dr Jayson Wang (Consultant Histopathologist) for extending his help & support as a co-editor for the book.

We are very thankful to the reviewers of all the chapters for their valuable suggestions. The reviewers include Dr Alastair Deery, Professor John Gosney, Dr John du Parcq, Dr Lorrette Ffolkes, Dr Ruth Nash, Dr Jayson Wang, Dr Valerie Thomas and Dr Limci Gupta.

We would like to thank trainee registrars Dr Caitlin Beggan (FRCPath) and Dr Afsheen Wasif (FRCPath part 1) in helping us with writing up sample chapters and to Dr Duaa Saeed (FRCPath part 1), Dr Raphael Oluwaseyi Opanuga (FRCPath part 1) and Dr Hira Mir (FRCPath part 1) for taking time to give trainee feedback and congratulations to Duaa and Raphael on passing the exam!

Thanks to Dr Duaa Saeed (FRCPath part 1), Dr Jasmin Lee (FRCPath part 1), Dr Amanda Brown, Dr Emily Scott and Dr Leslie Cheng for proof-reading and formatting.

We also thank Aminur Rahman (BMS) for IT assistance to facilitate publication.

Dr Valerie Thomas & Dr Limci Gupta
Editors-in-chief

Table of contents

1) Diagnostic cytopathology 1
   Dr Valerie Thomas *(St George's Hospital, London)*

2) Cervical Cytology 11
   Dr Valerie Thomas *(St George's Hospital, London)*

3) Breast pathology 21
   Dr Valerie Thomas *(St George's Hospital, London)*

4) Female genital tract pathology 34
   Dr Limci Gupta *(St George's Hospital, London)*

5) Gastro-intestinal pathology 48
   Dr Jayson Wang *(St George's Hospital, London)*

6) Liver & Pancreatic pathology 64
   Dr Jayson Wang *(St George's Hospital, London)*

7) Cardiovascular pathology 76
   Dr Fiona Scott *(Hemel Hempstead General Hospital, Herts, UK)*

8) Oral and nasal pathology 89
   Dr Miguel Perez *(Royal Free Hospital, London)*

9) Thoracic pathology 102
   Dr Brendan Tinwell *(St George's Hospital, London)*

10) Dermatopathology 117
    Dr Limci Gupta *(St George's Hospital, London)*

11) Osteo-articular and soft tissue pathology 129
    Dr Rukma Doshi & Dr Jayson Wang *(St George's Hospital, London)*

12) Urinary tract pathology 144
    Dr Brendan Tinwell & Dr Jayson Wang *(St George's Hospital, London)*

13) Male genital tract pathology 158
    Dr Limci Gupta & Dr Jayson Wang *(St George's Hospital, London)*

14) Endocrine pathology 169
    Dr Rashpal Flora *(Hammersmith Hospital, London)*

15) Lymphoreticular pathology 180
    Dr Richard Attanoos & Dr Matthew Pugh *(University Hospital of Wales, Cardiff, UK)*

16) Nervous system and neuromuscular pathology.. 196
    Dr Leslie Bridges *(St George's Hospital, London)*

17) Paediatric and placental pathology... 210
    Dr Iona Jeffrey & Dr Ruth Nash *(St George's Hospital, London)*

18) General pathology.. 230
    Dr Charanjit Kaur *(St George's Hospital, London)*

19) Autopsy & Forensic pathology... 245
    Dr Fiona Scott *(Hemel Hempstead General Hospital, Herts, UK)*
    Dr Charanjit Kaur *(St George's Hospital, London)*

20) Management issues in cellular pathology... 263-273
    Dr Val Thomas *(St George's Hospital, London)*
    Dr Martin Young *(Royal Free Hospital, London)*

CHAPTER 1

DIAGNOSTIC CYTOLOGY
Dr Valerie Thomas

1. Which one of the following statements is correct?

    a) Carcinoma of the bronchus can be diagnosed on sputum cytology.
    b) Sputum cytology is a good screening test for lung cancer.
    c) It is not possible to subtype tumours of the bronchus on cytology.
    d) Curschmann's spirals are associated with small cell carcinoma.
    e) Tuberculosis in sputum is usually recognised the presence of caseating granulomas.

2. Which one of the following is true about a good bronchiolo-alveolar lavage specimen?

    a) Bronchial epithelial cells are the predominant cell type present.
    b) Sarcoidosis is revealed by the presence of plasma cells.
    c) An increase in small lymphocytes favours a diagnosis of non-Hodgkin's lymphoma.
    d) Alveolar macrophages are the predominant cell population.
    e) Ciliocytophthoria raises the possibility of a diagnosis of Psittacosis.

3. In a bronchial wash specimen, which one of the following statements is correct?

    a) Mucolysis should be avoided in the preparation of the sample.
    b) A differential cell count reflects underlying pulmonary disease.
    c) It is not possible to identify small cell carcinoma.
    d) Bronchial washings have a sensitivity for the detection of malignancy as good as a good deep cough sputum.
    e) Fungal hyphae are a frequent finding.

Scenario for two questions 4 & 5: A 60 year old woman with a history of sarcoidosis is becoming increasingly breathless. A chest X-ray shows mediastinal lymphadenopathy and a solitary new centrally necrotic lesion in the left upper lobe of the lung. She undergoes further investigations.
Bronchial washings and brushings were negative for malignant cells.

4. Which one of the following statements is true?

    a) Mediastinal lymphadenopathy indicates metastatic carcinoma.
    b) A new centrally necrotic mass is most likely due to sarcoidosis.
    c) A new centrally necrotic mass may be due to a primary bronchogenic carcinoma.
    d) The yield of bronchial washing samples on microbiological testing is lower than the yield from sputum samples.
    e) The lungs are rarely affected by sarcoidosis.

The patient goes on to have fine needle aspiration of the lung mass and endobronchial ultrasound (EBUS) with fine needle aspiration of the mediastinal nodes. The cytology of the nodes shows granulomatous inflammation whilst that of the necrotic mass shows dichotomously branching, narrow, occasionally septate hyphal structures.

5. Which one of the following statements is true?

   a) The patient has an Aspergilloma.
   b) The patient has Candidal infection.
   c) The patient's sample is contaminated with Mucor from the environment.
   d) The patient has tuberculosis.
   e) The patient has Mycobacterium avium intracellulare.

6. Which one of the following is **not** a normal cellular constituent in urine?

   a) Neutrophils
   b) Eosinophils
   c) Squamous cells
   d) Transitional cells
   e) Umbrella cells

7. Urine cytology is most effective in which one of the following clinical settings?

   a) Follow-up of high grade TCC.
   b) Screening for in situ bladder cancer.
   c) Testing urine from urodynamic patients.
   d) Detection of renal cancer.
   e) Detection of low grade TCC.

Scenario for two questions 8 & 9: A voided urine preparation from a 70 year old man with a history of high grade transitional cell carcinoma shows scattered cells as shown in the micrograph.

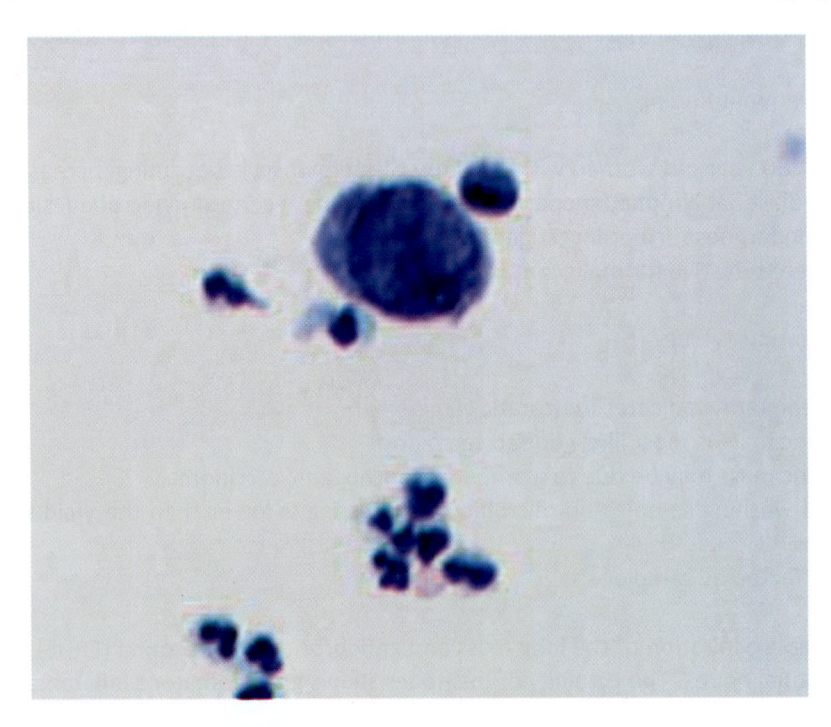

8. What is this appearance?

    a) High grade invasive TCC.
    b) Low grade invasive TCC.
    c) Carcinoma in situ.
    d) Polyoma virus infection.
    e) Cytomegalovirus infection.

9. What clinical pathway should the patient follow?

    a) Flexible cystoscopy.
    b) On-going cytological surveillance.
    c) Virological investigation.
    d) Rigid cystoscopy.
    e) Bladder washings should be taken.

10. In the CSF which of the following is correct?

    a) Usual cell populations include 'left shifted' neutrophils.
    b) Fragments of metachromatic cartilage may be seen.
    c) Small lymphocytes are not usually seen.
    d) Monocytoid cells are an unusual population.
    e) A 1 ml sample is suitable for the assessment of tumour.

11. In fine needle aspiration cytology of the breast, which of the following is true?

    a) Invasive carcinoma is identified with the same sensitivity as core biopsy.
    b) Microcalcification can be identified with ease.
    c) Malignant cells are not easily detected due to the quantity of benign breast elements in normal breast tissue.
    d) Aspirates from invasive lobular carcinoma are routinely more cellular than those from ductal carcinomas.
    e) Aspirates from fibroadenomas are a known source of potential false positive diagnosis.

Scenario for questions 12 & 13: A 18 year old woman attends her GP with a breast lump. The mass is smooth and rounded. The GP refers the woman to a breast unit for assessment.

12. The breast unit would be expected to carry out which one of the following options:

    a) Clinical examination and fine needle aspiration (FNA)
    b) Clinical examination, FNA and targeted ultrasound examination of the lesion.
    c) Clinical examination, FNA and ultrasound examination of the whole breast.
    d) Clinical examination, core biopsy and ultrasound of the whole breast
    e) Clinical examination, core biopsy and MRI.

13. A fine needle aspirate was performed which shows the following:

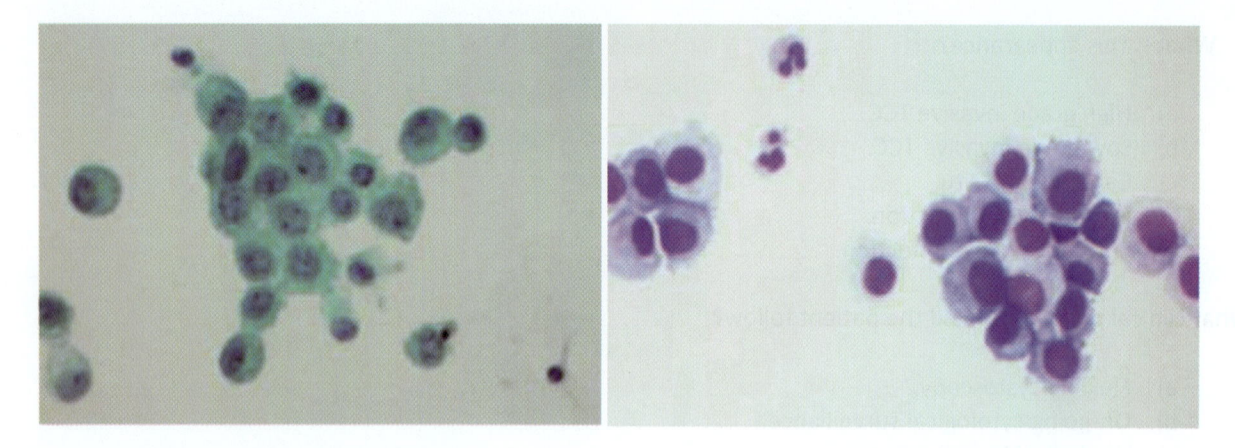

What does this fine needle aspirate show?

    a) High grade invasive ductal carcinoma.
    b) Low grade malignancy with ductal features.
    c) High grade malignancy with ductal features.
    d) Fibroadenoma with ductal hyperplasia of usual type.
    e) Fibroadenoma with atypical ductal hyperplasia.

Scenario for two questions 14 & 15: A fifty-five year old man goes to his GP with a painless lump in his neck which has slowly increased in size over the past 6 months. The GP examines the patient and palpates a 2cm level 2 mass. He refers the man to the rapid access clinic for ultrasound and FNA.

14. Which one of the following is the most likely diagnosis statistically?

    a) Metastatic papillary carcinoma of thyroid.
    b) Pleomorphic salivary adenoma.
    c) Metastatic squamous cell carcinoma.
    d) Adenoid cystic carcinoma.
    e) Non-Hodgkin's lymphoma.

15. On aspiration the mass yields 1.5ml of thick opaque yellow fluid. Which of the following is the most likely diagnosis?

    a) Metastatic papillary carcinoma of thyroid.
    b) Pleomorphic salivary adenoma.
    c) Metastatic squamous cell carcinoma.
    d) Adenoid cystic carcinoma.
    e) Non-Hodgkin's lymphoma.

16. A pleural fluid tap from a 75 year old female living in the UK reveals a dominant population of small lymphocytes. What is the single most likely diagnosis from the following options?

    a) Lymphoma.
    b) Tuberculosis.
    c) Filariasis.
    d) Meig's syndrome.
    e) Carcinoma.

Scenario for two questions 17 & 18: A 48 year old woman presents to the GP with a history of increasing shortness of breath. On direct questioning she admits to episodes of tarry black stools. When examined, she is found to have ascites and has a BMI of 18. On investigation she has a haemoglobin of 70g/l with a macrocytic anaemia and an albumin of 20 g/l.

17. Which of the following options is the single most likely underlying disease?

    a) Metastatic carcinoma.
    b) Gastric ulcer.
    c) Alcoholic liver disease.
    d) Leukaemia.
    e) Anorexia nervosa.

18. She is referred to hospital where the ascites is tapped. Which one of the following is the most likely cytological finding?

    a) A lymphohistiocytic effusion.
    b) A xanthochromic effusion.
    c) An effusion containing malignant glandular cells.
    d) An effusion containing atypical mesothelial cells.
    e) A purulent effusion.

19. A 25 year old man presents with a lump in his neck of acute onset and persistent for 6 weeks. On examination he has a solitary node in left level V, measuring 2cm in maximum diameter. An FNA shows a cellular aspirate predominantly comprising small lymphocytes with some crushed follicle fragments, a few small loose aggregates of macrophages and scattered lymphoid blasts. Which one of the following is the most likely diagnosis?

    a) Infectious mononucleosis.
    b) B cell non-Hodgkin's lymphoma, follicular subtype.
    c) Hodgkin's disease.
    d) Toxoplasmosis.
    e) Persistent generalised lymphadenopathy.

Scenario for two questions 20 & 21: A 65 year old male engineer presents with shortness of breath and a unilateral pleural effusion. He admits to smoking 20-30 cigarettes and drinking up to 3 pints of beer per day. He has lost 10 kg of weight in the past three months.
The fluid is drained and submitted for cytology.

20. Which one of the following options is the most likely cytological appearance?

    a) Purulent effusion.
    b) Lymphohistiocytic effusion with benign mesothelial cells.
    c) Malignant effusion.
    d) Eosinophilic effusion.
    e) Chylous effusion.

21. On chest X-ray, after drainage, pleural plaques are visible together with a mass in the lung parenchyma. The mass lesion is most likely to be:

a) Pulmonary pseudotumour.
b) Mesothelioma.
c) Secondary carcinoma.
d) Hamartoma.
e) Primary pulmonary carcinoma.

22. A 50 year old woman presents with an enlarged left neck lymph node. The FNA comprises necrotic cellular material. Which of the following is the **least** likely diagnosis?

a) Suppurative lymphadenitis
b) Benign reactive lymph node.
c) Tuberculosis.
d) Carcinoma.
e) Lymphoma.

23. A pericardial effusion **does not** usually include:

a) Macrophages.
b) Lymphocytes.
c) Mast cells.
d) Mesothelial cells.
e) Epithelial cells.

Correct Answers and Explanations:

1) Correct answer: a) Carcinoma of the bronchus can be diagnosed on sputum cytology.

Lung cancer can be identified in sputum but sputum cytology is not a good test for population screening.
In patients with subsequently proven cancer, the sensitivity of sputum testing for malignancy improves with repeated samples up to three when sensitivity is up to 80-85%, with little improvement on further sampling.
Whether the sample is sputum, bronchial washings, lavage or transthoracic or endobronchial ultrasound-guided FNA, tumours can be subtyped morphologically and if a liquid sample is obtained adjunctive immunochemistry can subtype adenocarcinoma into likely site of origin.
Curschmann's spirals are found in asthma and chronic obstructive pulmonary disease most commonly. Sputum samples from active pulmonary tuberculosis are usually purulent as the infection starts within lung parenchyma and ulcerates into bronchi with an acute inflammatory reaction. Granulomas are rarely seen.

2) Correct answer: d) Alveolar macrophages are the predominant cell population.

A lavage specimen is obtained by wedging the bronchoscope into the smallest possible bronchiole possible instilling and re-aspirating saline.
The cellular component should be largely alveolar macrophages. If bronchial epithelial cells are present the specimen is contaminated with bronchial elements.

An increase in small lymphocytes may favour a diagnosis of sarcoidosis or other inflammatory condition. An increase in plasma cells is not a feature of sarcoidosis. Ciliocytophthoria is a non-specific feature of reactive change with the loss of the ciliated part of the cell leaving residual cuboidal respiratory epithelial cells and separated ciliated fragments. It has no specific relationship with psittacosis.

3) Correct answer: d) Bronchial washings have a sensitivity for the detection of malignancy as good as a good deep cough sputum.

Bronchial washings can be used to detect tumour in patients without a productive cough and are often taken at the time of endoscopic bronchial biopsy to improve tumour yield.
Mucolysis allows excellent cellular liquid preparations to be made. A differential cell count on a wash specimen does not provide useful clinical information; cell counts may be useful on bronchiolo-alveolar lavage samples in relation to parenchymal disease.
Fungal hyphae can be seen in lavage samples but are not a frequent finding.

4) Correct answer: c) A new centrally necrotic mass may be due to a primary bronchogenic carcinoma.

Mediastinal lymphadenopathy can be seen in malignancy and in benign conditions. Symmetrical mediastinal lymphadenopathy is a characteristic feature of sarcoidosis; the lungs are involved in 90% of cases with peribronchial granulomas and in 5-15% with chronic pulmonary fibrosis. A new centrally necrotic mass is unlikely to be due to sarcoidosis.
The yield of bronchial washing samples on microbiological testing is at least as good as the yield from sputum samples.

5) Correct answer: a) The patient has an Aspergilloma.

Aspergillus can colonise abnormal lung tissue particularly in patients who are immunosupressed for example on steroid treatment.
Mucor shows wide irregularly branching fungal hyphae. Candida is a yeast that does not form septate hyphae.
The granulomas found in the node are likely to be due to sarcoidosis which is associated with small discrete granulomas in most cases. Tuberculosis usually causes large irregular areas of granulomatous inflammation with central necrosis. The diagnosis of simultaneous tuberculosis should be considered however and ZN stains performed; material from the EBUS sample should be sent for culture.
Mycobacterium avium intracellulare is an atypical Mycobacterium associated with infection in patients with severe immunosuppression usually related to HIV/AIDS.

6) Correct answer: b) Eosinophils.

Eosinophils in urine can be associated with parasites (e.g. Schistosomes); renal parenchymal disease (where they are seen with renal tubular cells and casts) and associated with tumour.
Neutrophils are seen commonly in female urine and often seen in small numbers in males. In increased numbers they may indicate urinary tract infection or be associated with stone.
Squamous cells, transitional cells and umbrella cells (superficial transitional cells) are normal constituents of urine.

7) Correct answer: a) Follow-up of high grade TCC.

Screening urine for in situ or invasive transitional cell carcinoma, even when focused on at risk populations (e.g. industrial workers in aniline dye industry) is not an effective screen. Similarly urine cytology in patients undergoing urodynamic investigations has a poor yield no better than screening well women. Renal cancer generally has to be very advanced to be detected by urine cytology.

Detection of low grade TCC is difficult on cytology as reactive processes can lead to indistinguishable appearances. Where there is a highly cellular urine containing papillary microbiopsies the diagnosis should be raised as a possibility having excluded infection, instrumentation of the urinary tract and stone disease.

8) Correct answer: d) Polyoma virus infection.

Polyoma virus infection can be asymptomatic or be associated with haematuria. The infected cells are large with a large rather featureless hyperchromatic nucleus with obliteration of chromatin detail.

High grade TCC can be difficult to distinguish but better preserved tumour cells will show nuclear envelope abnormality, chromatin detail and nucleolation; invasive disease cannot be accurately predicted on cytology but a single cell pattern would be more in keeping within carcinoma in situ. Low grade TCC would not show this degree of atypia and again invasion cannot be predicted.

CMV infection can be seen in the urine of immunosuppressed transplant patients but has characteristic intranuclear 'owl's eye' inclusions which do not obliterate the whole nucleus.

9) Correct answer: b) On-going cytological surveillance.

No additional investigations are required.

10)   Correct answer: b) Fragments of metachromatic cartilage may be seen.

Small lymphocytes and monocytes comprise the normal cell populations seen.

Fragments of cartilage from the intervertebral disc may be inadvertently sampled whilst procuring the CSF.

Neutrophils may be seen in blood stained otherwise normal CSF but will be mature hypersegmented constituents of the attendant blood.

Whilst a 1ml sample could yield a tumour diagnosis, studies have shown that a 3-5ml sample is more accurate for detection and, when negative, gives more assurance that no tumour is present.

11)   Correct answer: e) Aspirates from fibroadenomas are a known source of potential false positive diagnosis.

It is not possible to identify invasion within a cytological sample but can be seen in core biopsy. Microcalcification is more easily identified in core biopsy though multiple levels of the core may be required. Unless pregnant or lactating, there is very little glandular tissue in normal breast tissue so this does not obscure malignant cells.

Aspirates from invasive lobular carcinoma can be very poorly cellular and careful assessment is necessary to avoid a false negative interpretation.

12)   Correct answer: b) Clinical examination, fine needle aspiration and targeted ultrasound examination of the lesion.

The 'triple assessment' should be carried out comprising clinical examination, imaging and pathology. In the case of an 18 year old with a discrete lump the preferred imaging investigation is ultrasound as young women have relatively dense breast tissue which obscures mammography which itself carries the attendant risk associated with X-ray exposure. Targeted ultrasound of the lesion would be the preferred approach as screening the breast with ultrasound in the absence of an identified malignancy is unproductive. Though a core biopsy could be performed most clinicians would avoid this as being more 'invasive' than a fine needle aspirate. MRI is not appropriate as a diagnostic tool in this context.

13) Correct answer: c) High grade malignancy with ductal features.

Comparing the tumour cell size to attendant neutrophils, these are large cells with variable nuclei and little polarisation. They are loosely clustered and single cells present retain cytoplasm. A few apoptotic tumour cells are seen.
Cytology cannot accurately diagnose invasion except perhaps where destructive invasion of adipose tissue is seen (evidenced by fat necrosis) but even then great care must be taken to ensure superimposition of malignant cells and adipose microbiopsies is not giving a spurious impression of invasion.
There are no features typical of fibroadenoma here: no myoepithelial cells are seen either in the sheets or as bare bipolar nuclei in the background, no proteinaceous background or myxoid stroma is present; and the epithelial cells do not look benign.

14) Correct answer: c) Metastatic squamous cell carcinoma.

Given the age of the patient, the nodal presentation and the relative frequencies of these conditions none of the other options are appropriate.

15) Correct answer: c) Metastatic squamous cell carcinoma.

Metastatic squamous cell carcinoma, when necrotic, is aspirated as turbid yellow fluid.
None of the other options appear like this grossly.

16) Correct answer: e) Carcinoma.

Bit of a trick question, sorry.
A lymphocytic effusion is most likely to be due to chronic inflammation, however this was not a given option.
In the UK, the most likely underlying diagnosis is carcinoma but this may vary with tuberculosis being in the leading cause in endemic areas for example.

17) Correct answer*:* c) Alcoholic liver disease.

The combination of low albumin, macrocytic anaemia and episodic malaena points very strongly to liver disease and the low BMI is also in keeping.
Gastric ulcer may cause malaena but other features such as the very low albumin and a macrocytic rather than microcytic anaemia is less likely.
No features given lead clearly to other proffered diagnoses.

18) Correct answer*:* a) A lymphohistiocytic effusion.

Xanthochromia refers to a yellowish discolouration of CSF due to previous haemorrhage and breakdown of red cells. It can also be used in relation to yellowish skin patches.
Malignancy could be present but is less likely than the benign effusion generally seen in alcoholic liver disease. Hepatocellular carcinoma arising in cirrhosis consequent upon alcohol consumption rarely appears in ascites.
Mesothelial atypia is not seen in ascites due to alcoholic liver disease.

19) Correct answer: d) Toxoplasmosis.

The description given is a of classical lymph node appearances in Toxoplasmosis (Piringer-Kuchinka lymphadenitis).

Infectious mononucleosis generally involves multiple nodes with florid reactive changes prominent blasts, active germinal centres and cells which can cytologically resemble Hodgkin's disease.

The description does not fit with Hodgkin's disease with no Reed-Sternberg or variant multinucleate or mononuclear Hodgkin cells described. Follicular NHL would be unusual at this age and again the description does not correspond to a follicular lymphoma.

Persistent generalised lymphadenopathy involves multiple nodes (as its name suggests) and the description is not appropriate.

20) Correct answer: c) Malignant effusion.

The strong smoking history and weight loss should prompt a very likely diagnosis of malignancy.

Purulent effusions are seen in pneumonia but no acute symptoms are described. Eosinophilia within pleural fluid is due to air in the pleural space the most likely cause of which is previous fluid aspiration. Rupture of a bulla is also possible but no acute symptoms are given here.

A chylous effusion arises when lymphatic drainage is obstructed and is an unusual appearance in a pleural fluid so less likely than all other options.

21) Correct answer: e) Primary pulmonary carcinoma.

The patient is an engineer and depending on the exact nature of his work this could well have exposed him to asbestos which along with pleural plaques might lead to a diagnosis of mesothelioma. However, this mass is intrapulmonary and therefore a primary lung cancer. Other options are inappropriate.

It is of note that asbestos workers who are non-smokers have a relative risk of 5 of getting lung cancer compared with workers not exposed to asbestos. If the asbestos worker also smokes, the relative risk of developing lung cancer is 92 i.e. there is a massive potentiating effect between cigarette smoking and asbestos exposure in the development of lung cancer.

22) Correct answer: b) Benign reactive lymph node.

Sorry you are being asked for the **least** likely cause of a necrotising lymphadenitis. A non-specific benign reactive node unusually shows necrosis.

All other options regularly show necrosis.

23) Correct answer: e) Epithelial cells.

Epithelial cells are **not** a routine finding in pericardial fluid. All other options regularly occur within pericardial fluids.

CHAPTER 2

CERVICAL CYTOLOGY
Dr Valerie Thomas

1. Which one of the following statements is true of cervical cytology?

    a) A single test has a sensitivity of 35% for detection of dyskaryosis.
    b) A series of tests within an organised screening service reduces the risk of developing cervical cancer.
    c) The cervical screening test aims to detect invasive carcinoma.
    d) Cervical cytology results should always be available before referral of symptomatic patients to colposcopy.
    e) The first cervical cytology test should be carried out at the age of 18 in the context of the NHS CSP.

2. Which one of the following statements is true regarding liquid based cytology?

    a) The use of liquid based cytology results in fewer histologically proven adenocarcinomas being diagnosed.
    b) The specificity of liquid based cytology is higher than that seen in conventional (direct smears).
    c) When using liquid based cytology, a separate sample should be procured for HPV testing.
    d) The rate of inadequate cytology tests is equivalent in liquid based and direct (conventional) smears.
    e) The use of liquid based cytology results in fewer cytological diagnoses of abnormal glandular cells.

For questions 3 & 4: A 22 year old woman has her first cervical cytology test in the genito-urinary clinic.

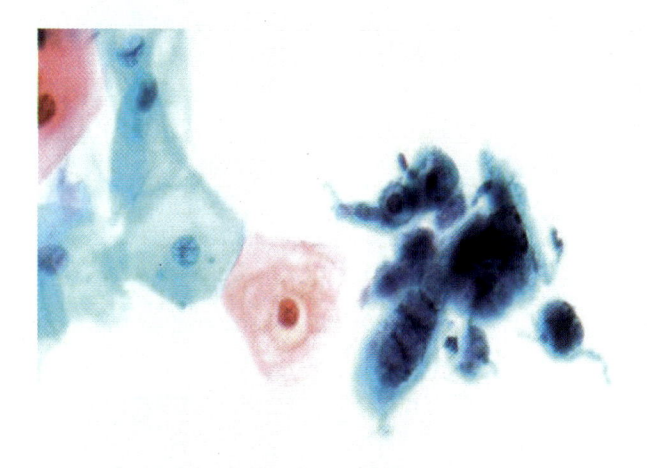

3. What abnormality is shown?

    a) HPV infection.

b) HPV infection with low grade dyskaryosis (mild dyskaryosis).
c) HSV infection.
d) HSV infection with low grade dyskaryosis (mild dyskaryosis).
e) Glandular neoplasia (CGIN or cervical in situ adenocarcinoma).

4. The patient also tests positive for Chlamydia trachomatis infection and negative for gonorrhoea in a cervical sample. Which one of the following statements is true of Chlamydial infection?

   a) Infection is symptomatic in 90% of infected males.
   b) Infection is associated with female infertility.
   c) Infection is not vertically transmitted.
   d) There is UK wide screening programme for Chlamydial infection of the cervix.
   e) Chlamydiae are small gram-positive obligate intracellular microorganisms

For questions 5 & 6: A 26 year old woman has her first cervical cytology test with the following appearance:

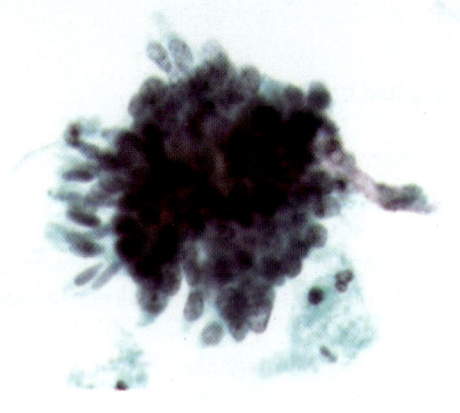

5. What abnormality is shown?

   a) An endocervical polyp.
   b) High grade squamous dyskaryosis.
   c) High grade cervical glandular intraepithelial neoplasia (adenocarcinoma in situ).
   d) Cervical ectropion.
   e) Lower segment endometrial cells.

6. What management would you advise within the context of the NHSCSP?

   a) Repeat cytology in 6 months.
   b) Repeat the cytology in 6 months ensuring the sample is from the second half of the menstrual cycle.
   c) Discharge to normal recall.
   d) Refer for colposcopy.
   e) Perform an HPV test.

7. A 40 year old patient has a cervical cytology test showing low grade dyskaryosis (LSIL). Which one of the following is true?

   a) An HPV test should not be carried out.
   b) Approximately 50% low grade cervical cytology tests result in a high grade biopsy or excision.

c) A negative HPV test would allow the patient to be returned to normal re-call.
d) The patient should be referred for colposcopic assessment prior to a LLETZ (large loop excision of the cervix).
e) The patient should have an examination under anaesthesia.

8. A 40 year old patient has a cervical cytology test showing high grade dyskaryosis (HSIL). Which one of the following is true within the context of the NHSCSP?

   a) The PPV for a high grade cervical cytology test being high grade on histopathology is approximately 60% nationally.
   b) High grade disease accounts for approximately 5% of all abnormalities detected.
   c) A negative HPV test would allow the patient to be returned to normal re-call.
   d) The patient should be referred for colposcopic assessment prior to a LLETZ (large loop excision of the cervix).
   e) The patient should have an examination under anaesthesia.

9. Which of the following factors are **not** associated with false positive reporting in cervical cytology?

   a) Use of an intra-uterine contraceptive device.
   b) Combined oral contraceptive pill.
   c) Tests taken during menstruation.
   d) Tests taken in the perimenopausal age group.
   e) Tests taken mid-cycle.

For questions 10 & 11: A 32 year old woman who uses an IUCD has a cervical cytology test on day 24 of her cycle which shows the following appearance:

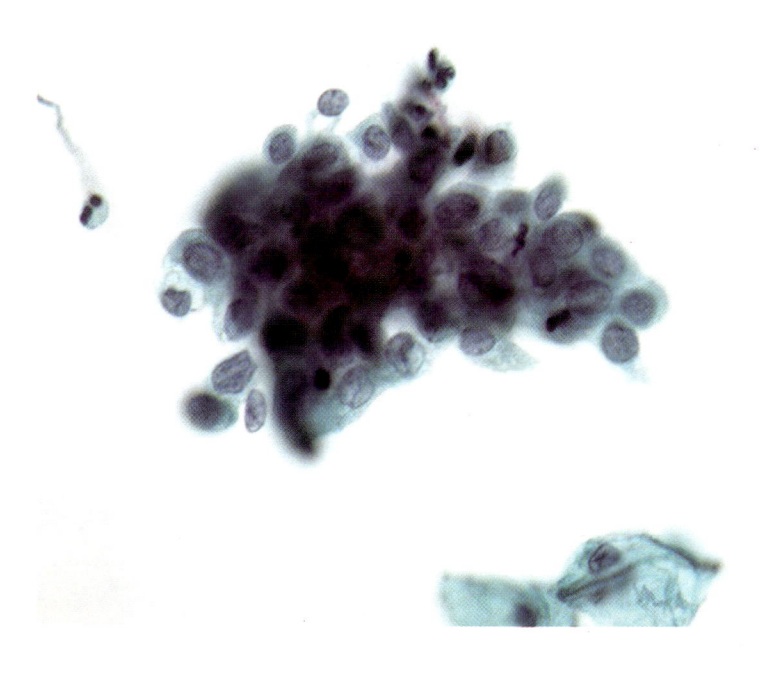

10. Which one of the following statements is correct?

a) The sample shows high grade glandular dyskaryosis.
   b) The sample shows reactive regenerative changes in squamous epithelium.
   c) The sample shows borderline changes.
   d) The sample shows high grade squamous dyskaryosis.
   e) The sample shows reactive regenerative changes in glandular cells.

11. The woman is referred to colposcopy. The colposcopy is incomplete (i.e. the complete transformation zone is not seen). A biopsy is taken which shows normal transformation zone tissue. An HPV test is positive for high risk HPV serotypes. Which one of the following would be appropriate management?

    a) Discharge to normal recall.
    b) Repeat the smear in 6 months.
    c) Repeat the smear and the HPV test in 6 months.
    d) Ablate the cervix.
    e) Review the case at a multidisciplinary meeting.

12. In relation to renal transplant patients, which one of the following statements is correct within the NHSCSP?

    a) All women who are immunosuppressed are recommended to have yearly cervical cytology screening.
    b) All women with renal transplants and a history of previous CIN should have increased cytology surveillance.
    c) All women with renal transplants should have yearly cervical cytology tests.
    d) All women with renal transplants should be routinely screened.
    e) All immunosuppressed women should have yearly (cervical) HPV tests.

For questions 13 & 14: A 26 year old woman presents with an unpleasant greenish smelly discharge to her GP practice. She has a history of normal cervical cytology tests with the last being one year ago. The nurse practitioner takes cervical cytology and sends a swab for microbiology. The micrograph shows the appearances of the smear. No abnormal bacteria are identified on culture.

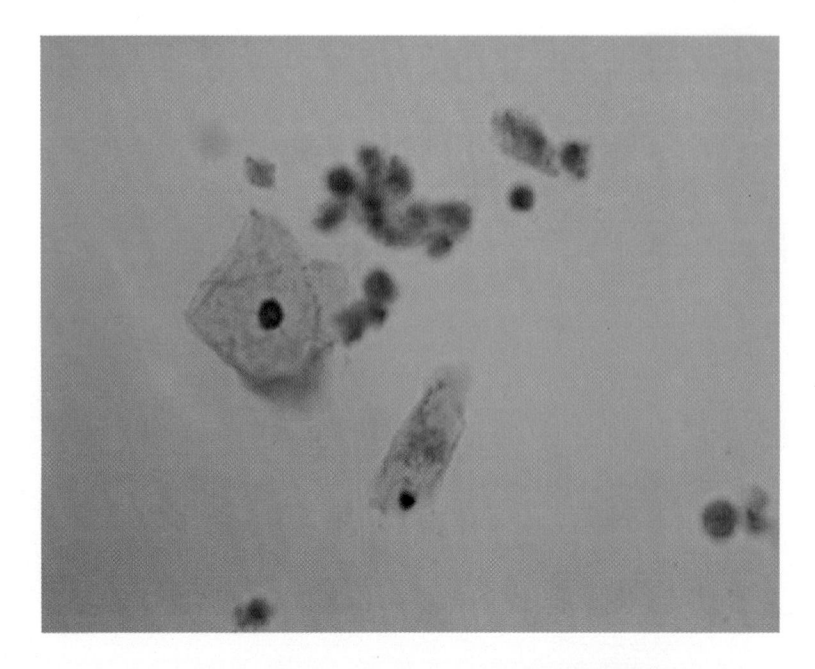

13. How would you report the cervical cytology test?

    a) Normal with a partly cytolytic smear pattern.
    b) Normal appearances.
    c) Normal appearances with Trichomonas vaginalis.
    d) Normal appearances with Amoebiasis.
    e) Normal appearances with degenerate cell debris.

14. The nurse practitioner has taken an 'out of programme' cytology test i.e. an early test. What might have been more appropriate?

    a) Reassurance and discussion of vaginal hygiene.
    b) Referral to a gynaecologist.
    c) Referral to a community health clinic.
    d) Referral to a genito-urinary clinic.
    e) Referral to an infectious disease service.

For questions 15 & 16: A 45 year old woman has a routine cervical cytology test taken on day 15 of her cycle which shows the following appearances:

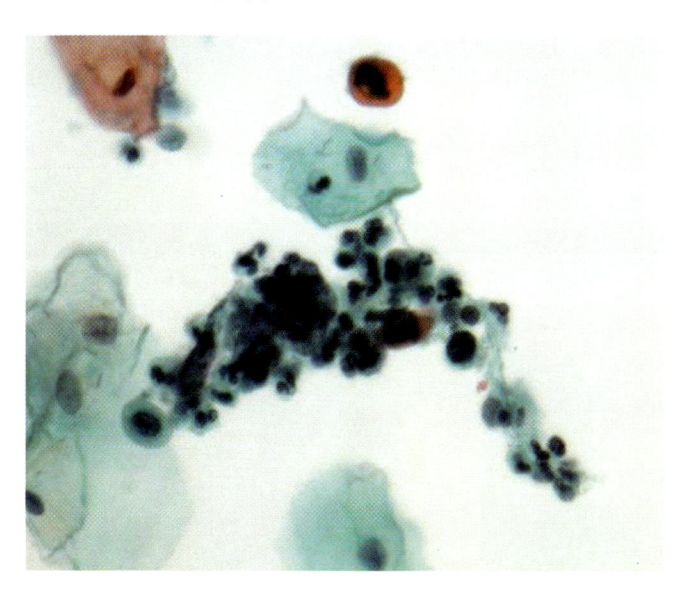

15. How would you report this test?

    a) Negative inflammatory changes.
    b) Negative menstrual smear.
    c) High grade squamous dyskaryosis (HSIL).
    d) High grade squamous dyskaryosis with features suggestive of invasive carcinoma.
    e) Negative endometritis.

16. What management would you advise?

    a) Normal recall.
    b) Gynaecological referral.
    c) Referral to colposcopy.

d) Repeat smear after treatment of any infection.
  e) Repeat smear in the latter part of the cycle.

17. Which of the following statements is correct concerning tubo-endometrioid metaplasia (TEM)?

    a) TEM is a pre-malignant condition.
    b) TEM is seen in the normal cervix.
    c) TEM is only seen after LLETZ or other cervical surgery.
    d) TEM is not seen in association with CIN or CGIN (adenocarcinoma in situ).
    e) TEM uniformly shows clear ciliation.

18. Which of the following are not a morphological feature of HPV infection?

    a) Nuclear integration of HPV DNA.
    b) Koilocytosis.
    c) Dyskeratosis.
    d) Parakeratosis.
    e) Hyperplasia.

For questions 19 & 20: A 60 year old woman presents with post-menopausal bleeding with three separate episodes. She is on Tamoxifen for breast cancer and has had frequent episodes of vaginal candidiasis.

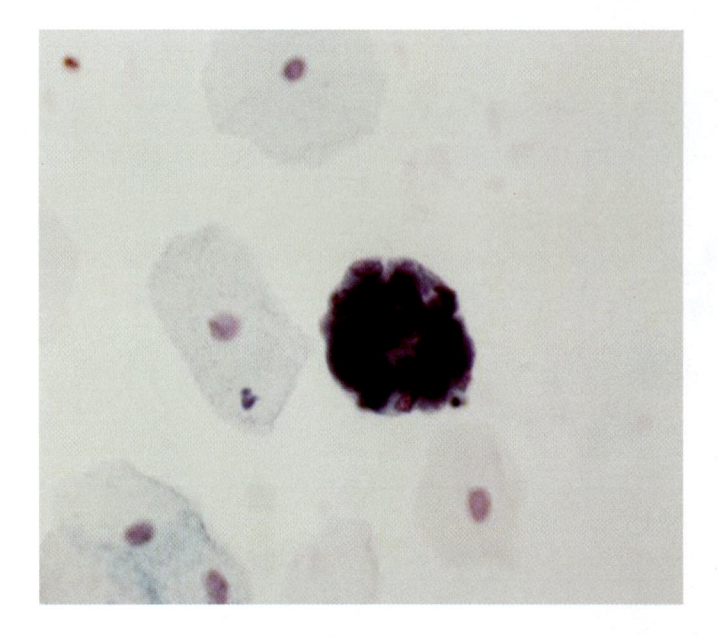

19. What does the cervical cytology show?

    a) Endocervical cells.
    b) Endometriosis.
    c) Endometrial adenocarcinoma.
    d) Normal endometrial cells.
    e) High grade CGIN (adenocarcinoma in situ).

20. What management would you suggest?

a) Refer to colposcopy.
b) Refer to gynaecology.
c) Repeat cervical smear test in 3 months.
d) Normal recall.
e) Treatment for candidal infection.

Correct Answers and Explanations:

1) Correct answer: b) A series of tests within an organised screening service reduces the risk of developing cervical cancer.

The aim of the NHSCSP is to detect CIN3 (high grade dyskaryosis; HSIL). A single smear test has a sensitivity of 60-80% for dyskaryosis. Most International advice is not to begin screening until the age of 21 and in England 25. The referral of a symptomatic woman should not await the result of cytology, indeed a cervical cytology is not advised in this group of patients.

2) Correct answer: e) The use of liquid based cytology results in fewer cytological diagnoses of abnormal glandular cells.

The use of liquid based cytology results in fewer cytological diagnoses of abnormal glandular cells, but the number of histologically proven adenocarcinomas remains unchanged. The specificity of liquid based cytology is equivalent to that seen in conventional (direct smears). Within the UK, since the introduction of liquid based cytology, the inadequate cervical cytology rate has dropped from around 9% to 1-2% of all tests taken.
The sample taken for cytology can be used to test HPV status.

3) Correct answer: c) HSV infection .

The micrograph shows HSV infection with reactive (inflammatory) changes in the adjacent squamous cells which shows a reactive 'halo' of clearance around its nucleus. The nucleus is small and does not show dyskaryosis.

4) Correct answer: b) Infection is associated with female infertility.

Repeated Chlamydial infection is associated with female infertility due to chronic inflammation with scarring of the fallopian tubes. Males are asymptomatic in approximately 50% of infections and females asymptomatic in 80% of infections. Chlamydiae are gram negative organisms. In England there is a Chlamydia Screening Programme which offers targeted screening to all sexually active under 25-year olds in England, annually or on change of sexual partner. Scotland and Wales do not have an organised screening programme for Chlamydia but offer opportunistic testing. Chlamydial genital infection is reported in 5%–30% of pregnant women, with a risk for vertical transmission during parturition to newborn infants of about 50%.

5) Correct answer: c) High grade cervical glandular intraepithelial neoplasia (adenocarcinoma in situ).

The micrograph shows a crowded hyperchromatic microbiopsy with discernable 'feathering' at the periphery and mitoses, indicating high grade CGIN (adenocarcinoma in situ).

6) Correct answer: d) Refer for colposcopy.

Hyperchromatic groups can be an area of diagnostic difficulty. In the context of screening the recognition of this lesion as high grade disease is the vital part of the screening cytology, discriminating between high grade squamous dyskaryosis vs glandular neoplasia is less critical though will affect subsequent management to a degree.

Early repeats of cytology are not part of normal recommendations unless the cytology is classed as inadequate. An HPV test would only be appropriate if this was considered a low grade or borderline lesion.

7) Correct answer: c) A negative HPV test would allow the patient to be returned to normal re-call.

HPV triage of low grade dyskaryosis is routine in the NHSCSP; if the test is negative the woman is returned to normal screening. Approximately 30% of high grade disease is seen after a low grade cervical cytology test. If HPV testing is positive, the patient should be referred for colposcopic assessment which may include biopsy and subsequent LLETZ. Unless persistent, low grade dyskaryosis is not an indication for LLETZ. Examination under anaesthesia is not part of the routine pathway for assessment or treatment for low grade dyskaryosis.

8) Correct answer: d) The patient should be referred for colposcopic assessment prior to a LLETZ (large loop excision of the cervix).

High grade disease (moderate dyskaryosis, severe dyskaryosis, glandular abnormality and invasive carcinoma) is seen in approximately 1% of cervical cytology tests within the NHSCSP. The PPV for high grade cytology being high grade on histopathology is approximately 80% nationally.
An HPV test would not routinely be carried out in a case of high grade abnormality within the NHSCSP; the cellular abnormality advises the referral. Examination under anaesthesia would not usually be appropriate unless colposcopy led to a suspicion of invasive carcinoma or specific patient related issues did not allow a colposcopic examination to be carried out without anaesthetic.

9) Correct answer: e) Tests taken mid-cycle.

Intra-uterine devices can cause reactive changes in endocervix and endometrial cells which can be misinterpreted as high grade dyskaryosis.
Endometrial cells seen in cytology tests taken during menstruation can present as hyperchromatic groups of cells and be mistakenly identified as high grade dyskaryosis.
Oral contraceptive pills can result in an immature smear pattern and irregular menstruation both of which can be misinterpreted as high grade disease. Similarly, atrophic smears from peri or post-menopausal women can lead to immature squamous metaplastic cells being misinterpreted as high grade dyskaryosis.
Mid-cycle testing should not present any of these problems; late in the cycle prior to onset of obvious menstruation degenerate endometrial cells may appear in the cytology.

10) Correct answer: d) The sample shows high grade squamous dyskaryosis.

These cells are large with an increased nuclear cytoplasmic ratio and convoluted nuclear membranes. The chromatin is stippled and numerous mitoses are present. The appearances are of high grade squamous dyskaryosis.

11) Correct answer: e) Review the case at a multidisciplinary meeting.

An MDM discussion should be held. Consideration of local positive predictive value for high grade cytology predicting high grade histology should be made. The cytology, current biopsy and any previous cervical pathology should be reviewed. Assuming the high grade cytology is assured a LLETZ should be undertaken. Ablation could be considered but is not normal practice without a confirmatory biopsy of high grade disease. Other options are inappropriate.

12) Correct answer: b) All women with renal transplants and a history of previous CIN should have increased cytology surveillance.

There is very specific guidance for cervical screening in immunosuppressed patients in various categories provided by NICE (last updated Feb 2015).
These recommendations are based on expert opinion in a guideline published by the NHS Cervical Screening Programme (NHSCSP) *Colposcopy and programme management, Guidelines for the NHS cervical screening programme*.
As further evidence arises, these guidelines may be changed.

In summary:
Patients with HIV should be screened annually.
Renal transplant patients should be screened at or around the time of diagnosis of end-stage renal failure and one year after commencement of immunosuppression post-transplant. Women with a history of CIN should have increased surveillance.
Women starting cytotoxic drugs for rheumatological conditions should have cervical screening if their screening history is incomplete.

13) Correct answer: c) Normal appearances with Trichomonas vaginalis.

Trichomonas vaginalis is a motile parasitic protozoan with a lemon or pear shaped appearance and with flagellae. It appears in cervical cytology tests and is often seen grouped on glycogenated squamous cells and may be somewhat degenerate.
The morphology is diagnostic.

14) Correct answer: d) Referral to a genito-urinary clinic.

At least at first presentation referral to genito-urinary services is the most appropriate patient pathway for a young woman with abnormal discharge so that this sexually transmitted disease can be rapidly diagnosed and appropriate advice and treatment given.

15) Correct answer: d) High grade squamous dyskaryosis with features suggestive of invasive carcinoma.

This test shows hard orangeophilic dyskeratosis in dyskaryotic cells together with spindled malignant nuclei admixed with neutrophils and necrotic debris (diathesis).
Other responses are inappropriate.

16) Correct answer: c) Referral to colposcopy.

The usual referral pathway would be to colposcopy as an urgent referral. Other options are inappropriate.

17) Correct answer: b) TEM is seen in the normal cervix.

TEM rather like respiratory epithelial cells can lose the top of the cell by degeneration (ciliacytopthoria) and present a cuboidal appearance. Recognition of terminal bars at the top of a columnar cell is another appearance of TEM.
TEM is not a pre-malignant condition and can be seen in routine cytology tests from cervixes of women who have not had any previous surgical procedure. TEM can be seen associated with normal cytology or any degree of squamous or glandular abnormality.

18) Correct answer: a) Nuclear integration of HPV DNA.

All of these are features of HPV infection however HPV DNA integration cannot be recognised morphologically.

19) Correct answer: d) Normal endometrial cells.

Normal endometrial cells in a cluster with admixed neutrophils.
All other options are inappropriate.

20) Correct answer: b) Refer to gynaecology.

Tamoxifen is a partial agonist at oestrogen receptors which has an anti-oestrogenic effect in ER positive breast cancer but stimulates oestrogen receptors in the genitourinary tract. This includes stimulation of endometrial cells to divide. It is associated with post-menopausal bleeding but also with a low risk of development of endometrial adenocarcinoma (approximately 1% per year). The patient requires endometrial assessment by hysteroscopy and/or endometrial curettage by a gynaecologist.
A cervical cytology test was not indicated and might have led to false positive cytology with consequent morbidity from inappropriate investigations.

CHAPTER 3

BREAST PATHOLOGY
Dr Valerie Thomas

1. When cutting up a wide local excision which one of the following is true?

    a) The margins should be inked.
    b) The specimen should be sliced into 10mm slices.
    c) The entire specimen should always be embedded.
    d) Gross assessment of margins is not necessary.
    e) Special care should be taken to look for grossly apparent lympho-vascular invasion.

2. When cutting up a mastectomy which one of the following is true?

    a) Preliminary slicing for fixation is unhelpful as it distorts the specimen.
    b) The tumours are not measured as microscopic assessment is required.
    c) If more than one tumour is present, the distance between the tumours should be measured.
    d) If more than one tumour is present margins should be measured only on the largest.
    e) Placing the specimen in formalin into the fridge before dissection improves preservation.

3. When cutting up an axillary dissection which one of the following is true?

    a) Each node should be cut at 2mm intervals and all embedded.
    b) Morphological landmarks on the dissection allow the pathologist to divide the nodes into levels 1-3.
    c) All identified nodes should be sampled.
    d) Frozen sections are routinely used to detect nodal disease
    e) The entire axillary contents should be embedded in megablocks to ensure an accurate and complete nodal count.

4. Which of the following is a risk factor for cancer?

    a) Columnar cell metaplasia
    b) Collagenous spherulosis
    c) Pseudoangiomatous stromal hyperplasia
    d) Radial scar
    e) Fibroadenomatoid hyperplasia

5. When grading invasive carcinoma according to the modified Bloom and Richardson system, which one of the following factors is assessed?

    a) Tubule formation
    b) Tumour necrosis
    c) Proliferation index (Ki67)
    d) Lymphovascular invasion
    e) Nucleo-cytoplasmic ratio

6. In assessment of tumour proliferation which one of these factors is assessed in the modified Bloom and Richardson system?

   a) Apoptotic count
   b) Mitotic count
   c) Ki67
   d) Number of nuclei in metaphase
   e) Nucleolar count per tumour cell

7. Using the American Joint Committee on Cancer 7th edition breast cancer guidance (TNM staging), in the presence of multiple tumours which one of the following is true?

   a) The tumour size is the summation of the maximum diameter of all tumours present.
   b) The tumour size is the maximum diameter of the highest grade tumour present.
   c) The tumour size is the maximum diameter of the largest individual invasive carcinoma including surrounding DCIS.
   d) The tumour size is the diameter of the largest invasive tumour present.
   e) Each tumour is separately staged.

8. Which one of the following is true regarding lobular carcinoma of the breast?

   a) Invasive lobular carcinoma is not graded according to the modified Bloom and Richardson system.
   b) When staging a lobular carcinoma, LCIS is included in the overall measurement of the tumour.
   c) Invasive lobular carcinoma is a morphological diagnosis.
   d) All lobular carcinomas are negative for e-cadherin on immunochemistry.
   e) Invasive lobular carcinoma has a worse prognosis than invasive ductal carcinoma of the same grade and stage.

9. Which of the following statements regarding DCIS are true?

   a) Low grade DCIS is a risk factor for invasive carcinoma with 30% of patients developing invasive disease within 10 years of diagnosis if untreated.
   b) The extent of the DCIS is irrelevant to treatment.
   c) If high grade DCIS is incompletely excised, the risk of invasive recurrence is 90%.
   d) When DCIS is seen in an ultrasound guided 14G core biopsy, there is a 70% chance that the resection will show invasive carcinoma.
   e) The extent of comedo pattern high grade DCIS is very accurately mapped by mammographically-detected calcification.

10. Which of the following is true of the van Nuys index?

    a) It provides an indication of whether a patient with invasive ductal carcinoma will benefit from radiotherapy.
    b) It is based entirely on grading.
    c) The score is higher where necrosis is absent.
    d) It provides an indication of whether a patient with DCIS will benefit from radiotherapy after surgery.
    e) It provides an indication of whether a patient with DCIS will benefit from radiotherapy without the requirement for surgery.

11. Which one of the following statements regarding in situ lobular neoplasia (ISLN) is true?

    a) ISLN is a malignant condition coded as B5a within the screening programme
    b) Diagnosis of ISLN can be supported with CK5+/e-cadherin- immunochemistry
    c) Diagnosis of ISLN can be supported with CK5-/e-cadherin- immunochemistry
    d) Diagnosis of ISLN can be supported with CK5-/e-cadherin+ immunochemistry
    e) ISLN with necrosis and calcification is not a recognised variant.

A scenario for questions 12-14: A 49 year old woman has her first screening mammography which reveals a stellate mass. On ultrasound an irregular, hypoechoic density is noted corresponding to the mass. The mass is impalpable.

12. Which of the following is the most likely diagnosis?

    a) Invasive lobular carcinoma
    b) Invasive ductal carcinoma
    c) Mucinous carcinoma
    d) An intramammary lymph node
    e) Medullary carcinoma

13. Which one of the following investigations would contribute the most to a pre-operative diagnosis?

    a) FNA cytology of the mass.
    b) Diagnostic excision of the mass.
    c) MRI scan of both breasts.
    d) FNA cytology of the ipsilateral nodes.
    e) Core biopsy of the mass.

14. A core biopsy taken under US shows a fibroadenoma. Which of the following would you advise at the MDM discussion of the case?

    a) The histology is benign so no further action is required
    b) The biopsy does not correlate with the imaging; imaging should be repeated.
    c) The biopsy does not correlate with the imaging; biopsy should be repeated.
    d) The case should be reported as a serious incident due to non-correlation.
    e) The patient should have a diagnostic excision.

A scenario for questions 15-17: A 48 year old woman has her first screening mammogram which reveals a stellate lesion with surrounding slender spicules. On ultrasound a stellate distortion is seen in the corresponding area without a definite mass.

15. Which of the following is the likely diagnosis?

    a) Tubular carcinoma
    b) Fat necrosis
    c) Medullary carcinoma
    d) Invasive ductal carcinoma NST.
    e) Hamartoma

16. She has an ultrasound guided biopsy and then goes on to excision. The excision shows the following pathology:

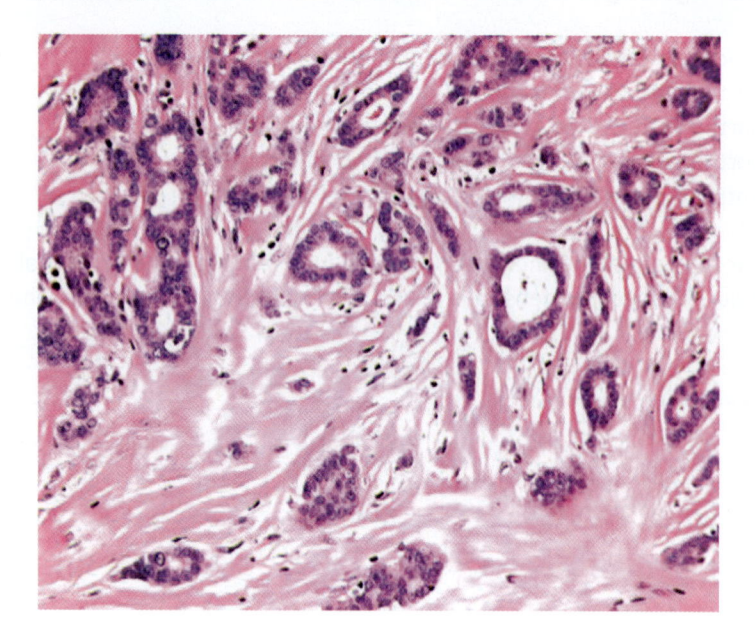

    What is the diagnosis?

    a) Tubular carcinoma
    b) Radial scar
    c) Medullary carcinoma
    d) Invasive ductal carcinoma NST.
    e) Fibroadenoma

17. What immunochemistry would you use to confirm the diagnosis?

    a) Collagen IV, CK5, p63
    b) ER, Her2
    c) ER, PgR, Her2
    d) CK5, CK7, CK20
    e) ER, Her2, e-cadherin

18. When submitting a core biopsy for reporting from the NHS breast screening programme, which one of the following should the radiologist submit as part of the minimum data set for biopsies?

    a) Breast density
    b) Weight of the sample
    c) Number of cores taken
    d) The core biopsy reporting category
    e) The cytology reporting category

19. Which of the following statements about microcalcification is true?

    a) Mammographically detected microcalcification is normally associated with malignancy.
    b) High grade comedo DCIS often shows calcification of the central necrotic areas.

c) Low grade DCIS is not associated with calcification.
   d) Fibroadenomas rarely show calcification.
   e) Radiology can detect smaller flecks of calcification than seen histologically.

20. A 48 year old woman has her first screening mammogram which reveals a 5mm cluster of variable calcifications classified as M3. Ultrasound of the corresponding area shows no lesion. The woman has a stereo-guided core biopsy with representative microcalcification noted in the specimen X-ray.

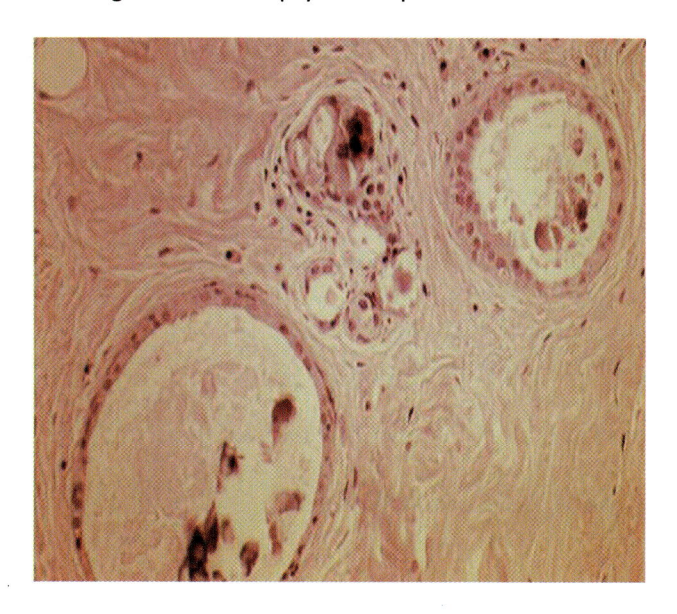

   Which diagnosis listed is the best fit?

   a) Fibrocystic changes
   b) Microcystic changes with calcification.
   c) Low grade DCIS with calcification
   d) Columnar cell changes with microcysts and calcification
   e) Flat epithelial atypia with calcification.

21. Which of the following statements is correct?

   a) Columnar cell metaplasia in microcysts is not usually associated with calcification
   b) Coarse calcification in a ductal distribution is pathognomonic for high grade DCIS.
   c) Mammary duct ectasia can show dense calcification of the duct wall
   d) Psammomatous calcification is often associated with malignancy.
   e) The use of polarised light is unhelpful in the detection of calcification.

22. Which of the following statements is correct?

   a) Immunochemistry routinely used for assessment of Her2 status stains intranuclear Her2 genes.
   b) Immunochemistry routinely used for assessment of Her2 status stains cytoplasmic Her2 genes.
   c) Immunochemistry routinely used for assessment of Her2 status stains Her2 genes on the cell membrane.
   d) Immunochemistry routinely used for assessment of Her2 status stains Her2 gene product located on the cell membrane.

e) Immunochemistry routinely used for assessment of Her2 status stains nuclear Her-2 gene product.

23. In relation to 'triple-negative' tumours of breast, which of the following is correct?

    a) Immunochemistry for CK7, CK5 and CK14 is negative in tumour cells
    b) Immunochemistry for EGFR-1 is negative.
    c) Immunochemistry for CK5 is usually positive.
    d) Immunochemistry for ER, Her2 and GCDP-15 is negative.
    e) Immunochemistry for ER, Her2 and EGFR-1 is negative.

24. In relation to 'triple-negative' tumours of breast, which of the following is correct?

    a) All basal-like carcinomas are triple negative.
    b) Many triple negative carcinomas are associated with amplification of Her2 gene.
    c) All triple negative carcinomas are basal-like carcinomas.
    d) Many triple negative carcinomas are associated with 1p19q deletion.
    e) Many triple negative carcinomas are associated with BRCA1 mutations

25. Which of the following statements is correct in relation to medullary carcinoma of the breast?

    a) Diagnosis relies on the gross appearance of a discrete rounded mass
    b) Diagnosis relies on the microscopic demonstration of admixed in situ and invasive carcinoma with a lymphoid infiltrate.
    c) Diagnosis requires the presence of osteoclast-like giant cells within the tumour.
    d) Up to 90% of cases are associated with BRCA 1 gene mutations
    e) Most of these cases are ER positive.

26. Which of the following is **not** a described 'special type' of breast cancer?

    a) serous papillary cystadenocarcinoma
    b) micropapillary carcinoma
    c) adenoid cystic carcinoma
    d) mucinous carcinoma
    e) tubulo-lobular carcinoma

27. Which one of the following statements is true?

    a) 'Special types' of breast cancer are defined largely by their radiological appearances.
    b) 'Special types' of breast cancer almost always have a better prognosis than invasive ductal carcinoma of no special type.
    c) In mucinous carcinoma, the tumour cells usually show intracytoplasmic mucin resembling goblet cells.
    d) In tubulo-lobular carcinoma the lobular component is always e-cadherin negative.
    e) Invasive lobular carcinoma is a 'special type' of breast cancer.

28. Which of the following statements about the current molecular classification of breast tumours is true?

a) It delineates 6 predominant breast cancer subtypes: luminal A, luminal B, ERBB2-enriched, basal, claudin-low and normal-like.
b) The molecular groupings directly correspond with morphological subtypes.
c) Lobular carcinoma directly corresponds to a luminal A subtype.
d) The ERB2 enriched group corresponds to 'triple negative' immunophenotype.
e) The 6 predominant groups have few differences in their therapeutic responses.

Scenario for questions 29-30: A 30 year old woman presented in the symptomatic clinic with a mass in the upper central part of her left breast. She was currently breast feeding her 6 month old baby. She initially had an ultrasound and then a mammogram to assess the mass. The mass was measured as 50mm maximum diameter.
A biopsy was taken and an ER negative, Her2 positive, grade 3 invasive ductal carcinoma of no special type diagnosed. FNA of the ipsilateral axillary nodes showed metastatic ductal carcinoma.
Treatment options were discussed and the patient opted for neoadjuvant chemotherapy followed by mastectomy. Her tumour responded well to chemo-therapy and she subsequently had the planned mastectomy and axillary node clearance.

29. Which of the following describes the most likely gross appearances of her mastectomy?

a) Normal.
b) Presence of a hard irregular tumour mass measuring 50mm.
c) Presence of an irregular area of soft fibrosis at the site of the previous tumour.
d) Presence of small islands of lymphocytes and haemosiderin-laden macrophages in an area of fibrosis.
e) Presence of a hard area of scarring at the site of previous tumour.

30. Which of the following does **not** require microscopic assessment?

a) The size of any residual tumour.
b) The size of the tumour bed.
c) The percentage cellularity of the residual tumour.
d) The number of nodes involved.
e) The area of any necrosis within the tumour.

Correct answers and explanations:

1) Correct answer: a) The margins should be inked.

Cut up of a wide local excision should be compliant with gathering data for the minimum data set. In order to assess marginal involvement the specimen should be differentially inked in multiple colours and a record of the inking pattern recorded in the macroscopic description.
Where there is an established screening programme, many of the lesions will measure less than 10mm and therefore potentially contained within a single slice if sliced at 10mm. Initially slicing should be at 5mm or less; it may be difficult to slice large fatty wide local excisions thinly but, pragmatically, slicing at 3mm will allow slices to be directly blocked. It is not necessary to embed the whole specimen though in small excisions this may be the most straight forward approach. Where there is no discernible macroscopic lesion it may be

necessary to embed the whole specimen. Where there is a guide wire in situ and clinico-radiological information that the lesion is close to the needle tip, initial blocks may be focussed around the needle tip with the proviso that the orientation of the sample must be maintained in order to take additional blocks if necessary. Gross assessment of margins is important particularly if the specimen is large; blocks should be taken as a minimum from the areas where tumour most closely approaches the margin. Gross assessment for lymphovascular invasion is not appropriate.

2) Correct answer: c) If more than one tumour is present; the distance between the tumours should be measured.

Slicing particularly of large mastectomies can help optimise fixation; care should be taken to maintain the orientation of the slices e.g. by keeping the overlying skin intact but taking care to slice all the way through the breast tissue to skin.
Measuring the tumours grossly is a key element of the data set being critical for staging. Tumours should be assessed in three dimensions – often the largest dimension is across the slices. Stage 1 tumours must also be assessed microscopically. Where there are multiple tumours the distance between tumours should be measured as pathologically those lying less than 10mm apart are considered to be part of a single mass; this may alter the tumour stage. Margins of all the tumours should be assessed.
Formalin fixation is a chemical process; putting the specimen in formalin and then refrigerating will slow down the process of fixation and impair fixation. To optimise fixation large specimens should be sliced and placed in a large volume of neutral buffered formalin (ideally 10x the volume of the specimen). If formalin is heavily blood-stained on receipt of the specimen, it should be changed for fresh solution as the initial chemically reactive formalin will have been consumed by reacting with blood.

3) Correct answer: c) All identified nodes should be sampled.

When sampling an axillary dissection all nodes and all nodular areas of possible nodal nature should be sampled. This may result in slides including nodules of brown fat but improves the overall lymph node yield. The expected minimum number of nodes you should identify within a level 1-3 axillary clearance is 15; if the axilla has been resected after neoadjuvant chemotherapy fewer nodes are usually found. There are no morphological land marks on an axillary dissection. Surgeons may send the different levels in different pots or identify the position of the highest node with a suture. If clearly orientated the axilla can be systematically dissected starting from the highest point. Every node should be sampled.
Slicing at 2mm and embedding all slices is part of the protocol for assessing sentinel lymph nodes and is not applied to axillary dissections.
Frozen sections are not routinely used to identify involved nodes within an axillary dissection, however some units use frozen sections in the intra-operative assessment of sentinel lymph nodes.

4) Correct answer: d) Radial scar.

Radial scar on biopsy is associated with an increased risk of the patient developing breast cancer. This is a complex area with a lot of variably good studies available. Retrospective outcome reviews give a figure around 10% of subsequent cancer development. A large study as part of the Nurses Health Study (USA) published in 2014 supports a x2 risk of developing cancer (in situ or invasive) with a 3 fold increase in risk for ER/PR negative cancers compared with other tumour types. The cancer does not necessarily develop within the area of the radial scar nor necessarily within the same breast. Women with multiple radial scars have a higher risk than those with a single radial scar.
None of the other conditions listed have a documented associated increased risk of cancer.

5) Correct answer: a) Tubule formation.

The Modified Bloom and Richardson grading system for breast cancer which is endorsed by the NHSBSP is based on three parameters: tubule formation, nuclear pleomorphism and mitotic rate each given a score which is summed to give a grade for the tumour (see NHSBSP publications). None of the other factors listed are part of this grading system.

6) Correct answer: b) Mitotic count.

Although all the factors listed give an indication of proliferation and turnover of the tumour, only the mitotic count or mitotic rate per 10 high power fields forms part of modified Bloom and Richardson grading.
It is important to assess your own high power field area and check the algorithm provided in the NHS breast screening programme publications to establish the cut-off points in scoring the mitotic rate as these will vary according to the specific objective lens you use.

7) Correct answer: d) The tumour size is the diameter of the largest invasive tumour present.

The largest focus of invasive carcinoma is used to stage the tumour. All the other options are incorrect.

8) Correct answer: c) Invasive lobular carcinoma is a morphological diagnosis.

All subtypes of invasive breast cancer, including invasive lobular carcinoma, are graded using the modified Bloom and Richardson grading system.
ISLN (in situ lobular neoplasia up to and including lobular carcinoma in situ) is considered a risk marker within the NHSBSP and is not included in the whole tumour size. DCIS associated with an invasive lobular carcinoma is included in the whole tumour size.
A percentage of invasive lobular carcinoma is positive for e-cadherin (particularly pleomorphic subtype). The diagnosis of invasive lobular carcinoma remains primarily a morphological process.
Conventional invasive lobular carcinoma has a better prognosis than invasive ductal carcinoma of the same grade and stage.

9) Correct answer: a) Low grade DCIS is a risk factor for invasive carcinoma with 30% of patients developing invasive disease within 10 years of diagnosis if untreated.

Low grade DCIS is a risk factor for invasive carcinoma as stated. More extensive disease is associated with an increased rate of recurrent disease and also is more likely to harbour invasive ductal carcinoma in the resection. When assessing where to biopsy an extensive area of DCIS, radiologists look for potential mass lesions within the area with ultrasound to optimise detection of invasive disease and/or target dense clusters of calcification. The chance of finding invasive carcinoma in the resection after diagnosis of DCIS is a parameter assessed within the NHSBSP; finding invasive cancer pre-operatively allows axillary surgery to be planned as part of a single operation rather than a second operation if found post-operatively. The likelihood of this occurring has been examined in a large USA based meta-analysis and found to vary according to sampling method and needle width (ECRI Institute for the Agency for Healthcare Research and Quality). For ultrasound guided cores 25% of the cores containing DCIS were found to contain invasive carcinoma. Some studies show a higher conversion rate to invasive carcinoma than this.
Extensive areas of DCIS are treated by mastectomy; smaller areas may be treated by local excision. The role of radiotherapy is discussed in question 10.
Although calcification is a guide to the extent of DCIS is underestimates DCIS size and the DCIS is not usually calcified at the periphery and is hence more extensive on histology than predicted radiologically.

10) Correct answer: d) It provides an indication of whether a patient with DCIS will benefit from radiotherapy after surgery.

The van Nuys index allows assessment of whether a patient with DCIS would benefit from post-excisional radiotherapy; more recent studies (2014) include prognostic data based on 12 years of follow-up.
The parameters assessed on the excisional specimen are: the type of DCIS (non-high nuclear grade without necrosis; non-high nuclear grade with necrosis; high nuclear grade with or without necrosis); size, margins and patient's age.

11) Correct answer: c) Diagnosis of ISLN can be supported with CK5-/e-cadherin- immunochemistry.

ISLN comprises in situ lobular neoplasia up to and including lobular carcinoma in situ. It is considered a risk factor for the subsequent development of invasive carcinoma and is categorised as a B3 lesion within the NHSBSP. The variant with high grade cellular appearances and comedo necrosis is classified as B5a within the screening programme.
ILSN is negative for CK5 on immunohistochemistry as is most DCIS.
ISLN is also negative for e-cadherin, whereas low grade DCIS will be positive for e-cadherin.
It is important to confirm whether a low grade in situ proliferation is lobular or ductal in nature as these are classified and treated differently:
Low grade DCIS is categorised as B5a and excised; conventional ISLN is categorised as B3 and the client put under surveillance.

12) Correct answer: b) Invasive ductal carcinoma.

You may think this is a radiology question but radiology unsurprisingly reflects the gross morphology in the pathology. We have to correlate our biopsy findings routinely with the imaging, working with our radiology colleagues to ensure we have representative samples of the radiological lesion.
The imaging points to a tumour with a stellate outline.
Medullary and mucinous carcinomas are two malignant lesions which can be underestimated on imaging due to their smooth outline. This is of course reflected in the histology where grossly a rounded mass is present. Benign lymph nodes also present a rounded profile.
Invasive ductal carcinoma accounts for approximately 85% of invasive cancers with lobular at approximately 12%. Both can present as stellate masses, however the question asks for the most likely diagnosis which is then 8:1 in favour of invasive ductal carcinoma.

13) Correct answer: e) Core biopsy of the mass.

The 'triple approach' is applied to breast disease i.e. there should be clinical, radiological and pathological data contributing to the pre-operative diagnosis to ensure the correct treatment is initiated.
FNA, whilst excellent at diagnosing malignant cells, cannot differentiate between in situ and invasive carcinoma as the architecture is not apparent. FNA is often used in symptomatic clinics as part of the triple assessment as it is quick and simpler to perform than a core biopsy. A core biopsy is the preferred option in this case which is radiologically malignant and if confirmed will require assessment of the presence of invasion to direct surgery and immunochemistry for steroid receptors and Her2.

14) Correct answer: c) The biopsy does not correlate with the imaging; biopsy should be repeated.

The case should be re-discussed at MDM, if all findings confirmed, in the first instance, the biopsy should be repeated ensuring that the correct lesion has been sampled.

If a second biopsy of the stellate area is also benign then the lesion could potentially sampled by vacuum biopsy giving a larger sample or diagnostic excision performed.

Within the NHSBSP there is an expectation that >90% of patients will have a non-operative diagnosis of cancer prior to excision. Carrying out a diagnostic excision followed by a formal wide local excision or mastectomy is not an acceptable routine pathway in the UK.

15) Correct answer: a) Tubular carcinoma.

Stellate lesions comprise radial scars, complex sclerosing lesions, invasive ductal carcinoma NST, tubular carcinoma and sometimes areas of fat necrosis.
Stellate lesions without an associated mass are most likely to be radial scar or tubular carcinoma.
Medullary carcinoma and hamartoma are masses and usually rounded.

16) Correct answer: a) Tubular carcinoma.

The lesion shows angulated ductal profiles lined by a single layer of atypical ductal cells characteristic of tubular carcinoma. There is no apparent radial arrangement of epithelial elements to support a diagnosis of radial scar.
Whilst tubular carcinoma is a type of invasive ductal carcinoma, it is important to recognise specific subtypes which impact on prognosis.
Other answers are inappropriate.

17) Correct answer: a) Collagen IV, CK5, p63.

You are being asked to confirm the diagnosis. CK5 and p63 as basal markers should be negative in tubular carcinoma aiding the diagnosis of this low grade malignancy. Collagen IV (or laminin) staining should be absent around invasive carcinoma.
ER, PgR and Her2 give prognostic information. Other options are inappropriate.

Note: ER staining pattern may be helpful in identifying a 'clonal' pattern of staining particularly in atypical intraductal epithelial proliferations but was not offered in this context in the list of options.

18) Correct answer: c) Number of cores taken.

Data items that you would expect to receive on the pathology request include the side, size and nature of the lesion. If the lesion is calcification the radiologist should indicate whether there is representative calcification within the X-ray of the specimen. You should also be told how many cores of what core width have been sent.
Breast density is an experimental measure currently.
Where vacuum assisted excision is performed, the specimen is weighed by the pathologist. Core biopsy and cytology reporting categories are assigned by the pathologist when reporting.

19) Correct answer: b) High grade comedo DCIS often shows calcification of the central necrotic areas.

Mammographic calcification is very common in fibrocystic changes and other benign conditions such as fat necrosis or fibroadenoma. Much of this calcification is categorised as benign (M2) and not biopsied.
DCIS of any grade is often associated with calcification.
Radiology does not detect individual flecks of calcification <100 micrometres; histology picks up extremely small flecks which are not seen on radiology. Careful correlation of the mammogram with the specimen X-ray and the histology is required to ensure representative calcification has been seen.

20) Correct answer: d) Columnar cell changes with microcysts and calcification.

The question asks for the diagnosis which 'best fits' the image.
Whereas options a, b and d are all correct, option d best describes the appearances. There is no evidence of atypia or malignancy in the sample.

21) Correct answer: c) Mammary duct ectasia can show dense calcification of the duct wall.

Microcysts are often associated with calcification and columnar cell lined cysts are no exception. The calcification in clusters of microcysts is often given an indeterminate (M3) grading by radiologists and is therefore a common biopsy appearance for screening pathologists. Specific morphological types of calcification can be seen in benign and malignant lesions but none are specific for a benign or malignant lesion. Coarse calcification in a ductal pattern is seen in comedo DCIS but also duct ectasia and sometimes fibroadenomatoid changes,
Polarisation of the slide can reveal Weddellite-type calcification. Weddellite is calcium oxalate composed of angulated crystals of pale amber colour or colourless on microscopy. Haematoxyphilic (dark purple) calcification is calcium phosphate (also known as hydroxyapatite) clearly visible under conventional light microscopy.

22) Correct answer: d) Immunochemistry routinely used for assessment of Her2 status, stains Her2 gene product located on the cell membrane.

Immunochemistry demonstrates the Her2 gene product on the cell membrane. In situ hybridisation is required to demonstrate any amplification of the Her2 gene.

23) Correct answer: c) Immunochemistry for CK5 is usually positive.

'Triple negative' tumours are invasive ductal carcinomas which are negative on immunochemistry for ER, PgR and Her2 over expression. They are often basal-like and express CK5 (71- 91% reported) and EGFR-1 is often expressed.

24) Correct answer: e) Many triple negative carcinomas are associated with BRCA1 mutations.

There is a large (approximately 70%) overlap between 'triple negative' tumours and basal-like tumours but they are not identical. There are subsets of basal-like tumours which are not triple negative and of triple negative tumours that are not basal-like. Many of the 'triple negative' tumours are associated with BRCA1 gene mutations.
Her2 amplification would lead to over expression of Her-2 product – the tumour would not be 'triple negative'. 1p19q deletion is described as a good prognostic feature in oligodendrogliomas.

25) Correct answer: d) Up to 90% of cases are associated with BRCA 1 gene mutations.

Although medullary carcinomas present a rounded outline, this is not exclusive to medullary carcinomas, it is the microscopy which is distinctive. Neither giant cells nor in situ carcinoma are a requirement for diagnosis. Most are ER negative.

26) Correct answer: a) Serous papillary cystadenocarcinoma.

This is the typical descriptor for a subtype of ovarian carcinoma.

27) Correct answer: e) Invasive lobular carcinoma is a 'special type' of breast cancer.

Special types of breast cancer are largely morphologically defined. Special types include tumours of different prognostic types. Mucinous carcinomas usually comprise small tumour acini floating in pools of extracellular mucin. In tubulo-lobular carcinoma the lobular component is always positive for e-cadherin, in contrast to pure lobular carcinoma which is usually negative.

28) Correct answer: a) Delineates 6 predominant breast cancer subtypes: luminal A, luminal B, ERBB2-enriched, basal, claudin-low and normal-like.

The molecular classification does not correspond to the morphological correlation e.g. most but not all lobular carcinomas are luminal A as are ER positive ductal carcinomas.
ERB-2 enriched tumours are Her2 positive. The six different groups have strikingly different therapeutic responses.

29) Correct answer: c) Presence of an irregular area of soft fibrosis at the site of the previous tumour.

The question asks for the most likely answer. The clinical data states the tumour has responded well to chemotherapy.
Although a breast can appear grossly normal after neo-adjuvant chemotherapy this is not the most usual appearance which is of a soft fibrous area of 'tumour bed' often with a residual irregular margin. Persistence of a 50mm hard tumour mass is also unlikely given the good response to chemotherapy; hard scarring is not a usual outcome.
The 'presence of small islands of lymphocytes and haemosiderin-laden macrophages in an area of fibrosis' is a microscopic description of the typical findings in the tumour bed after neoadjuvant chemotherapy.

30) Correct answer: e) The area of any necrosis within the tumour.

Necrosis may be noted but does not require measurement. All the other features listed are required to be assessed. When assessing nodes, it is also important to record the number of nodes showing partial or total obliteration by fibrosis which reflects nodes previously involved by metastatic carcinoma.

CHAPTER 4

FEMALE GENITAL TRACT PATHOLOGY
Dr Limci Gupta

1) A 58 yr old lady with known psoriasis and 6 months history of perianal and vulval erythema, scales and erosions presents to her GP. Clinical diagnosis was ? psoriasis ?lichen simplex. A vulval biopsy was performed which showed the following picture:

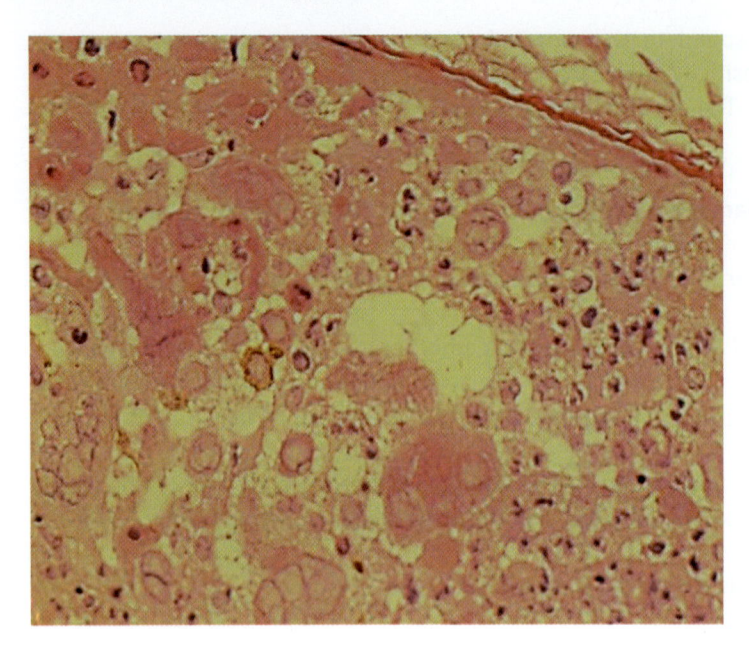

What is the diagnosis?

a) Psoriasis
b) Lichen simplex
c) Herpes simplex infection
d) Cytomegalovirus infection

2-6) Extended matching item questions:

A) Hidradenoma papilliferum
B) Endometriosis
C) Extramammary Paget's disease
D) Vulval intraepithelial neoplasia
E) Trichoepithelioma
F) Deep 'aggressive' angiomyxoma
G) Liposarcoma
H) Angiomyofibroblastoma
I) Bartholin's cyst
J) Fibro-epithelial polyp

For each of the following microscopic descriptions of a **vulval mass**, select the most appropriate diagnosis from the list of options above. Each option may be used once, more than once, or not at all.

2) Poorly marginated infiltrative hypocellular lesion with bland spindled cells and perivascular condensation of collagen.

3) Well demarcated lesion composed of plump round and spindle cells in an oedematous stroma with alternating hypercellular and hypocellular areas. The hypercellular stroma is condensed around prominent vascular spaces.

4) Polypoid mass with fibrovascular core with stellate multinucleate stromal cells in a variably cellular stroma.

5) Intradermal papillary proliferation composed of fibrovascular cores lined by a double layer comprising both columnar and myoepthelial cells.

6) Intradermal fibrosis with columnar glands in a spindled, CD10 positive stroma.

7-10) Extended matching item questions:

    A) Contact dermatitis
    B) Lichen planus
    C) Erythema multiforme
    D) Hidradenitis suppurativa
    E) Syphilis
    F) Granuloma inguinale
    G) Varicella Zoster
    H) Herpes simplex
    I) Lichen Sclerosis
    J) Lymphogranuloma venereum
    K) Impetigo
    L) Crohn's disease
    M) Pyoderma gangrenosum

For each of the following microscopic descriptions of features identified on a **vulval biopsy**, select the most appropriate diagnosis from the list of options above. Each option may be used once, more than once, or not at all.

7) Vulval skin with epidermal basal layer hydropic degeneration, Civatte bodies and dermal band-like lymphoplasmacytic inflammation with pigment incontinence.

8) Painful ulceration with follicular centric chronic inflammation, apocrine gland destruction, abscess and sinus formation. Similar findings are noted in the axillary tissue.

9) Compact orthokeratosis, thin epidermis with vacuolar alteration at the dermo-epidermal junction, homogenized sclerotic collagen in the papillary and upper reticular dermis and moderately dense perivascular lymphocytic inflammation.

10) Ulcer with lymphocytes and plasma cells in the ulcer base. Silver stains demonstrate spirochetes.

11) A 2 year old girl, presents with vaginal bleeding and a soft tumour like mass protruding from the introitus. Microscopically the tumour demonstrates superficial squamous epithelium with underlying myxoid hypocellular stroma containing stellate cells. A cambium layer is present below the squamous epithelium. Which of the following is the most appropriate diagnosis?

   a) Fibroepithelial polyp
   b) Alveolar rhabdomyosarcoma
   c) Embryonal rhabdomyosarcoma
   d) Endometriosis

12) A 55 year old female presents with abnormal vaginal bleeding. Hysteroscopy reveals a tumour, located in the lower uterine segment and endocervical canal. Microscopically, the tumour is a moderately differentiated adenocarcinoma. The tumour shows positive staining with progesterone receptor and vimentin. CEA and p16 are negative. What is the most appropriate diagnosis?

   a) Endometrioid adenocarcinoma of the endometrium
   b) Endometrioid adenocarcinoma of the ovary
   c) Endocervical adenocarcinoma
   d) Serous carcinoma of the endometrium

13) Cervical carcinogenesis is most commonly associated with which of the following infectious agents:

   a) HHV-8
   b) HPV 6 and 11
   c) HPV 16 and 18
   d) HSV

14) A 37 year old lady presents with irregular P/V spotting. This is the microscopic picture following an endometrial biopsy.

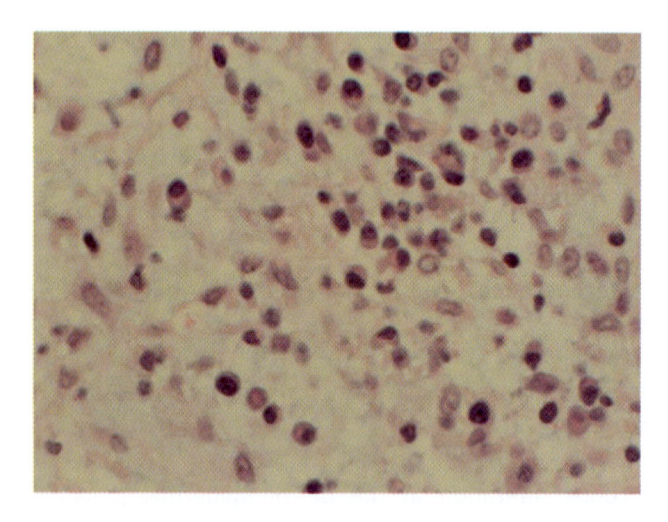

What is the diagnosis commonly associated with?

   a) Carcinoma
   b) Leiomyoma
   c) Intra-uterine contraceptive device
   d) Plasma cell myeloma

15) A 56 year old lady has a Grade 2 endometrioid type endometrial carcinoma involving the cervical stroma. The parametrial tissue and adnexa are free from tumour. Peritoneal washings and lymph nodes are negative for malignant cells. Abdominal/pelvic CT scan show no tumour metastasis. What is her FIGO stage?

   a) 1B
   b) 2
   c) 3
   d) 4

16) Which of the following is NOT an important mandatory core parameter to record in a microscopic report of endometrial carcinoma?

   a) Distance to serosal surface
   b) Microscopic involvement of cervical stroma & appendages
   c) Lymphovascular (LV) invasion
   d) Immature squamous metaplasia in endocervix

17) The International Federation of Gynaecology and Obstetrics (FIGO), as recommended by WHO, advise histological grading of which of the following endometrial adenocarcinomas?

   a) Serous carcinoma of endometrium
   b) Endometrioid adenocarcinoma
   c) Clear cell carcinoma of endometrium
   d) Carcinosarcoma

18-20) Extended matching questions:

   A) Leiomyoma
   B) Symplastic/atypical leiomyoma
   C) Cellular leiomyoma
   D) Leiomyosarcoma
   E) Epithelioid leiomyoma
   F) Intravascular leiomyomatosis
   G) Myxoid leiomyosarcoma
   H) Leiomyoma treated with GNRH analogues
   I) Mitotically active leiomyoma
   J) Myxoid leiomyoma

18) A 5cm smooth muscle tumour. Microscopic examination demonstrates a bland smooth muscle proliferation with up to 10 mitoses per 10 high power fields in the most active areas. No coagulative necrosis is present.

19) Smooth muscle tumour with diffuse large atypical smooth muscle cell nuclei and multinucleated cells. Mitotic figures identified (3 per 10 high power fields in the most active areas). No coagulative necrosis present.

20) Smooth muscle tumour with diffuse atypical smooth muscle cell nuclei and abrupt coagulative necrosis.

21) A 49 year old lady presented with peritoneal disease and ascites. An ultrasound guided peritoneal biopsy was done which showed a poorly differentiated malignant tumour. Immunohistochemistry showed WT1+, CK7+, CK20-, ER+, CA125+, CDX2- and GCDFP15- tumour cells. What is the likely diagnosis?

   a) Poorly differentiated adenocarcinoma of most likely ovarian origin
   b) Poorly differentiated adenocarcinoma of breast origin
   c) Poorly differentiated adenocarcinoma of colorectal origin
   d) Mesothelioma

22) A 46 year old lady has an ovarian cyst measuring 10 cm in diameter. The outer surface is smooth and shiny with no papillary excrescences and there is no capsular rupture. The inner surface shows papillary excrescences in 4% of the entire surface area. Microscopically, this is a serous tumour with epithelial tufting and pseudo-stratification in 4% surface area. There is also mild cytological atypia but no stromal invasion is seen. What is the diagnosis?

   a) Benign serous cystadenoma
   b) Borderline serous cystadenoma
   c) Serous cystadenofibroma
   d) Serous cystadenocarcinoma

23) A 70 year old female is diagnosed with a serous borderline tumour of the ovary. On examination of the submitted peritoneum, a 1.5cm firm cream nodule is identified. Histologically this comprises of a papillary proliferation of cells with cytological appearances similar to those seen in the ovarian borderline tumour. They are arranged in sub-peritoneal papillae with smooth contours and there is no associated stromal reaction. Which of the following best describes the peritoneal lesion?

   a) Endometriosis
   b) Invasive peritoneal implant
   c) Endosalpingiosis
   d) Non invasive peritoneal implant

Scenario for questions 24 -27: A 54 year old lady was diagnosed with a large 30cm ovarian cyst, clinically ?mucinous carcinoma. Macroscopically, it was a ovary with a partly solid and a partly cystic ovarian tumour filled with haemorrhagic fluid. Microscopy showed the following picture:

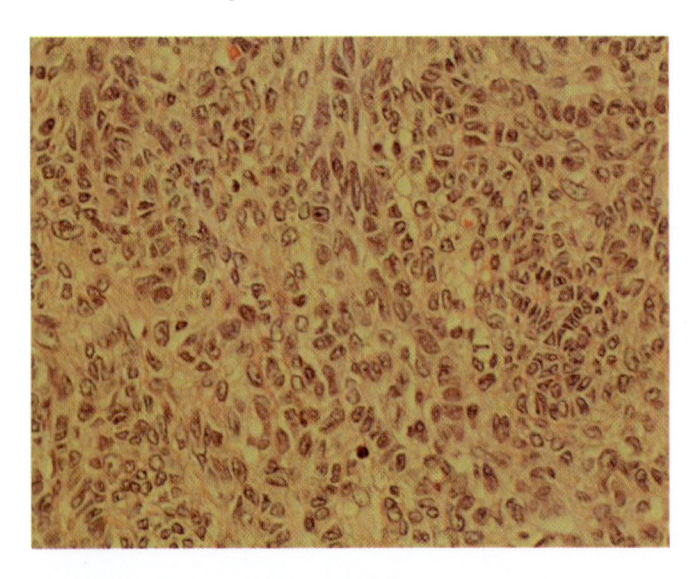

24) What is the likely diagnosis?

   a) Adult granulosa cell tumour
   b) Fibroma
   c) Adenocarcinoma
   d) Germ cell tumour

25) If you had a choice of only two immunostains, which ones would you do choose to confirm the diagnosis?

   a) AE1/3 and Oct 3/4
   b) Inhibin and calretinin
   c) CK7 and CK20
   d) WT1 and P53

26) Which is the best special stain for this case which would confirm the entity?

   a) EVG
   b) Reticulin
   c) PAS
   d) ZN

27) Which factors are poor prognostic factors for this tumour?

   a) Size > 2cm
   b) 3 mitoses/10HPF
   c) 35 yrs of age
   d) Capsular rupture

Scenario for questions 28 & 29: A 52 year old lady has unilateral ovarian and also an endometrial tumour.

28) Which of the following features favour metastases from endometrium to ovary?

   a) Different morphology of tumours
   b) Ovary showing multiple subcapsular nodules
   c) Low grade malignancy
   d) Unilateral ovarian tumour

29) Which of the following features favour two separate synchronous tumours?

   a) Involvement of uterine serosa
   b) One mass involving the inner aspect of ovary
   c) Multiple subcapsular nodules in ovary
   d) Lymphovascular invasion

30) The grading of immature teratoma of the ovary is based on:

   a) Size of tumour
   b) Amount of the immature neuro-epithelial component
   c) Differentiation of the immature component

d) Presence of necrosis

31) A unilocular, 10 cm ovarian cyst was received for frozen section. It was filled with haemorrhagic fluid. No solid areas or necrosis were identified. The cyst wall was focally thickened and measured 3-4 mm in maximum thickness. No papillary excrescences were seen. What is the likely diagnosis on macroscopic examination?

a) Lymphoma
b) Endometrioma
c) Carcinoma
d) Sex-cord stromal tumour

32) A 5 cm ovarian mass is received for frozen section diagnosis. There are solid brownish areas with a few cystic spaces containing dark inspissated secretions. The frozen sections show follicles lined by non-mucinous cuboidal cells and filled with colloid-like secretions. What should you tell the surgeon?

a) Mucinous neoplasm
b) Teratoma, most likely benign but needs to be sampled thoroughly to exclude immature components.
c) Immature teratoma
d) No answer available; defer the diagnosis to the paraffin sections

33) Which immunostain can be used to differentiate between a complete mole and a partial mole?

a) P53
b) P16
c) P57
d) CEA

34) A partial hydatidiform mole may display which of the following karyotypes?

a) 46XY
b) 32XY
c) 46XX
d) 69XXY

Correct answers and explanations:

1) Correct answer: c) Herpes simplex infection

In this picture, there are intra-epidermal vesicles with pale ground-glass / basophilic nuclear inclusions. This is a classic example and there should be no differentials for this.
Sometimes, beginners can confuse this with cytomegalovirus (CMV) infection. In CMV, there will be reddish-purplish nuclear as well as cytoplasmic 'owl-eye' inclusions with haloes around the inclusions.

2) Correct answer: F) Deep 'aggressive' angiomyxoma

This tumour is locally aggressive and can recur. These are poorly circumscribed, pauci-cellular lesions with bland spindled to round cells in abundant myxoid stroma. Medium-sized thick walled blood vessels with collagen condensation around vascular space is characteristic.

3) Correct answer: H) Angiomyofibroblastoma

Uncommon benign tumour which occurs in the soft tissue of the vulva and vagina in women of reproductive age. There is a well demarcated proliferation of benign spindle cells in oedematous stroma with perivascular hypercellularity. Vascular component is prominent in the stroma with numerous small vessels.
Differential included aggressive angiomyxoma but is differentiated from this by its well demarcated margin, superficial nature rather than deep and with hypercellular areas.

4) Correct answer: J) Fibroepithelial polyp

This is typically a polypoid lesion with normal squamous epithelium overlying an oedematous central fibrovascular core. The usual finding is that of stellate or multinucleate stromal cells in the fibrovascular core.

5) Correct answer: A) Hidradenoma papilliferum

The described features are classic for hidradenoma papilliferum, the most commonly described adnexal tumour seen in the vulva and perineum. It is an unencapsulated but well demarcated tumour with a papillary pattern. There are branching papillae with true fibrovascular cores lined by a double layer of epithelium; an outer myoepithelial layer and an inner columnar or cuboidal layer.

6) Correct answer: B) Endometriosis

Endometriosis describes the condition in which endometrial tissue is identified outside the uterine cavity. It can occur at various sites including the vulva. Ovarian involvement by endometriosis gives rise to an endometriotic (chocolate) cyst.
A diagnosis of endometriosis can be made when two of the three features listed below are identified histologically:
- Endometrial glands
- Endometrial stroma
- Evidence of old haemorrhage i.e. haemosiderin laden macrophages (fresh haemorrhage is not enough)

A focus of endometriosis undergoes the same hormone dependent cyclical changes as uterine endometrial tissue. Additionally, hyperplastic and neoplastic proliferation which may affect the endometrium can also occur in foci of endometriosis.

7) Correct answer: B) Lichen planus

Lichen planus is an autoimmune chronic inflammatory disease which can affect both skin and mucous membranes.
The classical features allowing diagnosis of lichen planus include:
- Basal layer hydropic degeneration with Civatte body/colloid body formation (rounded homogenous eosinophilic masses seen in the basal layer of the epidermis).
- Lichenoid (band like) chronic lymphohistiocytic inflammation in the upper dermis.
- Pigment incontinence with melanophages in the upper dermis.

Differential diagnosis includes lichenoid drug reaction. Consider this if there are eosinophils associated with the inflammatory infiltrate.

8) Correct answer: D) Hidradenitis suppurativa

Hidradenitis suppurativa is an inflammatory condition affecting the terminal hair follicles in apocrine gland bearing skin. It classically involves the axilla, inguinal folds, perineum, genitalia and peri-areolar region. The histological appearances are not specific. However, the diagnosis is suggested by an acute and chronic inflammatory process centred on dilated apocrine glands and hair follicles. Suppuration, abscess formation and sinus tract formation are common. A foreign body giant cell reaction may be seen in response to ruptured apocrine secretions or hair.

9) Correct answer: I) Lichen Sclerosis

This is one of the most common chronic inflammatory dermatoses of the ano-genital region. It generally affects middle aged and elderly women. Clinically, it appears as flat white papules which coalesce to form plaques.
The microscopic features are:
- Thin epidermis with vacuolar degeneration at the dermo-epidermal junction (DEJ)
- Homogenisation of the papillary and upper reticular dermis – the dermis appears eosinophilic and sclerosed.
- Pigment incontinence with melanophages.
- Moderate perivascular lymphocytic inflammation with admixed plasma cells and histiocytes surrounding the superficial vascular plexus which appears displaced downwards by the sclerosed collagen.

10) Correct answer: E) Syphilis

Syphilis is a sexually transmitted infection caused by the spirochete *Treponema pallidum.*
Microscopy shows a perivascular plasma cell infiltrate and endothelial cell proliferation. Neutrophils may be seen in early lesions. Granulomas may also be present. Secondary syphilis may mimic other inflammatory dermatoses and may exhibit spongiotic and lichenoid type patterns of inflammation. The dominance of **plasma cells** in the inflammatory infiltrate is a helpful clue to the diagnosis.
The spirochetes can be seen on a Warthin-Starry or other silver impregnation special stains.

11) Correct answer : c) Embryonal rhabdomyosarcoma

This is the most common vaginal neoplasm in children and young adults, although on the whole it is a rare tumour. The mean age at diagnosis is 2 years. It generally presents with vaginal bleeding.
The tumour comprises normal squamous mucosa overlying thickened or polypoid submucosa. The tumour diffusely involves the submucosa and lamina propria, appearing hypocellular and myxoid with a **classic cambium layer** (condensed layer of stromal cells just under the normal surface squamous epithelium). A characteristic finding is that of stellate cells or cells with elongated eosinophilic cytoplasm, some of which may show cross-striations.
MSA, desmin and myogenin are positive. This profile indicates a tumour of skeletal muscle origin.
Alveolar rhabdomyosarcoma histologically consists of small round blue cells with no cambium layer.

12) Correct answer: a) Endometrioid adenocarcinoma of the endometrium

A panel including PR, Vimentin, CEA and p16 is used to distinguish endometrioid adenocarcinoma of the endometrium from endocervical adenocarcinoma.

Endometrioid adenocarcinoma will be positive for PR and Vimentin. CEA and p16 are usually negative, although occasionally focal and weak positivity may be seen on CEA & p16. (Please note that interpretation of immunohistochemistry should be on the glandular component alone and squamous morules should not be considered).

Endocervical adenocarcinoma is diffusely and strongly positive for p16 and CEA. PR and Vimentin are generally negative.

13) Correct answer: c) HPV 16 and 18

Of the over 100 types of HPV, up to 13 are cancer causing, and thus designated as high risk. HPV types 16 and 18 are classical high risk HPV types that are implicated in the carcinogenesis of the majority of cervical carcinoma.

Low risk HPV such as types 6 and 11 cause genital warts, but are not implicated in cervical cancer oncogenesis.

14) Correct answer: c) Intra-uterine contraceptive device

This picture shows chronic endometritis which is a common association with an intra-uterine contraceptive device.

15) Correct answer: b) 2

Endometrioid carcinoma involving cervical stroma is classified as FIGO stage 2.

FIGO staging for endometrial carcinomas:
- IA   Tumour confined to the uterus, no or inner half of myometrial invasion
- IB   Tumour confined to the uterus, outer half of myometrial invasion
- II   Cervical stromal invasion, but not beyond uterus
- IIIA   Tumour invades serosa or adnexa
- IIIB   Vaginal and/or parametrial involvement
- IIIC1   Pelvic node involvement
- IIIC2   Para-aortic involvement
- IVA   Tumour invasion bladder and/or bowel mucosa
- IVB   Distant metastases including abdominal metastases and/or inguinal lymph nodes

16) Correct answer: d) Immature squamous metaplasia in endocervix

In the Royal College of Pathologists dataset for endometrial cancers, the following are core items to be recorded in the microscopy section:

- tumour type
- tumour grade
- myometrial invasion
- tumour free distance to serosa
- lymphovascular invasion
- cervical stromal invasion
- vaginal involvement
- uterine serosal involvement

- parametrial involvement
- adnexal involvement
- lymph node involvement
- omental involvement
- provisional FIGO stage

17) Correct answer: b) Endometrioid adenocarcinoma

Endometrial adenocarcinoma is divided into two types, Type I (endometrioid type) and Type II (non endometrioid type).

Endometrioid type adenocarcinoma (type I) accounts for the majority (80%) of endometrial carcinomas. The carcinogenesis is oestrogen dependent and is frequently associated with inactivation of PTEN tumour suppressor gene, K-ras mutations, and abnormalities in Beta Catenin. These tumours are associated with complex atypical hyperplasia and have a favourable prognosis with less myometrial invasion, vascular invasion and lymphatic spread than type II tumours.

FIGO and WHO recommend the grading of endometrioid tumours (type I) as it infers prognostic significance.
Grading is based on the proportion of non- squamous solid component of the tumour:

Grade 1 (well differentiated):   <5% solid growth
Grade 2(moderately differentiated):   6-50% solid growth
Grade3 (poorly differentiated):   >50% solid growth

If severe cytological atypia is present, the grade should be raised by one level.

Non endometrioid type adenocarcinomas (type II) in contrast tend to occur in older women and are not associated with oestrogen stimulation. They are classically associated with altered p53.

By definition type II tumours are high grade, and as such grading is not recommended. They have a poor prognosis with rapid progression, deep myometrial invasion and frequent lymphovascular spread.

Type II tumours include
- Serous carcinoma of endometrium
- Clear cell carcinoma of endometrium
- Carcinosarcoma
- Undifferentiated carcinoma

18) Correct answer: I) Mitotically active leiomyoma

Mitotically active leiomyomas have more than 5 mitotic figures per 10 HPF. Upto 20 mitotic figures / 10HPF can be seen. Cytological atypia or coagulative tumour cell necrosis are absent.

19) Correct answer: B) Symplastic/atypical leiomyoma

Symplastic leiomyoma is the historic name given to the entity now referred to as atypical leiomyoma.
These are characterised by pleomorphic atypical cells with multinucleated nuclei. Mitotic count should be below 5/10HPF and no coagulative necrosis should be seen.

20) Correct answer: D) Leiomyosarcoma

Leiomyosarcoma is a malignant smooth muscle tumour. Grossly they tend to have a more infiltrative border than leiomyoma and may appear paler than leiomyoma or yellow on cut section. Areas of haemorrhage or necrosis are commonly seen at cut up.

Microscopically the diagnosis of leiomyosarcoma is based on the presence of 2 out of 3 of the following criteria:
- Increased cellularity & cellular atypia
- Increased mitotic rate >10/10HPF with atypical mitoses
- Coagulative tumour necrosis – shows abrupt transition from viable to necrotic tissue without associated granulation tissue or inflammation (associated granulation tissue or inflammation would be present in hyaline necrosis). Ghost outlines of cells may be seen with tumour preservation around vascular spaces. Coagulative tumour necrosis is often termed 'geographical' necrosis due to the complex contour of the necrotic focus.

21) Correct answer: a) Poorly differentiated adenocarcinoma of female genital tract origin, most likely ovarian.

In this case, the malignant cells are positive for CA125, WT1, ER and CK7 (this combination indicates female genital tract origin, likely ovarian origin).
The malignant cells are negative for CK20 and CDX2 which would have been positive in colorectal carcinomas. However, this can be positive in mucinous carcinomas from ovary / breast but a combination of well selected immunostains and clinico-radiological correlation will help in reaching the correct diagnosis. They are negative for GCDFP 15 (breast marker).
Mesothelioma would be WT1 and CK7 positive and usually negative for the rest of the immunostains in question.

22) Correct answer : a) Benign serous cystadenoma

To call it a borderline tumour, we need to have 10% surface involvement by papillary areas with mild nuclear atypia.
High nuclear grade changes would make it in-situ carcinoma.

23) Correct answer: d) Non- invasive peritoneal implant

Peritoneal implants are identified in 30-40% of serous borderline tumours. They are classified as non-invasive or invasive.
Invasive implants show a disordered infiltration of the peritoneum with irregular contours and with associated desmoplasia.
Non invasive implants have well defined contours with no infiltration into surrounding tissue.
Separation of implants into invasive and non- invasive forms has a prognostic significance. Invasive implants carry a worse prognosis.
Endosalpingiosis is a benign peritoneal lesion with demonstrates cystically dilated glands lined by fallopian tube type epithelium. Psammomatous calcification is occasionally seen.

24) Correct answer : a) Adult granulosa cell tumour

This is a classical example. There are polyhedral cells with angular nuclei and nuclear grooves (coffee bean shaped nuclei). Call-Exner bodies with eosinophilic secretions may or may not be present.
A fibroma would have more oval to spindled cells with a storiform pattern.
Adenocarcinoma would be associated with glandular formation, necrosis and pleomorphic nuclei.
Each germ cell tumour has a characteristic picture (discussed in MGT chapter).
Poorly differentiated tumours have to be accompanied by immunohistochemical and special stains in order to reach a correct diagnosis.

25) Correct answer: b) Inhibin and calretinin

These are the optimal immunohistochemical stains for a granulosa cell tumour (positive in sex cord stromal tumours).

Oct 3/4 is a germ cell tumour marker.
P53 and WT1 are positive in serous adenocarcinomas of ovary.
CK7 and CK20 are used for indicating the origin of adenocarcinoma.

26) Correct answer : b) Reticulin

This is a brilliant special stain for differentiating granulosa cell tumour from fibroma/fibrosarcoma. It will stain around the nests in granulosa cell tumour and individual cells in fibroma/fibrosarcoma.

27) Correct answer: d) Capsular rupture

Poor prognostic features in granulosa cell tumour
- Size > 10 cm
- Mitotic count > 3/10HPFs
- Atypia
- Necrosis
- Capsular rupture
- Bilateral tumours
- Nuclear atypia
- Age > 40 years

28) Correct answer: b) Ovary showing multiple subcapsular nodules

Features favouring metastasis from endometrial carcinoma to ovary:
- Similar morphology of tumours
- Ovary shows multiple nodules (usually capsular/subcapsular)
- High grade
- Lymphovascular invasion
- Uterus – Ca in outer half of myometrium / serosa / parametrial tissue / adnexal structures
- Bilateral ovarian involvement with co-existing endometrial carcinoma

29) Correct answers : b) One mass involving the inner aspect of ovary

Features favouring synchronous primary tumours:
- Different morphologies but may have similar morphologies... These are usually low grade
- One mass in ovary rather than multiple nodules
- Mass involving the inner aspect of ovary as well, rather than just subcapsular
- No lymphovascular invasion (variable)
- Carcinoma confined to endometrium or inner half of myometrium
- Low grade
- Serosa / parametrial tissue not involved

30) Correct answer: b) The amount of the immature neuro-epithelial component

Immature teratoma is a germ cell tumour demonstrating elements differentiated from one or more embryonic germ cell layers, commonly in the form of immature neuro-epithelial tissue. Immature teratomas are graded using the Norris grading system which is based on the proportion of immature elements present with grade 1 tumours showing the least amount immature tissue and grade 3 the most.

Norris grading system for immature teratoma:
Grade 1: rare foci of immature neuro-epithelium, in no greater than 1 low power field in maximum size
Grade 2: immature neuro-epithelium in no more than 3 low power field foci
Grade 3: immature neuro-epithelium in more than 4 low power field foci

31) Correct answer: b) Endometrioma

Lymphoma, carcinoma and sex-cord stromal tumours would have solid areas.
Malignant tumours can have necrotic areas.

32) Correct answer: b) Teratoma, most likely benign but needs to be sampled thoroughly to exclude immature components.

Correct sampling is very important in ovarian tumours. All representative and suspicious areas should be taken.
1 section per cm of the tumour should be sampled.

33) Correct answer : c) P57

P57 is positive in cytotrophoblast in partial moles and in normal POCs. It is negative in cytotrophoblast in complete moles.

34) Correct answer : d) 69XXY

Partial hydatidiform moles are formed by fertilisation of the normal haploid maternal chromosome set in the ovum (23X) by an extra haploid set of paternal genome (i.e. two sperm) (23X or 23Y) leading to a **triploid genome** composed of **both** parental and maternal chromosomes.
Thus the possible karyotypes for partial mole include
- 69XXY
- 69XXX
- 69XYY

In contrast, complete hydatidiform moles are formed by fertilisation of an empty ovum by either one or two sperms, resulting in a diploid genome.
All chromosomes are derived from sperms, usually 46XX, less commonly 46XY. Therefore, foetal parts are absent in them.

CHAPTER 5

GASTRO-INTESTINAL TRACT
Dr Jayson Wang

1) Achalasia of the oesophagus may be caused by all of the following, except:

   a) Amyloidosis
   b) Aspiration pneumonia
   c) Diabetes mellitus
   d) Malignancy
   e) Sarcoidosis

2) A patient presents with anaemia, swollen and red tongue and lips with ulceration. Endoscopy showed an upper oesophageal web. The underlying condition is associated with deficiency of which nutrient?

   a) Calcium
   b) Iron
   c) Magnesium
   d) Potassium
   e) Zinc

3) A 65 year old man presents with indigestion and acid taste in the mouth. An endoscopy shows red patches in the lower oesophagus near the gastro-oesophageal junction. Oesophageal biopsies show squamous epithelium, with intervening columnar epithelium and crypts, with oesophageal glands in the submucosa. What is the most correct diagnosis based on the information?

   a) Consistent with Barrett's oesophagus
   b) Hiatus hernia
   c) Insufficient information for diagnosis
   d) Oesophageal adenocarcinoma
   e) Reflux oesophagitis

4) All of the following are associated with Barrett's oesophagus, except:

   a) Alcohol
   b) HPV infection
   c) Male
   d) Obesity
   e) Smoking

5) A 42 year old woman presents with dysphagia. Endoscopy shows a smooth nodule in the upper oesophagus. A biopsy shows an ill-defined lesion in the submucosa, composed of nests of polygonal cells with abundant eosinophilic cytoplasm and small bland nuclei. The cells are positive for S100 and DPAS. What is the most likely diagnosis?

   a) Adenocarcinoma
   b) Carcinoid tumour

c) Epithelioid schwannoma
d) Granular cell tumour
e) Melanoma

6) A 39 year old man presents with dyspepsia. The gastric biopsy shows moderate to severe acute and chronic inflammation in the lamina propria with focal lymphoid follicle formation. There are no atrophic gastric glands identified. Which of the following is the most likely cause of the gastritis?

   a) Alcohol excess
   b) Autoimmune disease
   c) Crohn's disease
   d) Helicobacter pylori infection
   e) Trauma

7) Fundic gland polyp may be associated with which of the following drugs?

   a) Diclofenac
   b) Iron
   c) Methyl-dopa
   d) Omeprazole
   e) Penicillin

8) In a well-circumscribed mural tumour in the small bowel, the histology shows mildly atypical spindled cells and epithelioid cells in fascicles. Which of the following immunostains, if positive, would help in the diagnosis of gastro-intestinal stromal tumour?

   a) AE1/3 and CAM5.2
   b) DOG1 and CD117
   c) S100 and MelanA
   d) SMA and h-caldesmon
   e) Synaptophysin and NSE

9) Which mutation sites are characterised by the most sensitivity to imatinib in gastro-intestinal stromal tumours?

   a) Exon 11 of c-kit gene
   b) Exon 12 of PDGFR gene
   c) Exon 14 of PDGFR gene
   d) Exon 9 of c-kit gene
   e) Wild-type for c-kit and PDGFR

10) Which of the following is not a prognostic factor for gastro-intestinal stromal tumour?

   a) Mitotic activity
   b) Necrosis
   c) Presence of metastasis
   d) Site of tumour
   e) Size of tumour

11) A 42 year old woman presents with indigestion and weight loss. A gastric biopsy shows a dense infiltrate of eosinophils in the lamina propria as well as infiltrating muscularis propria. The subsequent diagnosis is known to be associated with which of the following diseases?

    a) Barrett's oesophagus
    b) Coeliac disease
    c) HSV infection
    d) Squamous cell carcinoma of oesophagus
    e) Ulcerative colitis

12-15) Extended matching item questions: Match the following descriptions with the appropriate diseases of the small bowel:

    A. Acute bulbitis
    B. Adenoma
    C. Giardiasis
    D. Whipple's disease

12) Widened villi containing macrophages with PAS-positive granules.

13) Gastric foveolar metaplasia with neutrophils in the lamina propria.

14) Presence of basophilic binucleate organisms on the surface of the lumen.

15) Pedunculated lesion with pseudostratified epithelium.

16) Viral causes of diarrhoea include which of the following viruses?

    a) Calicivirus
    b) Coxsackievirus
    c) Paramyxovirus
    d) Rotavirus
    e) All of the above

17) A 50 year old woman presents with profuse diarrhoea. Endoscopy was normal. The colonic biopsy shows numerous intraepithelial lymphocytes in the surface epithelium and crypts which have a normal architecture, as well as a subepithelial band of hyalinised material within the lamina propria. Which is the most likely diagnosis?

    a) Collagenous colitis
    b) Lymphoma
    c) Pseudomembranous colitis
    d) Ulcerative colitis
    e) Whipple's disease

18) A patient with previous surgery for colonic cancer has rectal biopsies performed, which showed mild to moderate chronic inflammation with crypt abscesses, as well as prominent lymphoid follicles with germinal centre formation. Which is the most likely diagnosis?

    a) Colon cancer

b) Diversion colitis
c) Ischaemic colitis
d) Pseudomembranous colitis
e) Ulcerative colitis

19) A 28 year old HIV positive man develops watery diarrhoea. The colonic biopsies show colonic mucosa with normal architecture and mild chronic inflammation. The brush border of the surface epithelium appears thickened and basophilic. Which further stain would you perform?

a) Diastase Periodic acid Schiff
b) Gram
c) Grocott
d) Warthin-Starry
e) Ziehl-Neelsen stain

20) A 34 year old HIV positive man develops diarrhoea and upper abdominal pain. He has no respiratory symptoms. He has an endoscopy and duodenal biopsies are taken. The Ziehl-Neelsen stain of the biopsies stains organisms in the cytoplasm of the enterocytes. What is the likely diagnosis?

a) Cytomegalovirus
b) Histoplasmosis
c) Mycobacterium Avium
d) Mycobacterium Tuberculosis
e) Sarcoidosis

21) A newborn baby presents with delay in passing meconium. Rectal biopsies show a lack of ganglion cells. There are hypertrophic nerve fibres which are highlighted by acetylcholine esterase stain. The condition is caused by mutations of which of the following genes?

a) APC
b) KRAS
c) RET
d) TP53
e) VHL

22) A 20 year old man presents with anaemia and weight loss. A duodenal biopsy shows flattening of the villi with numerous intra-epithelial lymphocytes, counted at 50 per 100 enterocytes. No pathogens are identified. Which serological test would you advise the gastroenterologist to perform?

a) Anti-endomysium antibody
b) Anti-nuclear antibody
c) Anti-smooth muscle antibody
d) Rheumatoid factor
e) Serum IgE

23-27) Extended matching item questions: Match the following descriptions with the appropriate diseases of the large bowel:

A. Diverticular disease
B. Ischaemic colitis

C. Necrotising enterocolitis
D. Pseudomembranous colitis
E. Ulcerative colitis

23) Young man with bloody diarrhoea. Colon biopsies show moderate crypt architecture distortion with Paneth cell metaplasia, cryptitis and crypt abscesses extending along whole length of large bowel.

24) A 78 year old man with bloody diarrhoea. Colonic resection shows segments of necrotic ulcerated mucosa with haemorrhage, with adjacent normal mucosa.

25) A 69 year old female inpatient with profuse diarrhoea. Colonic biopsies show plaque-like fibrinous necrotic material adherent to ulcerated mucosa.

26) An 81 year old woman with acute abdominal pain. Colonic resection shows out-pouchings of colonic mucosa with surrounding hypertrophic muscularis propria.

27) A 3 month old baby in distress undergoes bowel resection showing transmural necrosis and infarction with submucosal gas bubbles.

28) Which of the following gross features favours a diagnosis of ulcerative colitis over Crohn's disease?

a) Discontinuous (skip) lesions
b) Fat wrapping
c) Hose-pipe thickening of the colon
d) Involvement of small bowel
e) Involvement of the rectum

29) Which of the following microscopic features favours a diagnosis of Crohn's disease over ulcerative colitis?

a) Crypt abscesses
b) Granulomas
c) Inflammation limited to submucosa
d) Marked crypt distortion
e) Paneth cell metaplasia

30) Which of the following types of colonic polyps are associated with Cronkite-Canada syndrome?

a) Inflammatory pseudopolyp
b) Juvenile polyp
c) Peutz-Jegher's polyp
d) Submucosal ganglioneuroma
e) Tubulo-villous adenoma

31) During routine reporting of colonic biopsies from an 80 year old man with rectal bleeding of recent onset and no other symptoms, it is noted that the biopsies are normal apart from a very tiny detached fragment of a poorly differentiated carcinoma. Which is the most appropriate immediate action to be taken?

a) Cut levels and check against other specimens of the day
b) Perform tumour marker immunohistochemistry and DNA analysis
c) Report as normal colonic biopsies and ignore the malignant fragment

d) Report as normal colonic biopsies with poorly differentiated carcinoma
e) Speak to sender and discuss at the multidisciplinary team meeting

32) Sections of a transverse colectomy specimen for colon cancer show that the tumour is seen to extend through the muscularis propria and into the peri-colonic fat. The serosa is not involved. Three local lymph nodes are infiltrated by carcinoma. What would be the appropriate TNM stage?

   a) pT2 N1
   b) pT2 N2
   c) pT3 N1
   d) pT3 N2
   e) pT4 N1

33) In the adenoma-carcinoma sequence of colorectal carcinogenesis, which gene is commonly the first to be altered?

   a) APC
   b) DCC
   c) KRAS
   d) SMAD4
   e) TP53

34) Mutations of which of the following genes have potential treatment implications in colorectal cancer currently?

   a) APC
   b) PMS2
   c) KRAS
   d) MSH2
   e) TP53

35) The Haggitt staging system is used for which of the following cancers?

   a) All colorectal cancers
   b) Lymph node positive cancers
   c) Metastatic cancers
   d) Sessile polyp cancers
   e) Stalked polyp cancers

36) Colorectal carcinoma associated with microsatellite instability pathway is associated with which of the following features?

   a) Hundreds of adenomatous polyps in small and large intestine
   b) Involves about 50% of sporadic cancers
   c) Right sided colon tumours
   d) Mutations on the APC gene

37) An adolescent boy presents with abdominal pain. On CT, he was found to have a large mass in small bowel. The resection specimen shows a tumour composed of a monotonous sheet of medium sized

lymphoid cells with interspersed macrophages. The cells are positive for CD20, CD79a, bcl6 and CD10, and are negative for CD3, CD5, CD23 and cyclin D1. The Ki67 proliferative index is >95%.
Which of the following is the most likely diagnosis?

    a) Burkitt lymphoma
    b) Diffuse large B-cell lymphoma
    c) Enteropathy-associated T-cell lymphoma
    d) Follicular lymphoma
    e) MALT-type marginal zone lymphoma

38) The commonest primary lymphoma of the gastrointestinal tract in the Western countries is:

    a) Burkitt lymphoma
    b) Diffuse large B-cell lymphoma
    c) Enteropathy-associated T-cell lymphoma
    d) Follicular lymphoma
    e) MALT-type marginal zone lymphoma

39) Which of the following positive immunostains are diagnostic for neuroendocrine tumour in the small bowel?

    a) CD117 and DOG1
    b) CD3 and CD5
    c) Chromogranin and CD56
    d) CK7 and CK20
    e) S100 and MelanA

40) Carcinoid syndrome is associated with which hormone expression?

    a) Gastrin
    b) Histamine
    c) Insulin
    d) Serotonin
    e) Vasoactive intestinal peptide

41) A 56 year old man presents with acute appendicitis. Sections of the appendicectomy specimen reveal infiltrative small nests of cells forming glands. With some cells showing a signet ring morphology. The cells are positive for AE1/3, synaptophysin, NSE, chromogranin and CD56. Which of the following is the diagnosis?

    a) Goblet cell carcinoid
    b) Mucinous cystadenocarcinoma
    c) Mucinous cystadenoma
    d) Mucocoele

42) Anal gland carcinomas are associated with which immunoprofile?

    a) CK7+/CK20-
    b) CK7+/CK20+
    c) CK7-/CK20+

d) CK7-/CK20-

43) A young woman presents with numerous polypoid lesions around the anal margin. Histologically, they are composed of hyperplastic squamous epithelium with a papillomatous growth pattern. There is parakeratosis, hypergranulosis and occasional cells with pyknotic nuclei and perinuclear clearing. Which is the correct diagnosis?

   a) Anal fibroepithelial polyps
   b) Condylomata acuminatum
   c) Haemorrhoids
   d) Rectal prolapse
   e) Squamous cell carcinoma

44) A 49 year old woman presents with nausea and vomiting. On endoscopy, she had a submucosal lesion found in the jejunum. This was excised and shows a lesion as shown below. The lesion is positive for DOG1, CD117 and CD34, and focally for SMA. Which of the following is the diagnosis?

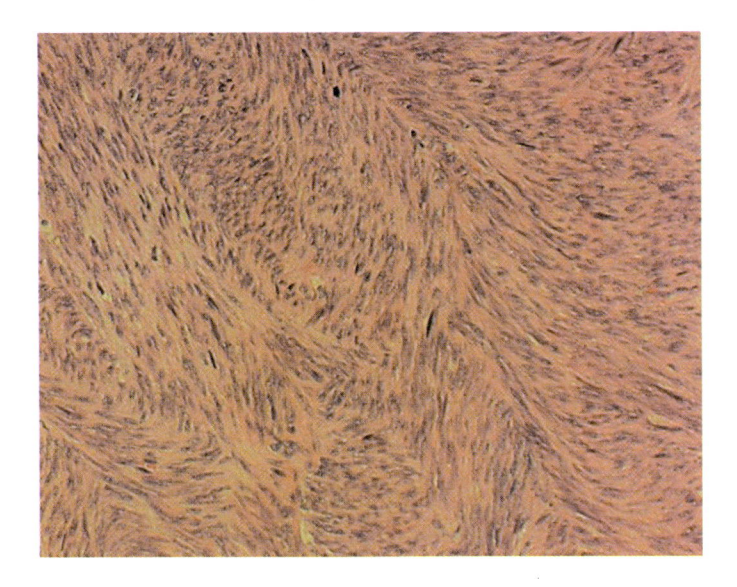

   a) Gastro-intestinal stromal tumour
   b) Inflammatory fibroid polyp
   c) Leiomyoma
   d) Schwannoma
   e) Spindle cell melanoma

45) A 36 year old man presents with lethargy and weight loss. On endoscopy, no obvious lesions were identified. Biopsies of the duodenum were taken, and the histology is shown below. Which of the following is the most likely diagnosis?

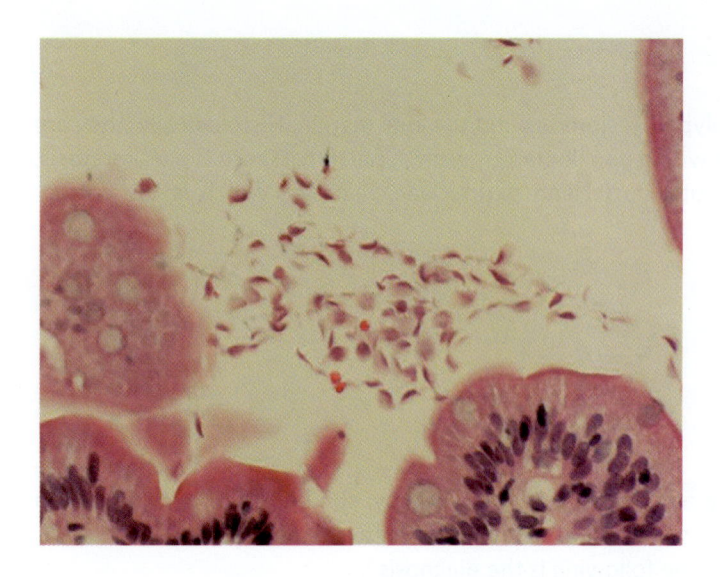

a) Acute bulbitis
b) Coeliac disease
c) Cryptosporidiasis
d) Giardiasis
e) Whipple's disease

Correct answers and explanations:

1) Correct answer: b) Aspiration pneumonia.

Achalasia is a functional abnormality of the oesophagus characterised by the contraction of the lower oesophageal sphincter and aperistalsis, resulting in dilatation of the oesophagus. It is caused by degeneration of the nerves fibres or roots controlling the oesophagus, including infiltrative processes. Causes therefore include amyloidosis, sarcoidosis and malignancy, as well as denervation caused by diabetes mellitus. Achalasia itself causes reflux symptoms and aspiration pneumonia.

2) Correct answer: b) Iron.

Plummer-Vinson syndrome is caused by iron deficiency, and results in glossitis, mouth ulcers and oesophageal strictures.
Potassium deficiency causes nausea, constipation, neurological and muscular symptoms, particularly cardiac arrhythmias.
Calcium and magnesium deficiency causes neurological, muscular and cardiac abnormalities.
Zinc deficiency causes a distinctive rash (acrodermatitis enteropathica) as well as diarrhoea and mental and immune function abnormalities.

3) Correct answer: a) Consistent with Barrett's oesophagus.

Barrett's oesophagus is the presence of columnar metaplasia in native oesophagus, which is confirmed by the presence of oesophageal gland ducts underlying squamous epithelium. The presence alone of glandular epithelium is insufficient for the diagnosis, as this can be seen in a gastro-oesophageal junction biopsy.
Previously, the presence of intestinal metaplasia was necessary for the diagnosis, but this in now not a requisite in the UK.
Reflux oesophagitis is inflammation, with basal cell hyperplasia and reactive changes in squamous epithelium, which is often a precursor to Barrett's oesophagus.
A hiatus hernia is a radiological or endoscopic diagnosis of stomach herniating above the diaphragm.
In adenocarcinomas, the malignant glands are infiltrative with the glandular cells displaying obvious cytological atypia.

4) Correct answer: b) HPV infection.

Barrett's oesophagus is associated with chronic reflux, which is in turn caused by a variety of conditions, including the presence of hiatus hernia, obesity, alcohol excess and smoking. It is commoner in men.
HPV infections are postulated to be associated with oesophageal squamous cell carcinoma.

5) Correct answer: d) Granular cell tumour.

Granular cell tumour is an often benign tumour affecting skin and mucosa, particularly in the head and neck region. It is likely of neural origin and is S100 positive. The granules in the cytoplasm of these cells are positive for DPAS.
While melanomas and schwannomas are positive for S100, they are negative for DPAS. Melanoma cells are often atypical and pleomorphic, but can be deceptively bland.
Adenocarcinomas and carcinoid tumours, while morphologically similar to granular cell tumours, are negative for S100.

6) Correct answer: d) Helicobacter pylori infection.

The features of mixed inflammation with lymphoid follicle formation is characteristic of H pylori associated gastritis.
Alcohol abuse usually results in acute gastritis and erosions, as does trauma, shock and sepsis.
Autoimmune or atrophic gastritis are due to anti-intrinsic factor autoantibodies and results in atrophic glands.
Crohn's disease can have similar picture to H pylori infection but is rarely manifest in the stomach.

7) Correct answer: d) Omeprazole.

Omeprazole is a proton pump inhibitor. It is used for the treatment of hyperacidity in the stomach. Its use results in the increased stimulation of histamine which results in hyperplasia of the gastric glands. Prolonged use can result in fundic gland polyps, which have dilated gastric glands.

8) Correct answer: b) DOG1 and CD117.

CD117 and DOG1, as well as CD34 positivity are diagnostic for a gastrointestinal stromal tumour.
AE1/3 and CAM5.2 cytokeratin stains are positive in carcinomas.
SMA and h-caldesmon staining are for smooth muscle tumours, such as leiomyomas and leiomyosarcomas (but can be positive in some GISTs).
S100 and MelanA positivity indicate a melanoma (or less likely clear cell sarcoma).

Neuroendocrine tumours are positive for chromogranin, CD56, synaptophysin and NSE, as well as having a dot-like cytoplasmic staining for CK8/18.

9) Correct answer: a) Exon 11 of c-kit gene.

Imatinib is a tyrosine kinase inhibitor, which targets the c-kit protein. Mutations involving exon 11 of the c-kit gene is associated with the most sensitivity to imatinib in gastro-intestinal stromal tumours.
In contrast, mutations involving exon 9 or 12 are less responsive to imatinib, while mutations involving PDGFR or wild-type GIST are usually resistant to the drug.

10) Correct answer: b) Necrosis.

Miettenin et al published a large series of GIST, which showed that GIST can be risk stratified according to the site (with rectal and small intestinal tumour having high risk of progression compared with gastric lesions), size (grouped into <2cm, 2-5cm, 5-10cm and >10cm), and mitotic activity (=<5 mitosis/high power fields or >5/hpf). This has been used internationally (see explanation for question 44).
The presence of metastasis usually confers inoperable status and much poorer prognosis.
The presence of tumour necrosis is not a recognised prognostic factor.

11) Correct answer : b) Coeliac disease.

The histological diagnosis is that of eosinophilic gastritis. The presence of eosinophils may be associated with parasites or an allergic reaction, with subsequent raised serum IgE levels. Although this is a rarely reported condition, it is known to be associated with coeliac disease.

12) Correct answer: D) Whipple's disease.

Whipple's disease is caused by the Tropheryma whippelii, a Gram-positive bacteria, which infects the bowel, joints and central nervous system. In the bowel, it infects the macrophages which accumulate in the lamina propria of the villi. It results in malabsorption and diarrhoea, as well as neurological and joint symptoms.

13) Correct answer : A) Acute bulbitis.

Acute bulbitis is acute inflammation affecting the first part of the duodenum. It is usually associated with hyperacidity in the stomach which affects the proximal duodenum, resulting in gastric metaplasia.

14) Correct answer: C) Giardiasis.

Giardia lamblia is a pear or kite shaped binucleate organism which attaches and infects the villous surface of the small bowel, resulting in malabsorption syndromes and diarrhoea. The villous architecture may be preserved, or there may be flattening of the villi. Infections are common and spread via contaminated water or food, with outbreaks seen in institutions such as hospitals.

15) Correct answer: B) Adenoma.

Small bowel adenomas are rare, but are morphologically similar to the large bowel counterparts. They are most commonly seen in the duodenum, around the ampulla of Vater.

16) Correct answer: e) All of the above.

All the above viral organisms are known to cause diarrhoea. In particular, Rotavirus and Calicivirus (such as the Norwalk-like virus) cause seasonal outbreaks.
Coxsackievirus cause a variety of diseases including Hand-Foot-Mouth disease.
Paramyxovirus causes mumps, which can have diarrhoea as a symptom.

17) Correct answer: a) Collagenous colitis.

Collagenous colitis is an idiopathic condition affecting usually middle aged women, who develop watery diarrhoea. The endoscopic findings are normal, and biopsy findings are as described, with a thickened band of collagen in the lamina propria. It is associated with autoimmune conditions, including coeliac disease and thyroiditis.

18) Correct answer: b) Diversion colitis.

Diversion colitis is an inflammatory condition involving patients post-bowel surgery who have a rectal stump or pouch. Due to the lack of short chain fatty acids from the normal bowel contents, the enterocytes develop nutrient deprivation, and develop features as described.

19) Correct answer: d) Warthin-Starry.

The morphological features and clinical setting is most in keeping with spirochaetosis, in which spirochaetes attach to and infect the brush border of the enterocytes in immunosuppressed patients. The organisms are detected by Warthin-Starry stain.
ZN stain is for Mycobacterium species, D-PAS and Grocott stains are for fungal organisms and Gram stain is for bacterial organisms.

20) Correct answer: c) Myobacterium Avium.

In an immunosuppressed patient, the intestinal tract is susceptible to a variety of infectious agents. The ZN stain is specific for Mycobacterium species. Although Mycobacterium Tuberculosis is a possibility, this is rarer in the GI tract, and Mycobacterium Avium Intracellulare is a more likely agent in this clinical setting.

21) Correct answer: c) RET.

Hirschsprung's disease is a rare congenital disease affecting 1:5000 neonates. It is caused by mutations in the RET gene, which results in the arrest of migration of neural crest cells, resulting in lack of ganglion cells in the rectum extending proximally over a variable length. There is secondary hypertrophy of the extrinsic nerve fibres of the aganglionic segment of Hirschsprung's disease, which shows aberrant staining for acetylcholine esterase. Patients present with obstructive constipation.

22) Correct answer: a) Anti-endomysium antibody.

Coeliac disease is an allergic or hypersensitivity condition to gluten in the diet. Patients develop antibodies to endomysium, which are the reticulum of connective tissue, as well as to peptides found in gluten, such as gliadin. Previously anti-gliadin antibodies were tested for coeliac disease, but more recently anti-endomysium antibodies were shown to be more accurate.

23) Correct answer: E) Ulcerative colitis.

Ulcerative colitis is one of the inflammatory bowel diseases. It involves the rectum and extends proximally, affecting the small bowel rarely. The inflammation is usually limited to the submucosa, but there is extensive ulceration, crypt architecture distortion, as well as more severe crypt inflammation and distortion. The features of crypt architecture abnormalities and mucin depletion help distinguish chronic inflammatory bowel disease from acute colitis, such as infectious causes.

24) Correct answer: B) Ischaemic colitis.

Ischaemic colitis occurs when the blood vessels (arterial or venous) supplying a segment of the bowel are occluded by thrombus or emboli. Due to the non-anastomosing end arteries, there is segmental infarction of a part of the bowel with necrosis, ulceration and subsequent haemorrhage.

25) Correct answer: D) Pseudomembranous colitis.

Pseudomembranous colitis results from the usage of broad spectrum antibiotics, in particular cephalosporins, which reduce the population of the normal gut flora. This results in the overgrowth of Clostridium difficile, an obligate anaerobe, which causes this infection, and which is characterised clinically by severe diarrhoea. Histologically, there is plaque-like fibrinous necrotic material adherent to ulcerated mucosa with a mushroom-like architecture, the so-called pseudomembrane.

26) Correct answer: A) Diverticular disease.

Diverticular disease is the presence of one or more out-pouchings of the bowel mucosa through the muscle wall. It can be congenital or acquired, the latter is due to increased intraluminal pressure often as a result of constipation. It is most common in middle aged to elderly population. Secondary inflammation or perforation leads to clinically symptomatic diverticulitis.

27) Correct answer: C) Necrotising enterocolitis.

Necrotising enterocolitis commonly affects neonates especially if preterm. It is of uncertain aetiology but often occurs after oral feeding has commenced. There is a necrotising inflammation of the small and large bowel with bacterial colonisation, transmural necrosis and gas formation.

28) Correct answer: e) Involvement of the rectum.
29) Correct answer: b) Granulomas.

Explanations for 28 & 29: Crohn's disease and ulcerative colitis are both inflammatory bowel diseases which have some similarities but are often clinically and pathologically distinct.
Crohn's disease can involve the whole GI tract (including anus and mouth) but with skip lesions. It often affects the entire thickness of the bowel wall with granulomas and lymphoid follicle formation. Grossly, it leads to "hosepipe" thickening of the bowel, with a cobblestone appearance of the mucosa.
Ulcerative colitis invariably involves the rectum and extends proximally, affecting the small bowel rarely. The inflammation is usually limited to the submucosa but there is extensive ulceration, as well as more severe crypt inflammation and distortion. Granulomas are not a feature though granulomas associated with crypt rupture can be present.

30) Correct answer: b) Juvenile polyp.

Juvenile polyps are hamartomatous polyps, which are seen in Cowden syndrome and Cronkhite-Canada syndrome.

Peutz-Jaegher polyp is also a hamartomatous polyp which is associated with Peutz-Jaegher syndrome.
Inflammatory pseudopolyps are associated with inflammatory bowel disease.
Tubulo-villous adenomas are often sporadic but may be associated with familial adenomatous polyposis.
Submucosal ganglioneuroma are associated with neurofibromatosis.

31) Correct answer: a) Cut levels and check against other specimens of the day.

The histological findings should raise the suspicion that there is a carry-over artefact. The first action should be to investigate this possibility by checking the specimen or performing levels to determine the plane of the malignant fragment.
Reporting the biopsy without further investigative work would be negligent.
While performing immunohistochemical tests or molecular tests or further discussion with clinical team might be necessary, this would not be the first course of action.

32) Correct answer: c) pT3 N1.

This extension of the tumour into subserosal fat without serosal involvement is pT3. Involvement by 1-3 lymph nodes is pTN1. Reporting colonic resection specimens for colorectal cancer should include UICC/AJCC/TNM staging information, which would guide prognostication and adjuvant treatment options.

Table showing brief overview of the TNM staging for colorectal cancers.

Stage	Level of involvement
Tumour	
T1	Limited to mucosa and submucosa
T2	Extension into (but not through) muscularis propria
T3	Invasion through muscularis propria into subserosal fat
T4	Direct invasion of other organs or perforates visceral peritoneum
Nodes	
N0	No lymph nodes involved
N1	1-3 lymph nodes involved
N2	4 or more lymph nodes involved
Metastasis	
M0	No metastasis
M1	Distant metastasis

33) Correct answer: a) APC.

The adenoma-carcinoma sequence is a model of colorectal carcinogenesis postulated by Fearon and Vogelstein, in which the majority of colorectal cancers accumulate genetic abnormalities in a stepwise sequence, as they progress from adenomas with low grade dysplasia through high grade dysplasia and finally invasive malignancy. The sequence is APC, KRAS, SMAD and TP53 followed by others. This has been proven by studies, although alternative pathways also exist, such as the microsatellite instability pathway involving mismatch repair genes.

34) Correct answer: c) KRAS.

All 5 genes are implicated in the pathogenesis of colorectal cancers (APC, KRAS and TP53 by the adenoma-carcinoma sequence, and MLH1, PMS2, MSH2 and MSH6 via the microsatellite instability pathway). KRAS is involved in the EGFR pathway of cancer growth. Currently, only KRAS has implications for determining targeted therapy sensitivity. Mutation of the KRAS gene has been shown to confer resistance to anti-EGFR antibody therapy in colorectal cancers.

35) Correct answer: e) Stalked polyp cancers.

The Haggitt staging system is used for stalked polyp cancers limited to the submucosa.
The Kikuchi staging system is used for sessile polyp cancers limited to the submucosa.
All colorectal cancers can be staged with the TNM/AJCC system. The Duke's or Astor-Collier systems are older systems which can also be used on all colorectal cancers.

36) Correct answer: c) Right-sided colon tumours.

The microsatellite instability pathway is an alternative pathway for the pathogenesis of colorectal cancers. It is caused by mutation of one of the 4 mismatch repair genes (MSH2, MSH6, PMS2 and MLH1). It is involved in 10-15% of sporadic cancers, and in the Hereditary Non-polyposis Colorectal Cancer (Lynch) syndrome. It is associated with right sided hyperplastic polyps and large right sided tumours, but not large numbers of adenomas. Apart from colorectal cancers, patients with the Lynch syndrome also have other cancers, including in the endometrium.

37) Correct answer: a) Burkitt lymphoma.

Enteropathy-associated T-cell lymphoma is a T-cell lymphoma associated with coeliac disease, which is expected to be positive for CD3 and CD5.
Although the rest are B cell lymphomas which are CD20 and CD79a positive, MALT lymphoma does not have a germinal centre immunophenotype (bcl6 and CD10). Follicular lymphoma has the immunoprofile as described, but is a low grade lymphoma, with a low proliferation index and a mixture of centrocytes and centroblasts. The distinction between a diffuse large B-cell lymphoma and Burkitt lymphoma can be difficult, but a Burkitt lymphoma is favoured due to the starry-sky morphology, the size of the lymphoid cells and the higher Ki67 index. If necessary, molecular studies for c-myc translocation would confirm the diagnosis.

38) Correct answer: e) MALT-type marginal zone lymphoma.

In the Western world, the commonest primary gastrointestinal lymphoma is MALT-type lymphoma.

39) Correct answer: c) Chromogranin and CD56.

Neuroendocrine tumours are positive for chromogranin, CD56, synaptophysin and NSE, as well as having a dot-like cytoplasmic staining for CK8/18.
S100 and MelanA positivity indicate a melanoma (or less likely clear cell sarcoma).
CD3 and CD5 stains are for T lymphocytes.
CD117 and DOG1 positivity are diagnostic for a gastrointestinal stromal tumour.
CK7 and CK20 cytokeratin stains are positive in carcinomas, and the differential positivity is useful for distinguishing tumours from different organs.

40) Correct answer: d) Serotonin.
Neuroendocrine tumours from the gastrointestinal tract may secrete a number of hormones. However, some (previously termed carcinoid tumours) particularly from the small bowel secrete serotonin, which is usually metabolised in the liver.

Metastatic tumours in the liver however release serotonin into the systemic circulation, producing symptoms of flushing, diarrhoea, wheezing and breathlessness, termed the Carcinoid syndrome.

41) Correct answer: a) Goblet cell carcinoid.

The goblet cell carcinoid, also known as crypt cell carcinoma and neuroendocrine tumour with goblet cell differentiation, is a gastrointestinal tumour that is most commonly found in the appendix. It consists of a neuroendocrine component and a conventional carcinoma, thought to arise from Paneth cells. It is therefore positive immunohistochemically for both neuroendocrine and cytokeratin markers.

42) Correct answer: a) CK7+/CK20-.

Anal gland carcinomas are morphologically and immunohistochemically different from rectal carcinomas. They are typically CK7+/CK20-, compared with colorectal cancers which are CK7-/CK20+.
Other cancers which are CK7+/CK20- include breast, lung, oesophageal, gastric, pancreatic and some gynaecological carcinomas.

43) Correct answer: b) Condylomata acuminatum.

The description fits a genital viral wart (Condylomata acuminatum) caused by the human papilloma virus (HPV).
Fibroepithelial polyps and haemorrhoids are usually non-hyperplastic and do not have koilocytosis (perinuclear clearing due to viral inclusions). Haemorrhoids in addition have numerous dilated blood vessels in the connective tissue.
Rectal prolapses are lined by hyperplastic columnar epithelium.
Squamous cell carcinoma is an invasive malignant tumour, which may also be caused by HPV.

44) Correct answer: a) Gastrointestinal stromal tumour.

Gastrointestinal stromal tumour is a tumour thought to arise from the interstitial cells of Cajal, and can develop along the gastrointestinal tract, with stomach being the commonest site. These tumours express DOG-1, CD117 and CD34, as well as having variably smooth muscle differentiation. They can behave in a variable fashion, which can be predicted based on several factors, such as mitotic activity, tumour size and site.

Table showing risk of aggressive behaviour in GIST (from Fletcher et al, Human Pathology. 2002; 33(5):459-65).

Risk of aggressive behaviour	Size of tumour	Mitotic Count
Very low risk	<2cm	<5/50hpf
Low risk	2-5cm	<5/50hpf
Intermediate risk	<5cm	6-10/50hpf
Intermediate risk	5-10cm	<5/50hpf
High risk	>5cm	>5/50hpf
High risk	>10cm	Any mitotic rate

45) Correct answer: d) Giardiasis.

See answer to question 14.

CHAPTER 6

HEPATOBILIARY & PANCREATIC PATHOLOGY
Dr Jayson Wang

Extended matching questions for questions 1 to 6: Match the following descriptions with the appropriate diseases of the liver:

- A) Autoimmune hepatitis
- B) Haemochromatosis
- C) Hepatitis B infection
- D) Alpha-1 antitrypsin deficiency
- E) Primary biliary cirrhosis
- F) Primary sclerosing cholangitis

1) Female of East Asian origin with chronic hepatitis, showing presence of "ground-glass" hepatocytes.

2) Middle aged man with deposition of Perl's stain positive particles within hepatocytes.

3) Young male with cytoplasmic inclusions which are D-PAS positive in hepatocytes.

4) Middle aged woman with jaundice. Liver biopsy shows concentric layers of fibrosis within portal tracts.

5) Young woman with prominence of plasma cells and lymphocytes in chronic hepatitis.

6) Middle aged woman, with jaundice, presence of granulomas in the portal tracts.

7) A woman undergoes a anterior resection for a colonic adenocarcinoma. On laparotomy, a small nodule was noted on the surface of the liver. The nodule was excised and sent for frozen section diagnosis. What is the most likely diagnosis?

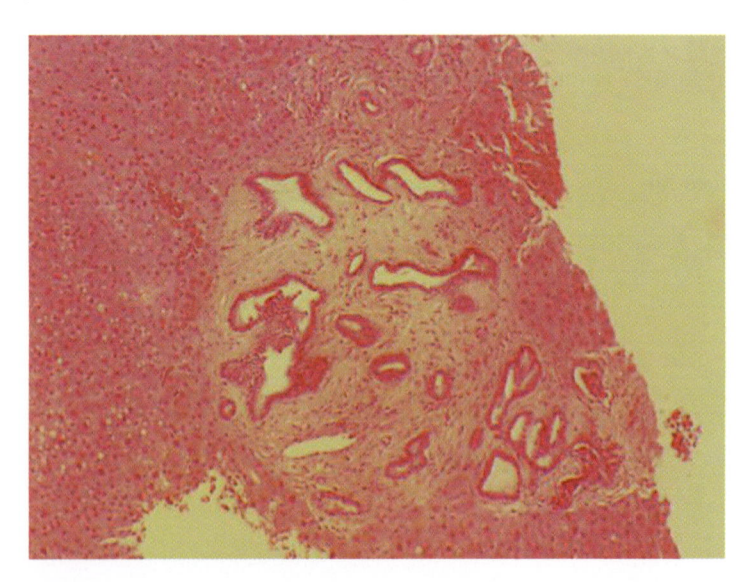

a) Cholangiocarcinoma
   b) Hepatocellular carcinoma
   c) Bile duct hamartoma
   d) Metastatic carcinoma from colon primary
   e) Primary sclerosing cholangitis

8) All of the following drugs can cause liver cirrhosis, except:

   a) Methotrexate
   b) Amiodarone
   c) Phenytoin
   d) Isoniazid
   e) All of the above can cause cirrhosis

9) A middle aged man presents with jaundice and encephalopathy. A liver biopsy shows the presence of small and large lipid droplets as well as eosinophilic inclusions in the cytoplasm of the hepatocytes. There are portal and parenchymal spotty neutrophilic infiltrates, as well as fibrous bridges between portal tracts. Which of the following is the most common aetiology?

   a) Alcohol
   b) Wilson's disease
   c) Aflatoxin ingestion
   d) Primary biliary cirrhosis
   e) Chronic autoimmune hepatitis

10) All the following are associated with non-alcoholic steatohepatitis, except:

   a) Diabetes mellitus
   b) Obesity
   c) Hypertension
   d) Paracetamol overdose
   e) Hypercholesterolaemia

11) Which of the following is a DNA virus?

   a) Hepatitis A virus
   b) Hepatitis B virus
   c) Hepatitis C virus
   d) Hepatitis D virus
   e) Hepatitis E virus

12) Haemochromatosis is caused by a defect in which gene?

   a) ATP17B
   b) HFE
   c) HLA-DR
   d) NF1
   e) G6PD

13) Which of the following organisms is/are common cause(s) of cholangiocarcinoma in parts of Asia?

a) Trypanosoma cruzi
b) Strongyloides stercoralis
c) Echinococcus granulosus
d) Schistosomiasis japonicum
e) Opisthorchis sinensis

14) A young woman presents with a mass in the liver. Macroscopically, there is a well-circumscribed brown lesion with a central pale stellate area of fibrosis. The rest of the liver is unremarkable. Microscopically, the lesion consists of hepatocytes with a thickened plate architecture. Which of the following is the most likely diagnosis?

a) Liver adenoma
b) Hepatocellular carcinoma
c) Focal nodular hyperplasia
d) Macronodular cirrhosis
e) Hepatoblastoma

15) Which of the following is associated with liver adenomas?

a) Male sex
b) Oral contraceptive use
c) Arterio-venous malformation
d) Cirrhosis
e) Alcohol excess

16) The commonest cause of hepatocellular carcinoma world-wide is :

a) Alcohol
b) Aflatoxin
c) Hepatitis B
d) Hepatitis C
e) Haemochromatosis

17) A 27 year old woman with no previous history of liver cirrhosis presents with a liver mass. The tumour was resected and shows a relatively circumscribed but unencapsulated tumour, with trabeculae of tumour cells and intervening layers of fibrosis. The cells have mild to moderate atypia and have moderate amounts of eosinophilic cytoplasm. What is the most likely diagnosis?

a) Hepatic adenoma
b) Nodular regenerative hyperplasia
c) Hepatoblastoma
d) Hepatocellular carcinoma, NOS
e) Fibrolamellar carcinoma

18) Which of the following positive immunostains favours the diagnosis of cholangiocarcinoma over hepatocellular carcinoma?

a) HepPar1
b) pCEA

c) AFP
   d) CK19
   e) CAM5.2

19) A 60 year old male undergoes a partial hepatectomy for liver metastasis. The histology shows an adenocarcinoma. Which of the following immunoprofiles would confirm a colorectal primary:

   a) CK7+/CK20-/CK19+
   b) CK7-/CK20-/AFP+
   c) CK7-/CK20-/PAX8+
   d) CK7-/CK20+/CDX2+
   e) CK7+/CK20-/TTF1+

20) An ill-defined tumour is present within the liver of a middle-aged male. The tumour is haemorrhagic and composed of highly pleomorphic spindled cells and numerous mitotic figures. The tumour cells are positive for ERG, CD34 and CD31. Which is the most likely diagnosis?

   a) Angiosarcoma
   b) Gastrointestinal stromal tumour
   c) Angiomyolipoma
   d) Solitary fibrous tumour
   e) Sarcomatoid carcinoma

21) A 3 month old baby presents with jaundice. A liver biopsy shows bile lakes with expanded oedematous portal tracts containing a proliferation of bile ductules. There are scattered giant hepatocytes. How frequent is the underlying condition?

   a) 1:10
   b) 1:100
   c) 1:1000
   d) 1:10000
   e) 1:100000

22) Which of the following are commonly associated with gallbladder carcinoma?

   a) Male sex
   b) Young adults
   c) Gallstones
   d) A good prognosis

23) A 55 year old man undergoes a subtotal pancreatectomy. Which of the following features favours chronic pancreatitis rather than pancreatic cancer?

   a) Fibrosis
   b) Lobular architecture
   c) Proliferation of small glands
   d) Nuclear pleomorphism
   e) Destruction of islets of Langerhans

24) Which of the following is NOT a cause of chronic pancreatitis?

a) Alcohol excess
b) Cystic fibrosis
c) IgG4-associated sclerosing disease
d) Renal calculi
e) Pancreatic cancer

25) Which of the following pancreatic cystic lesions are commoner in men?

a) Serous cystadenoma
b) Mucinous cystadenoma
c) Intraductal papillary mucinous neoplasm
d) Solid pseudopapillary tumour
e) Congenital cyst

26) Which of the following favours a diagnosis of intraductal papillary mucinous neoplasm over mucinous cystadenoma?

a) Located at tail of pancreas
b) Communicates with pancreatic duct
c) Female patient
d) Presence of surrounding ER-positive stroma
e) No invasive component

27) Which of the following is most commonly secreted by pancreatic neuroendocrine tumours?

a) Somatostatin
b) Insulin
c) Glucagon
d) Gastrin
e) VIP

28) Multiple Endocrine Neoplasia (MEN)-1 is associated with which of the following neoplasms, apart from pancreatic neuroendocrine tumours?

a) Pheochromocytoma
b) Medullary carcinoma of thyroid
c) Prolactinoma
d) Ganglioneuroma
e) Adrenocortical carcinoma

29) In a pancreatic neuroendocrine carcinoma, all of the following are suggestive of malignancy, except:

a) Size >2cm
b) Ki 67 index >10%
c) Lymphovascular invasion
d) Insulin secretion
e) Tumour necrosis

Extended matching questions for questions 30 till 34 : Match the following descriptions with the appropriate diseases of the pancreas:

- A) Pseudocyst
- B) Neuroendocrine tumour
- C) Microcystic cystadenoma
- D) Solid pseudopapillary pancreatic tumour
- E) Intraductal papillary mucinous neoplasm

30) Young woman with a tumour composed of sheets of cells with vague papillary and cystic formations. The cells are positive for CD10 and beta-catenin .

31) Middle aged man with history of alcohol excess presents with a mass behind the head of the pancreas. The cyst has a fibrous wall with no epithelial lining and contains necrotic haemorrhagic material.

32) Elderly man with a dilated ampulla of Vater with mucin extruded from the opening.

33) Young man with previous parathyroid adenomas, with a mass in the body of the pancreas, composed of polygonal cells with coarse chromatin pattern and granular eosinophilic nuclei.

34) Young woman with known VHL gene mutation, incidental finding of a mass in the tail of the pancreas, composed of aggregates of cysts lined by cuboidal cells with clear cytoplasm.

35) A 28 year old asymptomatic man has a liver biopsy performed. It shows mild portal and lobular chronic inflammation. The haematoxylin & eosin stain is shown below. The cytoplasm of some hepatocytes stain with Orcein.

What is the most likely diagnosis?

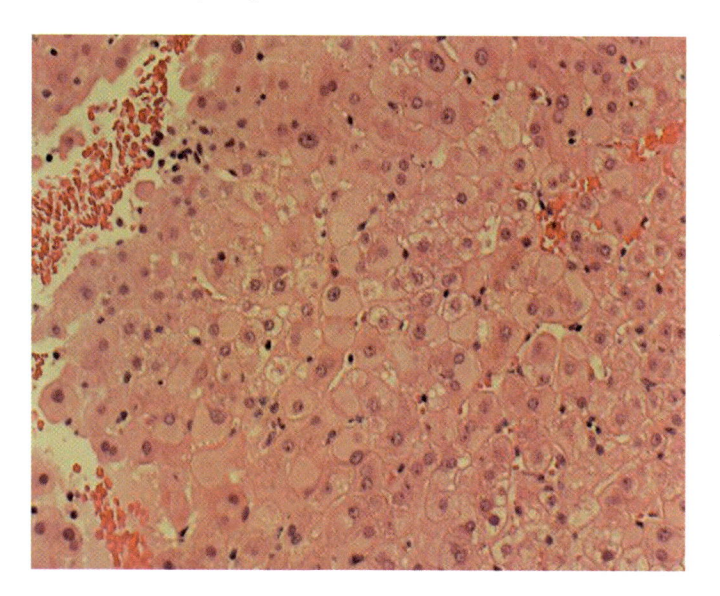

- a) Alcoholic liver disease
- b) Alpha-1 anti-trypsin deficiency
- c) Hepatitis B infection
- d) Hepatitis C infection
- e) Wilson's disease

36) A 72 year old woman presents with nausea and jaundice. An ultrasound scan showed a polyp in the gallbladder. No other lesions were identified on CT scan. A cholecystectomy was performed, and the histology of the polyp is shown below. Which is the most likely diagnosis?

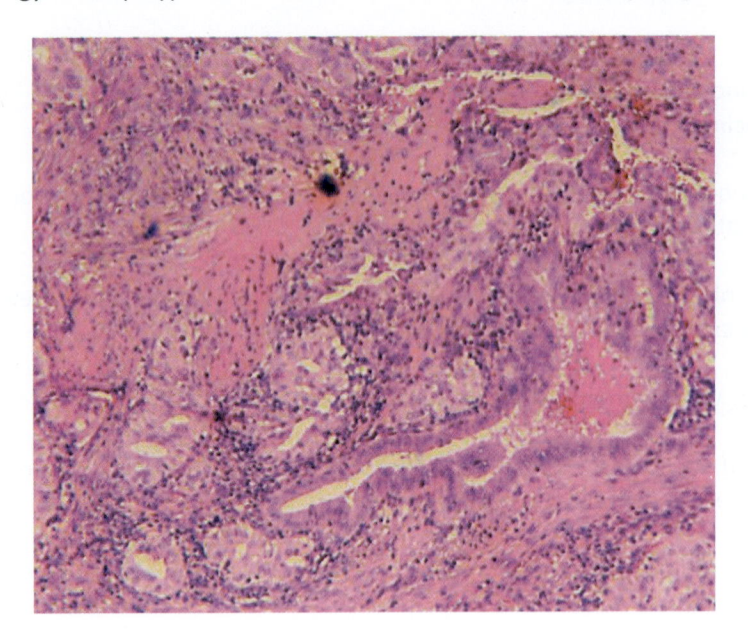

a) Adenoma of gallbladder
b) Adenocarcinoma of gallbladder
c) Follicular cholecystitis
d) Hepatocellular carcinoma
e) Metastatic gastric carcinoma

Correct answers and explanations:

1) Correct answer : C) Hepatitis B infection.

Hepatitis B chronic infection is common in East Asian countries. The viral organisms in hepatocytes give rise to a finely granular eosinophilic cytoplasm, termed "ground-glass" hepatocytes.

2) Correct answer : B) Haemochromatosis.

Perl's stain detects the presence of iron deposition. Iron deposits are seen secondary to many other hepatitic conditions as well as iron overload, but in the absence of these, a diagnosis of haemochromatosis should be considered.

3) Correct answer : D) Alpha-1 antitrypsin deficiency.

Alpha-1-antitrypsin deficiency is an autosomal recessive condition, marked by low serum levels of this enzyme inhibitor. The lack of this results in lung and liver disease in childhood and early adulthood. In this disease, abnormal antitrypsin protein is accumulated in the endoplasmic reticulum of cells including hepatocytes, which gives rise to red cytoplasmic granules with Diastase-Periodic acid Schiff staining.

4) Correct answer : F) Primary sclerosing cholangitis.

Primary sclerosing cirrhosis is a condition characterised by inflammation and fibrosis of the biliary tree, commonly associated with inflammatory bowel disease.
Histologically, there is concentric fibrosis around bile ducts with associated chronic inflammation.

5) Correct answer : A) Autoimmune hepatitis.

Autoimmune hepatitis is a chronic hepatitis associated with serum autoantibodies including antinuclear and anti-smooth muscle antibodies.
Histologically it gives rise to a chronic hepatitis picture but with prominent numbers of plasma cells.

6) Correct answer : E) Primary biliary cirrhosis.

Primary biliary cirrhosis is an autoimmune disease often characterised by the presence of serum anti-mitochondrial antibodies. It usually affects middle aged women, and shows portal tracts with mixed chronic inflammation, including non-necrotising granulomas. It often progresses to cirrhosis.

7) Correct answer : c) Bile duct hamartoma.

Bile duct hamartomas (von Meyenburg complexes) are small (<5mm), subcapsular, usually multiple nodules, often found incidentally at laparotomy. They are benign well-circumscribed lesions composed of small angulated bile ducts surrounded by loose fibrous stroma within portal tracts. The ductal cells are uniform and bland with no nuclear pleomorphism.
All these features help distinguish bile duct hamartomas from malignant lesions.

8) Correct answer : e) All of the above can cause cirrhosis.

The liver is the major organ for metabolising drugs, and is potentially susceptible to damage from a wide variety of medication and toxins. All the above medications are recognised to cause chronic hepatitis and subsequent cirrhosis.

9) Correct answer : a) Alcohol.

The histological picture is that of steatohepatitis with early cirrhosis. Worldwide, the commonest cause of this is chronic alcohol excess.
While a similar picture can arise from Wilson's disease or toxin ingestion, these are much rarer.
Primary biliary cirrhosis and autoimmune hepatitis are more likely to occur in women, not have fatty change, and show chronic rather than neutrophilic inflammation.

10) Correct answer : d) Paracetamol overdose.

Non-alcoholic steatohepatitis is a condition which histologically resembles alcoholic liver disease, but without a history of alcohol excess. It affects men and women equally, and is associated with obesity, hypercholesterolaemia, and insulin resistant states, including non-insulin dependent diabetes.

Paracetamol overdose results in an acute hepatitis picture with extensive necrosis.

11) Correct answer : b) Hepatitis B virus.

Hepatitis B is the only DNA virus, the others are all RNA viruses.

12) Correct answer : b) HFE.

Haemochromatosis is caused by a defect in the HFE gene located on chromosome 6. HLA-DR is also located on chromosome 6 and is a part of the major histocompatibility complex family of genes, which are implicated in a variety of autoimmune conditions including diabetes mellitus.
Defect in the ATP17B gene is responsible for Wilson's disease.
NF1 is the cause of neurofibromatosis type 1.
G6PD deficiency results in haemolytic anaemia and may lead to bile stones.

13) Correct answer : e) Opisthorchis sinensis.

Trypanosoma cruzi is a protozoa responsible for Chagas disease, resulting in myocardial infestation with myositis and arrhythmias.
Strongyloides stercoralis is a roundworm commonly named thread worm which infests the intestinal tract.
Echinococcus granulosus, called the hydatid worm or tapeworm also infests the intestinal tract, and also gives rise to hydatid cysts in the liver.
Schistosomiasis japonicum is a parasite found in East Asia which infests the intestinal tract, leading to intrahepatic portal vein obstruction and subsequent cirrhosis, portal hypertension and liver failure.
Opisthorchis sinensis is a liver fluke found in East Asian countries which infests the common bile duct. It is the commonest cause of cholangiocarcinoma in these countries.

14) Correct answer : c) Focal nodular hyperplasia.

Focal nodular hyperplasia is a benign lesion affecting young adults, more often in women. It is characterised by a central stellate scar macroscopically, which contains large blood vessels with fibrosis, chronic inflammation and bile duct proliferation. The non-scar area consists of normal hepatocytes with thickened plate architecture.

15) Correct answer : b) Oral contraceptive use.

Liver adenomas are associated with young women and with the use of oral contraceptives.
In contrast, arteriovenous malformations are associated with focal nodular hyperplasia, while cirrhosis and chronic hepatitis associated with alcohol excess are associated with hepatocellular carcinoma.

16) Correct answer : c) Hepatitis B.

Hepatocellular carcinoma is associated with three main aetiological agents : viral infections (Hepatitis B and C), alcohol excess and food contaminants, although chronic liver damage can also be associated with HCC pathogenesis.
The relative contribution to HCC varies in the world, with alcohol excess the main factor in the Western world, but Hepatitis B is postulated to be the commonest cause worldwide.

17) Correct answer : e) Fibrolamellar carcinoma.

Fibrolamellar hepatocellular carcinoma is a rare form of hepatocellular carcinoma which often affects young adults and is associated with a better prognosis.
Histologically, it is a relatively circumscribed tumour with trabeculae of tumour cells with intervening layers of fibrosis.
An adenoma will also have minimal pleomorphism, but have normal plate architecture, but no fibrosis of portal tracts.
Nodular regenerative hyperplasia has hypertrophied hepatocytes with adjacent atrophied cells but also normal plate architecture (with </= 3 cell plate thickness).
Hepatoblastoma usually occurs in early childhood and is composed of primitive cells in acini or tubules, some with mesenchymal differentiation.
Hepatocellular carcinoma, NOS usually occurs in older patients with established cirrhosis.

18) Correct answer : d) CK19.

Cholangiocarcinomas are adenocarcinomas arising from the biliary epithelium. It is often positive for CK19.
In contrast, hepatocellular carcinomas arise from hepatocytes and can be positive for HepPar1, AFP, CAM5.2 (CK8/18) and pCEA (the latter with a canalicular pattern).

19) Correct answer : d) CK7-/CK20+/CDX2+.

A colorectal cancer is CK7-/CK20+/CDX2+, with the pattern of CK7- and CK20+ rarely seen in other tumours apart from Merkel cell carcinomas. CDX2 is a nuclear protein involved in large bowel differentiation.
Tumours which are CK7-/CK20- include hepatocellular carcinomas (AFP+), renal cell carcinomas (PAX8+) and prostatic carcinomas (PSA+).
Tumours which are CK7+ and CK20- include the upper gastrointestinal tract or pancreatico-biliary tract (CK19+) and lung (TTF1+).

20) Correct answer : a) Angiosarcoma.

Angiosarcomas are recognised in the liver and are associated with previous exposure to Thorotrast (radiology contrast material), arsenic or vinyl chloride. The malignant endothelial cells are positive for ERG, CD34 and CD31.
GIST is positive for CD34, CD117 and DOG1, but not ERG and CD31.
The blood vessels component of angiomyolipoma can be positive for CD34, CD31 and ERG, but is usually benign with low mitotic count.
SFT is also usually benign and is positive for CD34, but negative for CD31 and ERG.
Sarcomatoid carcinomas can have heterologous differentiation, but angiosarcomatous differentiation is very rare.

21) Correct answer : d) 1:10000.

Primary biliary atresia is defined as a complete obstruction of the extrahepatic biliary tree within the first 3 months of life. It is a rare condition occurring in 1:10000 of live births. One-third of babies with neonatal cholestasis have this condition.
The cause is either in utero malformation or damage from possible viral infections.
A hereditary component has also been suggested.

22) Correct answer : c) Gallstones.

The majority of gallbladder carcinomas are related to previous history of chronic cholecystitis and gallstones. They frequently occur in elderly women and are often advanced and inoperable at the time of diagnosis. They therefore have a poor prognosis with a 5 year survival of <5%.

23) Correct answer : b) Lobular architecture.

The histological features of chronic pancreatitis can mimic pancreatic cancer, including fibrosis mimicking desmoplastic stroma of cancer and proliferation of small branching ducts which mimics infiltrative adenocarcinoma.
However, chronic pancreatitis retains a lobular architecture with the histological changes.
There is also islet cell hyperplasia in chronic pancreatitis.
Marked nuclear pleomorphism with crenated nuclei favours malignancy, as does an infiltrative destructive pattern of growth.

24) Correct answer : d) Renal calculi.

Chronic pancreatitis is caused by a variety of factors including alcohol excess, cystic fibrosis, IgG4-associated sclerosing disease, biliary calculi and pancreatic cancer.

25) Correct answer : c) Intraductal papillary mucinous neoplasm.

Intraductal papillary mucinous neoplasm is commoner in men, while all the others are more common in women.

26) Correct answer : b) Communicates with pancreatic duct.

Intraductal papillary mucinous neoplasms are more common in men, and often affect the head of the pancreas involving the main duct or one of its branches and are often associated with an invasive component

This, as compared to mucinous cystadenomas, affect women more frequently, are found in the body or tail, usually not communicating with a main duct, and have an ovarian type stroma surrounding the glands.

27) Correct answer : b) Insulin.

Insulinomas are the commonest type of neuroendocrine tumour found in the pancreas. It is often benign.

28) Correct answer : c) Prolactinoma.

MEN1 (Werner syndrome) is caused by mutations in the MEN1 oncogene, and is characterised by pituitary tumours, parathyroid hyperplasia and adenomas, adrenal cortical hyperplasia and pancreatic islet tumours.
MEN2 is caused by the RET oncogene, and is associated with medullary thyroid carcinomas and phaeochromocytomas, as well as parathyroid adenomas.

29) Correct answer : d) Insulin secretion.

The malignant potential of pancreatic neuroendocrine tumours is difficult to determine morphologically. However, studies have shown that the risk of malignancy (metastasis or recurrence) is associated with a larger size, infiltrative margins or lymphovascular invasion, high mitotic index and tumour necrosis.
Gastrinomas are also more likely to be malignant. In contrast, insulinomas are often benign.

30) Correct answer : D) Solid pseudopapillary pancreatic tumour.

Solid pseudopapillary pancreatic tumour is a rare tumour of low malignant potential usually affecting young women. The morphology is as described. Apart from CD10 and beta-catenin, they are also positive for alpha-1-antitrypsin, vimentin, CD56 and PAS.

31) Correct answer : A) Pseudocyst.

Pseudocyst is usually a sequelae of an attack of acute pancreatitis. They are usually on the surface of the gland and represent a walled off area of acute inflammation.

32) Correct answer : E) Intraductal papillary mucinous neoplasm.

Intraductal papillary mucinous tumour is a papillary tumour with mucin producing columnar epithelium involving the head of the pancreas. It often fills the main pancreatic duct which it dilates. There is frequently an invasive component with features of a mucinous carcinoma.

33) Correct answer : B) Neuroendocrine tumour.

Neuroendocrine tumours of the pancreas can be associated with several syndromes including MEN1 (patients may also have parathyroid or pituitary tumours) and Zollinger-Ellison syndrome. The neuroendocrine cells are as described.

34) Correct answer : C) Microcystic cystadenoma.

Microcystic cystadenoma is a form of serous cystadenoma, which is often an incidental finding in the tail of the pancreas. It occurs sporadically in elderly patients but is also seen in VHL families, where there is an association with renal cell carcinomas.

35) Correct answer : c) Hepatitis B infection.

Hepatitis B chronic infection is common in East Asian countries.
The Hepatitis B surface antigen can be detected using the Orcein stain.
Orcein stain also detects copper accumulation in Wilson's disease, which can also be indolent, but in view of the "ground-glass" appearance of the cytoplasm of some hepatocytes, which also suggests the presence of viral organisms in the cytoplasm, Hepatitis B infection is the more likely diagnosis.

36) Correct answer : b) Adenocarcinoma of gallbladder.

Gallbladder adenocarcinomas are commoner in women and may have a preceding history of chronic cholecystitis. Most adenocarcinomas of the gallbladder are invasive and infiltrate through the muscularis layers, but some form papillary adenocarcinomas.
These may be distinguished from adenomas by architectural complexity, invasion into the lamina propria and high grade nuclear atypia.
In the absence of other lesions, a primary tumour is more likely than metastatic disease from other sites.

CHAPTER 7

CARDIOVASCULAR PATHOLOGY
Dr Fiona Scott

1) Fibromuscular dysplasia (FMD) is a condition which causes non-atheromatous vascular stenosis. Which of the following statements is **not** true?

   a) The majority of patients with FMD are women between the ages of 40 and 60.
   b) The coronary arteries are the most commonly affected.
   c) FMD in men has a higher rate of complications such as dissection or aneurysms.
   d) FMD has 3 subtypes; intimal fibroplasia, medial fibroplasia and adventitial fibroplasia, with the medial subtype being most common.

2) A 79 year old man with a long history of smoking presents with increasing breathlessness, palpitations and fatigue. Echocardiography shows a large pericardial effusion. He undergoes percutaneous drainage of the effusion. Cytology of this shows lymphocytes, macrophages and mesothelial cells as well as a population of discohesive cells with high nuclear-cytoplasmic ratio and nuclear moulding. Immunohistochemistry carried out on a cell block preparation shows these cells stain positively with CD56, CK7 and TTF-1. Choose from the following options the most likely diagnosis.

   a) Malignant mesothelioma.
   b) Metastatic gastric adenocarcinoma.
   c) Metastatic small cell lung carcinoma of lung.
   d) Metastatic adenocarcinoma of lung.

3) Histology shows a section of left ventricular myocardium from an 83 year old man who had a long history of congestive cardiac failure. At autopsy, the left ventricular myocardium was thickened with a waxy cut surface. The histological diagnosis was amyloidosis. Which of the following options is the most likely type of amyloid protein in this case?

   a) Kappa light chain.
   b) AA protein.
   c) Beta-2 microglobulin.
   d) Transthyretin.

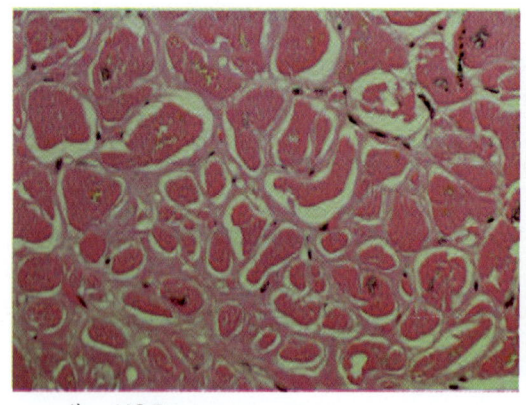

i) H&E

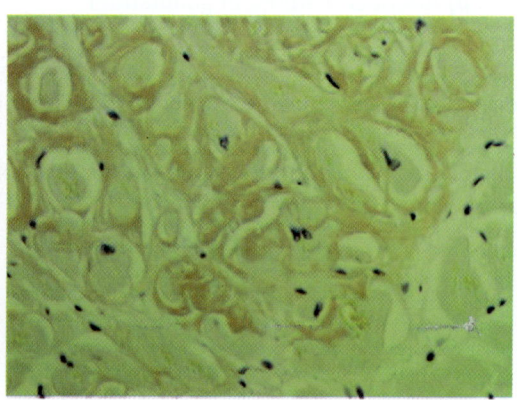

ii) Congo Red

4) A 25 year old female presents with a sudden collapse. The autopsy finding is of dissection of the thoracic aorta. Histology from the aorta shows small 'cystic' spaces in the tunica media and fragmentation of elastic tissue. What is the most likely diagnosis?

   a) Fibromuscular dysplasia.
   b) Marfan's Syndrome.
   c) Atherosclerosis.
   d) Polyarteritis nodosa.

5) Which of the following is true regarding Aschoff bodies?

   a) They are granulomatous foci characteristic of rheumatic heart disease.
   b) They are only found in the heart.
   c) They may only be seen microscopically..
   d) They occur in all three layers of the heart with the pericardium the least common site.

6) Which of the following immuno profiles would be the most characteristic of endocardial endothelial cells?

   a) CD34+ / CD31- / vWF+
   b) CD34+ / CD31+ / vWF-
   c) CD34+ / CD31+ / vWF+
   d) CD34- / CD31+ / vWF+

7) Which of the following is true regarding restrictive cardiomyopathy (RCM)?

   a) The commonest cause of RCM in the United Kingdom is amyloidosis.
   b) RCM is less common than dilated cardiomyopathy but more common than hypertrophic cardiomyopathy (HCM).
   c) Is defined by the WHO as 'a myocardial disease characterized by restrictive filling and reduced diastolic volume of either or both ventricles with abnormal systolic function and normal or near-normal wall thickness'.
   d) The differential diagnosis of RCM is a constrictive pericarditis and it is important to distinguish between the two, as constrictive pericarditis may be cured by surgical intervention.

8) A 35 year old male, heavy smoker, presents with a 2 year history of bilateral leg claudication with a new onset of rest pain. The left foot is found to be critically ischaemic and a below knee amputation is performed. Histology shows segmental thombosing inflammation of intermediate and small arteries as well as a superficial thrombophlebitis. Choose from the following options the most likely diagnosis.

   a) Buerger's disease.
   b) Atherosclerosis.
   c) Polyarteritis nodosa.
   d) Kawasaki disease.

9) Histology shows a section of temporal artery from a 76 year old female who presented with left sided headache. Which is the most likely diagnosis?

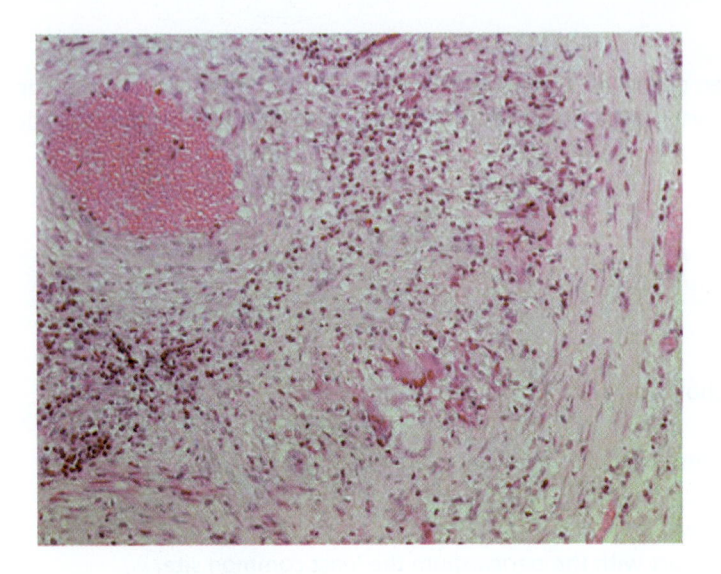

a) Atherosclerosis.
b) Hypertension.
c) Giant cell arteritis.
d) Takayasu's arteritis.

10) Which of the following facts regarding aortic dissection is **not** true?

a) Aortic dissection is uncommon in the presence of severe atherosclerosis.
b) Dissection of the aorta may extend retrogradely to involve the heart.
c) The dissection typically spreads along the plane between the intima and media.
d) Most aortic dissections occur in men with a history of hypertension.

11) A 50 year old man presents with rapidly progressive renal failure, shortness of breath, purulent nasal discharge and leg ulcers. A biopsy from a skin ulcer shows vascular changes including fibrinoid necrosis and a granulomatous inflammatory cell infiltrate including giant cells, eosinophils, neutrophils, lymphocytes and plasma cells. Serology shows the presence of cANCA. Choose from the following options the most likely diagnosis.

a) Chronic venous stasis with ulceration.
b) Leucocytoclastic vasculitis.
c) Sarcoidosis.
d) Wegener's granulomatosis.

12) Which one of the following histological features is **not** commonly seen in longstanding systemic hypertension?

a) Left ventricular myocyte enlargement with nuclear enlargement.
b) Reduplication of internal elastic lamina in arteries.
c) Fibrinoid necrosis.
d) Hyaline arteriolosclerosis.

13) Which of the following is **not** true of Anitschkow cells?

a) They are characteristically seen in rheumatic heart disease.

b) They are known as 'caterpillar' cells due to the shape of the cell.
c) They can be found in normal hearts.
d) They are of macrophage origin.

14) Endothelial cells provide a continuous lining of the cardiovascular system. Which of the following are produced by endothelial cells?

   a) Angiotensin-converting enzyme (ACE).
   b) Von Willebrand factor (VWF).
   c) Prostacyclin.
   d) Heparin-like molecules.
   e) All of the above.

15) Congenital heart disease is the commonest form of heart disease among children. Which of the following is the most frequent congenital malformation?

   a) Atrial septal defect.
   b) Ventricular septal defect.
   c) Coarctation of the aorta.
   d) Transposition of the great arteries.

16) Which of the following is **not** true of fatty streaks?

   a) Fatty streaks are composed of lipid-laden foamy macrophages within the intima.
   b) Can be seen in the aorta of children as young as 1 year old.
   c) All fatty streaks become atheromatous plaques.
   d) Fatty streaks are related to serum lipoprotein levels.

17) A 66 year old male overweight smoker presents with a history of chest pain on exertion, worse in cold weather. Coronary angiography shows severe stenosis of the left anterior descending coronary artery. What is the most likely cause?

   a) Kawasaki's disease.
   b) Myocarditis.
   c) Buerger's disease.
   d) Atherosclerosis.

18) A 23 year old man suffers a sudden collapse while playing football and cannot be revived. At autopsy, the left ventricular myocardium is found to be thickened and histology shows interstitial fibrosis and myocyte disarray. Choose from the following options the most likely diagnosis.

   a) Acute myocardial infarction.
   b) Hypertrophic obstructive cardiomyopathy (HCM).
   c) Viral myocarditis.
   d) Essential hypertension.

19) Which of the following statements is **not** true regarding the risk of congenital heart disease?

   a) It is increased by the use of ibuprofen in the first trimester of pregnancy.
   b) It is increased in Turner Syndrome.

c) It is increased with maternal diabetes.
d) It is increased with maternal flu infection in the second trimester.

20-22) Extended matching questions: Which of the following diagnoses best fits the description given?

A. Adriamycin-induced cardiotoxicity.
B. Radiation induced cardiotoxicity.
C. Cyclophosphamide cardiotoxicity.
D. Viral myocarditis.

20) This condition typically causes a haemorrhagic myocarditis with capillary microthrombi, interstitial fibrin deposition and myocyte necrosis.

21) This condition is a common cause of dilated cardiomyopathy.

22) In this condition, the earliest change seen histologically is vacuolation of cardiac myocytes followed by loss of cross-striations and myocytolysis. Inflammation is absent.

23) A baby is born at 36 weeks gestation and is immediately noted to be cyanosed by the midwife. Oxygen is administered but there is no improvement in the level of cyanosis. An urgent echocardiogram confirms the presence of a congenital heart defect. Which of the following is the most likely diagnosis?

a) Atrial septal defect.
b) Tetralogy of Fallot.
c) Ventricular septal defect.
d) Patent ductus arteriosus.

24) Which of the following statements regarding cardiac myxomas is **not** true?

a) Myxomas account for 50% of all primary cardiac tumours.
b) The peak incidence of cardiac myxoma is in men aged 20-40 years.
c) May show glandular differentiation histologically.
d) Immunohistochemical staining for calretinin is typically positive.

25-28) Extended matching questions: Which of the following best fits the histological appearances?

A. Infective endocarditis.
B. Rheumatic valvular disease.
C. Calcification of the mitral annulus.
D. Prolapsed mitral valve.
E. Systemic lupus erythematosus.

25) Histology of the valve shows myxoid degeneration.

26) Histology of the valve shows vegetations comprising fibrin, acute inflammatory cells, macrophages and bacterial colonies with relatively normal chordae tendineae.

27) Histology shows characteristic 'Libman-Sacks' vegetations comprising fibrin, histiocytes and lymphocytes on both the ventricular and atrial surface of the valve.

28) Histology shows calcium deposition along and beneath the mitral valve attachment with sparing of the anterior mitral valve leaflets.

29) Senile cardiac amyloidosis is a slowly progressive disease of the elderly. Which of the following statements regarding senile cardiac amyloidosis is **not** true?

   a) Predominantly affects elderly men.
   b) It is often missed because the symptoms are attributed to other more common forms of heart disease.
   c) It is most commonly the result of deposition of an abnormal protein called transthyretin.
   d) It is the most common cause of amyloid deposition in the heart.

30) A 25 year old man collapses while playing basketball and cannot be revived. During autopsy, the cut surface of the right ventricle appears yellow and histology shows fibro-fatty replacement of the right ventricular myocardium. Choose the most likely diagnosis from from the following options.

   a) Arrhythmogenic right ventricular dysplasia.
   b) Acute myocardial infarction.
   c) Viral myocarditis.
   d) Hypertrophic obstructive cardiomyopathy (HCM).

31) A 55 year old woman with beta thalassaemia is killed in a road traffic accident. The following histology (H&E stain) is taken from the left ventricular myocardium. What is the most likely diagnosis?

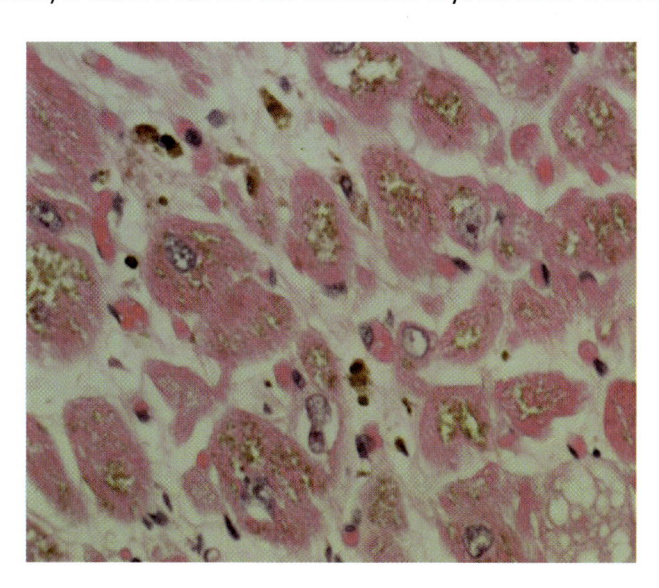

   a) Cardiac amyloidosis.
   b) Essential hypertension.
   c) Secondary haemochromatosis.
   d) Myocarditis.

32) Which of the following descriptions **best** fits with Takayasu's arteritis?

   a) This is a rare condition affecting predominantly young males aged 5 – 40 years from the Far East, Japan and Indian subcontinent.
   b) The condition affects large arteries mainly the aorta and its proximal large branches.

c) Histology initially shows an adventitial inflammatory infiltrate rich in eosinophils which progresses to medial fibrosis and intimal thickening.
d) Clinical presentation typically includes fever, malaise, Raynaud's phenomenon and limb claudication.

33) Which of the following are recognised cardiac complications of **both** type I and type II diabetes?

a) Congestive cardiac failure (CCF).
b) Coronary heart disease.
c) 'Silent' myocardial infarction.
d) Sudden cardiac death.
e) All of the above.

34) The figure below is a H&E stained section of a left ventricular myocardium taken from an autopsy of a 79 year old male smoker. Macroscopically, there was an area of yellow and tan softening with a hyperaemic rim in the free wall of the left ventricle.

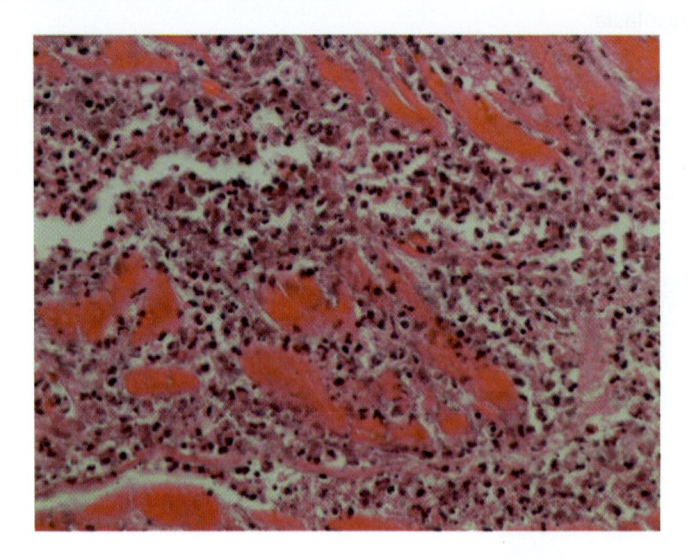

Choose the most likely diagnosis from the following options.

a) Acute myocardial infarction
b) Viral myocarditis
c) Sarcoidosis
d) Amyloidosis

35) Which of the following is the most common anatomical site for an atherosclerotic aneurysm?
a) Common iliac arteries.
b) Abdominal aorta above the level of the renal arteries.
c) Abdominal aorta below the level of the renal arteries.
d) Thoracic aorta.

36) Which of the following factors is **not** true regarding the UK screening programme for Abdominal Aortic Aneurysm (AAA)?

a) Screening involves an ultrasound scan that takes about 10 minutes.

b) Men aged 55 and over are eligible for screening.
c) Screening with subsequent repair reduces the mortality from aneurysm-related causes but has less effect on overall mortality from all causes.
d) The aim of the programme is to reduce the death rate from ruptured AAA in the screening population by up to 50%.

37) In the histological timeline of myocardial infarction, which of the following is **second** in the sequence of events?

a) Maximal neutrophil infiltrate.
b) Hypereosinophilia of cardiac myocytes.
c) Formation of granulation tissue.
d) Loss of myocyte cross striations and myocyte nuclear loss.

38) This is a cross section of the left anterior descending coronary artery taken at autopsy.

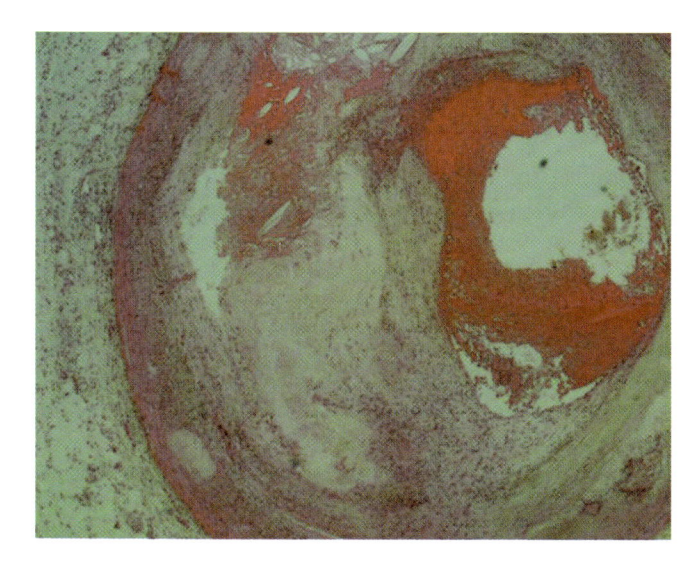

From the histology, which of the following options is the most likely clinical scenario?

a) A 15 year old Japanese boy with a several week history of low grade fever and myalgia.
b) A 49 year old man with longstanding essential hypertension.
c) A 65 year old man with a history of smoking and serum cholesterol: HDL ratio 7.5.
d) A 25 year old female with a history of facial rash, Raynaud's phenomenon and joint pains.

39) Left ventricular rupture is a rare complication post myocardial infarction. When does myocardial rupture **typically** occur after myocardial infarction?

a) Less than 1 hour.
b) 6-12 hours.
c) 3-5 days.
d) 12-24 days.

40) A 45 year old man presents with a short history of headache and blurred vision. Blood pressure was measured at 180/120 and ophthalmoscopy shows flame haemorrhages. Choose from the following options the most likely diagnosis.

    a) Essential hypertension.
    b) Chronic renal failure.
    c) Congestive cardiac failure.
    d) Malignant hypertension.

Correct answers and explanations:

1) Correct answer: b) The coronary arteries are the most commonly affected.

In fact, the carotid and renal arteries are most commonly affected, although the condition does affect the coronary arteries in some cases.

2) Correct answer: c) Metastatic small cell carcinoma of lung.

Pericardial effusions arising as a result of malignant disease are typically exudative. The commonest primary malignancies are lung, breast and leukaemia/lymphoma, as well as melanoma and mesothelioma.

3) Correct answer : d) Transthyretin.

The commonest type of amyloidosis affecting the heart is senile cardiac amyloidosis. In this condition, there is extracellular deposition of transthyretin protein. Cardiac amyloidosis is more common in men than in women. The disease is rare in people under the age of 40.

4) Correct answer: b) Marfan syndrome.

Marfan syndrome is a genetic disorder of connective tissue characterised by a variety of missense mutations in the gene encoding for fibrillin 1, an extracellular matrix glycoprotein essential for the formation of cellular microfibrils (FBN1 gene). Around two thirds of cases are due to familial transmission with the remainder the result of sporadic mutations.

5) Correct answer: a) Are granulomatous foci characteristic of rheumatic heart disease.

Aschoff bodies are a characteristic feature of rheumatic heart disease. Histology shows granulomata with central fibrinoid necrosis surrounded by lymphocytes, plasma cells and Anitschkow cells. They can measure up to 2mm in size (therefore, can be visible to naked eye also) and can be found in the proximal aorta in addition to the heart. They occur in all three layers of the heart with the most common site being the myocardium.

6) Correct answer: c) CD34+/CD31+/vWF+.

Both CD34 and CD31 are sensitive and specific for endothelial cells. vWF (Factor VIII related antigen) is less sensitive and shows variable staining of the sinusoids of liver, spleen and lymph nodes.

7) Correct answer: d) The differential diagnosis of RCM is a constrictive pericarditis and it is important to distinguish between the two, as constrictive pericarditis may be cured by surgical intervention.

Most cases of RCM are idiopathic. The WHO definition of RCM is 'a myocardial disease characterized by restrictive filling and reduced diastolic volume of either or both ventricles with normal systolic function and normal or near-normal wall thickness'.
RCM is less common than both dilated cardiomyopathy and hypertrophic cardiomyopathy (HCM).

8) Correct answer: a) Buerger's disease.

It is also known as thromboangiitis obliterans. This is a rare inflammatory vasculopathy in which there is segmental thombosing inflammation of medium and small sized arteries and veins of the lower and upper limbs. The ischaemic features are commonly accompanied by a migratory superficial thrombophlebitis. Cessation of smoking is the only measure found to prevent progression of the disease.

9) Correct answer: c) Giant cell arteritis.

Also known as temporal arteritis. Histology shows inflammation of the artery wall with frequent multinucleate giant cells. Up to half of people with temporal arteritis develop polymyalgia rheumatica. The temporal artery is most commonly affected but any artery can be involved.

10) Correct answer: c) The dissection typically spreads along the plane between the intima and media.

Typically, aortic dissection involves the plane between the middle and outer thirds of the media.

11) Correct answer: d) Wegener's granulomatosis (Granulomatosis with polyangiitis).

Wegener's granulomatosis is a necrotising granulomatous vasculitis which affects small and medium sized blood vessels. Serology typically shows positivity for c ANCA.

12) Correct answer: c) Fibrinoid necrosis.

This is characteristically seen in malignant hypertension. The other histological features are typically seen in longstanding essential hypertension.

13) Correct answer: b) Are known as 'caterpillar cells' due to the shape of the cell.

In fact, they are so called because of the condensed rod-like pattern of the nuclear chromatin in longitudinal section. Anitschkow cells are characteristically seen within Aschoff bodies in rheumatic heart disease. They are of macrophage origin and can be seen in normal hearts.

14) Correct answer: e) All of the above.

ACE is a vasoconstrictor molecule which modulates blood flow and vascular reactivity. VWF is a prothrombotic molecule. Prostacyclin is a vasodilator molecule. Heparin-like molecules act as anticoagulants.

15) Correct answer: b) Ventricular septal defects.

VSDs are present in 40-50% of all children with congenital heart disease. The symptoms and clinical presentation depend on the size of the defect.

16) Correct answer: c) All fatty streaks become atheromatous plaques.

In fact, fatty streaks are found in anatomical locations which are common for atheromatous plaques and also sites in which plaques are uncommon.

17) Correct answer: d) Atherosclerosis.

Male sex, high BMI and smoking are all risk factors for atherosclerosis.

18) Correct answer: b) Hypertrophic obstructive cardiomyopathy (HCM).

This is an inherited condition affecting approximately 1 in 500 people predominantly caused by mutations in the genes for beta-myosin heavy chain (MHC), myosin-binding protein C, cardiac troponin T, or tropomyosin. HCM causes thickening of the left ventricular myocardium and also disruption of the electrical function of the heart. HCM is the most common cause of sudden cardiac death in adolescents.

19) Correct answer: d) Is increased with maternal flu infection in the second trimester.

In fact, risk of congenital heart disease is increased twofold in women who get flu in the **first** trimester. The reasons for this are unclear.

20) Correct answer: C) Cyclophosphamide cardiotoxicity.

This condition typically produces a haemorrhagic myocarditis with widespread capillary thrombi, interstitial haemorrhage, fibrin deposition and myocyte necrosis.

21) Correct answer: D) Viral myocarditis.

Viral myocarditis is an important cause of dilated cardiomyopathy. Viral agents causing viral myocarditis include Parvovirus B19, HHV6, Coxsackie virus (A and B).

22) Correct answer: A) Adriamycin-induced cardiotoxicity.

Adriamycin (doxorubicin) is an antineoplastic agent which can give rise to a broad spectrum of early and late cardiac effects ranging from an acute pericarditis/myocarditis to congestive cardiac failure.

23) Correct answer: b) Tetralogy of Fallot.

This set of cardiac abnormalities results in a right-to-left shunt in early postnatal life with deoxygenated blood bypassing the lungs and entering the circulation. The other abnormalities result in a left-to-right shunting of blood.

24) Correct answer: b) The peak incidence in cardiac myxomas is in men aged 20 – 40.

In fact, the peak incidence is in middle aged females. Myxomas are the commonest primary cardiac tumours and may show glandular differentiation. Immunohistochemistry shows positivity for calretinin (75-100%), vimentin (50%) and variable positivity for S100, smooth muscle actin and desmin.

25) Correct answer: D) Mitral valve prolapse.

Histology shows increased extracellular matrix which stains with periodic acid-Schiff (PAS)/Alcian blue.

26) Correct answer: A) Infective endocarditis.

Common organisms include Staphylococcus aureus, Streptococcus viridans and coagulase negative Staphylococci.

27) Correct answer: E) Systemic lupus erythematosus.

Libman-Sacks endocarditis is a sterile endocarditis seen in systemic lupus erythematosus which typically affects the mitral valve and rarely embolises.

28) Correct answer: C) Calcification of the mitral annulus.

Although calcification of the mitral annulus does not usually involve the chordae tendinae or leaflets, advanced calcification may encroach on the posterior valve leaflet. However, the sparing of the anterior mitral valve leaflet distinguishes this from rheumatic valve disease.

29) Correct answer: c) Senile cardiac amyloidosis.

In this condition the amyloid is derived from transthyretin protein (TTP).

30) Correct answer: a) Arrhythmogenic right ventricular dysplasia (arrhythmogenic right ventricular cardiomyopathy).

his condition is an important cause of sudden cardiac death in young people as the result of life-threatening ventricular arrhythmias.

31) Correct answer: c) Secondary haemochromatosis.

This condition is due to iron overload. A spectrum of abnormalities has been described in iron overload including congestive cardiomyopathy, pericarditis, restrictive cardiomyopathy, and angina in the absence of coronary artery disease.

32) Correct answer: b) Takayasu's arteritis.

This condition predominantly affects young females. Histology shows a granulomatous vasculitis which can be indistinguishable histologically from giant cell arteritis. Claudication is common but Raynaud's phenomenon is rare.

33) Correct answer: e) All of the above.

CCF is unusually common in patients with diabetes (Diabetic cardiomyopathy) and atherosclerosis is accelerated. Silent myocardial infarction is a recognised complication, most likely due to cardiac autonomic neuropathy. Sudden cardiac death is also increased in diabetes due to a number of factors including accelerated coronary atherosclerosis, cardiac autonomic neuropathy with 'silent' ischaemia, blood glucose levels, a prothrombotic state and increased risk of arrhythmias.

34) Correct answer: a) Acute myocardial infarction.

Histology shows an interstitial neutrophilic infiltrate with degeneration of neutrophils and myocyte necrosis. The infarct is likely to be of 3 – 10 days duration.

35) Correct answer: c) Below the level of the renal arteries.

Atherosclerotic aneurysms are most common in the lower abdominal aorta proximal to the aortic bifurcation and below the level of the renal arteries.

36) Correct answer: b) Men aged 55 and over are eligible for screening.

In fact, the screening programme is open to men aged 65 and over.

37) Correct answer: d) Loss of myocyte cross striations and myocyte nuclear loss.

The earliest change is myocyte cytoplasmic hypereosinophilia followed by loss of cross striations and myocyte nuclei (4-12 hrs). Neutrophil infiltration is the next stage (1-3 days) followed by removal by necrotic myocytes by macrophages (3-10 days) and the laying down of granulation tissue (10 days – 4 weeks). Over time, increased collagen deposition results in a fibrous scar (6-8 weeks).

38) Correct answer: c) A 65 year old man with a history of smoking and a serum cholesterol: HDL ratio 7.5.

Histology shows an eccentrically placed atheromatous plaque. Male gender, smoking and raised cholesterol: HDL ratio (greater than 6) are all risk factors for atherosclerosis.

39) Correct answer: c) 3-5 days.

Left ventricular rupture occurs typically 3-5 days post myocardial infarction. It is a rare complication occurring in approximately 1% of cases of myocardial infarction. The incidence has decreased with the advent of emergency reperfusion therapies.

40) Correct answer: d) Malignant hypertension.

This occurs in approximately 1% of patients with hypertension. Blood pressure is typically 180/120 or over. An important cause is discontinuation of blood pressure medication. Other causes include collagen vascular disorders, renovascular disease, pre-eclampsia and drug use such as cocaine and amphetamines.

CHAPTER 8

NASAL & ORAL PATHOLOGY
Dr Miguel Perez-Machado

1) After a cold, a 22-year-old female notices the appearance of several clear vesicles on her lip. These small vesicles rupture, leaving painful ulcers that heal during the next 2 weeks. Several months later, during the exam period, this scenario is repeated. The most likely finding associated with these lesions is:

   a) Biopsy demonstrating fungal infection.
   b) Positive serology for herpes simplex virus type 1 (HSV-1).
   c) Abnormal increase of mast cells seen on the peripheral blood smear.
   d) Biopsy demonstrating a lesion composed of interlacing fascicles of slender spindle cells within a loose collagenous stroma.
   e) A predominantly lymphocytic inflammatory infiltrate on biopsy.

2-6) Extended matching questions:

   A) Gingivitis/Periodontitis.
   B) Aphthous ulcers (canker sores).
   C) Pyogenic granuloma (lobular capillary haemangioma).
   D) Palatal papillomatosis (inflammatory papillary hyperplasia).
   E) Peripheral giant cell tumour (epulis).

For each of the following explanations and descriptions of oral lesions, select the most appropriate answer from the list of options above.

2) Numerous multinucleated osteoclast-like foreign body giant cells with scant fibroangiomatous stroma.

3) Multiple foci of epithelial hyperplasia and pseudoepitheliomatous hyperplasia.

4) Inflammatory lesion composed of granulation tissue.

5) Single or multiple shallow, fibrin-coated ulcers with mononuclear underlying infiltrate. Painful.

6) Inflammation of soft tissue around teeth or surrounding alveolar bone, respectively.

For each of the clinical histories in questions 7, 8 and 9, match the most closely associated neoplastic and non-neoplastic process that may produce a mass lesion in the head and neck region:

   A) Pyogenic granuloma.
   B) Squamous cell carcinoma.
   C) Warthin tumour.
   D) Paraganglioma.
   E) Plasmacytoma.
   F) Pleomorphic adenoma.
   G) Mucoepidermoid carcinoma.
   H) Olfactory neuroblastoma.

I) Papilloma.
J) Angiofibroma.
K) Cholesteatoma.
L) Malignant lymphoma.

7) A 65-year-old female with a mass arising in a minor salivary gland located on the buccal mucosa, left side, beneath the tongue. The mass measures 2.2cm in diameter. It is excised, and histologically, the tumour is composed of varying proportions of mucous, epidermoid and intermediate-type cells.

8) A 2.5-cm, mobile, discrete, non tender mass is palpable on the right side of the face of a 65-year-old female. The mass has been slowly growing for several years and it is anterior to the ear and just superior to the mandible. Histological examination of the excised lesion demonstrates well-formed ductal structures with associated spindle cells in a myxoid stroma containing islands of chondroid formation.

9) A 35-year-old Caucasian male with a 3 cm mass arising in the left parotid region. The mass is painless with fluctuant swelling. Histological examination of the excised lesion demonstrates cystic spaces lined by papillary proliferation of bilayered oncocytic epithelium whose supporting stroma is composed of lymphoid tissue.

10) A 68-year-old male with history of chewing tobacco, present with some discrete white patches with a leathery surface on the lateral surface of the tongue. The lesion cannot be removed by rubbing. Histological examination reveals hyperkeratosis, parakeratosis, acanthosis, intranuclear inclusions and prominent ballooning of the squamous cells in the upper half of the epithelium. The underlying stroma shows mild chronic inflammation. This lesion is most likely to be:

   a) Oral candidiasis.
   b) Leukoplakia.
   c) Granular cell tumour.
   d) Ameloblastoma.

11) A 52-year-old male with a history of difficulty breathing through his nose for several months, presents with an enlarged cervical lymph node. A biopsy was performed that proved a metastatic carcinoma with squamous features; the tumour is composed of large epithelial cells having indistinct cell borders and prominent nuclei. There are mature lymphocytes scattered through this undifferentiated neoplasm. Immunostaining shows that the tumour cells are positive for Epstein-Barr virus (EBV). Which of the following sites is most likely to be primary tumour origin?

   a) Bronchus.
   b) Mouth.
   c) Nasopharynx.
   d) Skin.
   e) Oesophagus.

12) A 16-year-old female presents with a "lump" on her neck that she has noticed for several months. It does not seem to have increased in size appreciably during that time. On physical examination, there is a discrete, slightly movable nodule located in the midline just above the thyroid. The lesion is excised. Microscopic examination reveals a cystic mass lined by squamous and transitional / respiratory type epithelium. The underlying stroma contains mucus glands. Which of the following conditions is most likely to account for these findings?

a) Metastatic squamous cell carcinoma.
b) Thymoma.
c) Thyroglossal Cyst.
d) Branchial cleft cyst.
e) Warthin's tumour.

13) A 21-year-old female has difficulty breathing through her nose that has become progressively worse for the last 2 months. Physical examination reveals translucent, polypoid masses filling the nasal cavity bilaterally. Histological examination reveals respiratory mucosa overlying an oedematous stroma with scattered plasma cells and eosinophils. Which of the following conditions is most likely to account for these findings?

a) Allergic polyps.
b) Angiofibroma.
c) Rhinosporidiosis.
d) Sinonasal papilloma.
e) Meningioma.

14) In reporting mucosal malignancies of the oral cavity, all of the following are true, except:

a) The depth of invasion refers to the depth of greatest spread in presumed continuity below the top of the adjacent mucosa. For both nodular and ulcerated tumours, the line of the original mucosal surface is reconstructed to determine the true thickness.
b) The depth of invasion refers to the depth of greatest spread in presumed continuity below the top of the adjacent mucosa; only in nodular tumours.
c) Depth of invasion is greater than the tumour thickness in ulcerated tumours.
d) Depth of invasion is significantly related to nodal metastasis for oral T1/T2 carcinomas.
e) Reviews and a meta-analysis suggest that 4 mm is the optimal threshold for prediction of cervical node metastasis.

15) In reporting mucosal malignancies of the pharynx, all of the following are true, except:

a) Distance from invasive carcinoma to surgical margins. From a surgical point of view, >5 mm for both mucosal and deep margins is considered clear.
b) Distance from invasive carcinoma to surgical margins. From a surgical point of view, <10mm for both mucosal and deep margins is considered involved.
c) The presence of dysplasia at the margin is associated with a significantly increased risk of local recurrence.
d) The presence of carcinoma cells within an endothelial-lined space is the essential criterion and should be distinguished from retraction artefact.
e) It is necessary to distinguish between small lymphatics and venous channels.

16) A 72-year-old male has an ulcer on the side of his tongue for almost 8 months. Endoscopic examination reveals a lesion at the right base of the tongue. He underwent biopsy, followed by hemiglossectomy. Which of the following aetiological factors probably played the greatest role in the development of this lesion?

a) Human papillomavirus infection.
b) Allergic polyps.
c) Chewing tobacco.
d) Asbestos exposure.

e) HHV8 infection.

17) A patient undergoes a neck dissection as part of treatment for a metastatic SCC. This includes overlying skin as well as cervical lymph nodes (levels I–V). What type of neck dissection is this?

   a) Comprehensive neck dissection.
   b) Selective neck dissection.
   c) Extended neck dissection.
   d) Radical neck dissection.

18) Which of the following statements is true according to the Royal College of Pathologists Dataset for histopathology reporting of nodal excisions and neck dissection specimens associated with head and neck carcinomas, November 2013.

   a) For practical purposes, the critical factor influencing the use of adjuvant therapy is involvement of levels IV or V.
   b) Size of largest metastatic deposit is the same as the size of the largest node.
   c) Extracapsular spread of metastasis in a lymph node has no proven prognostic significance.
   d) Extracapsular spread (ECS) should be recorded separately in the 'Comments' section if desired.
   e) If the metastasis is 1 mm or less in diameter, it is considered a micrometastasis.

19) The grade of salivary carcinomas is related to the risk of local recurrence, regional and distant metastasis. Which of the following salivary gland malignancies is considered high grade?

   a) Pleomorphic adenoma.
   b) Carcinoma ex-pleomorphic adenoma.
   c) Acinic cell carcinoma.
   d) Basal cell adenocarcinoma.
   e) Cribriform carcinoma.

20) According to Royal College of Pathologists Dataset for histopathological reporting of salivary gland tumour, which TNM classification is applicable to a tumour that measures 39 mm with a focus of extraparenchymal extension?

   a) T1.
   b) T2.
   c) T3.
   d) T4a.
   e) T4b.

21) A 33yr old male, non – smoker, with a slow growing and asymptomatic mass in the posterior mandible. Radiologically the lesion is multi-loculated with "soap – bubble" radiolucency.
A biopsy was performed which showed the following picture:

What is the diagnosis?

   a) Oral squamous cell carcinoma.
   b) Metastatic squamous cell carcinoma of the lung.
   c) Ameloblastoma.
   d) Osteosarcoma.

22) A 64 year old lady with history of Sjogren's syndrome presents with ulceration of the oral mucosa. A biopsy was taken.
Positive immunostaining: CD19, CD20, CD22, CD79a, surface Ig, cytoplasmic Ig in plasma cell component, bcl-10, CD11c, bcl2, monoclonal light chain expression.
Negative staining: IgD, CD5, CD10, CD23, CD43 and bcl-1.
What is the diagnosis?

   a) CLL/SLL.
   b) Follicular lymphoma.
   c) Marginal zone lymphoma, MALT type.
   d) Mantle cell lymphoma.

23) A 45 year old man with a history of nasal obstruction presents with unilateral nasal obstruction, epistaxis, headaches, pain, excessive lacrimation, rhinorrhea, anosmia and visual disturbances. A biopsy was performed which showed the following picture:

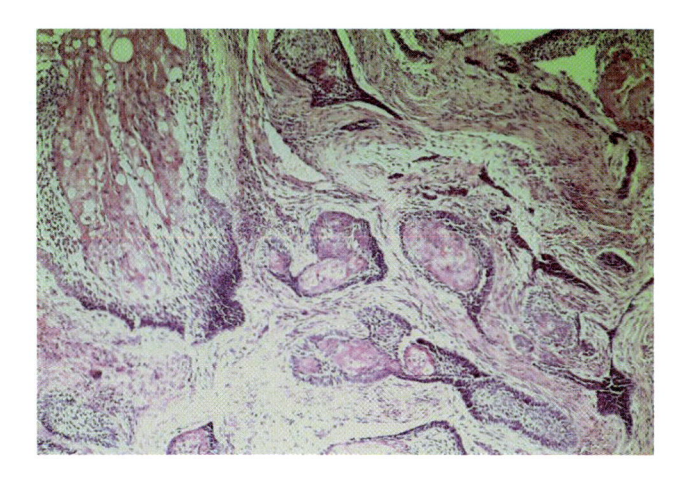

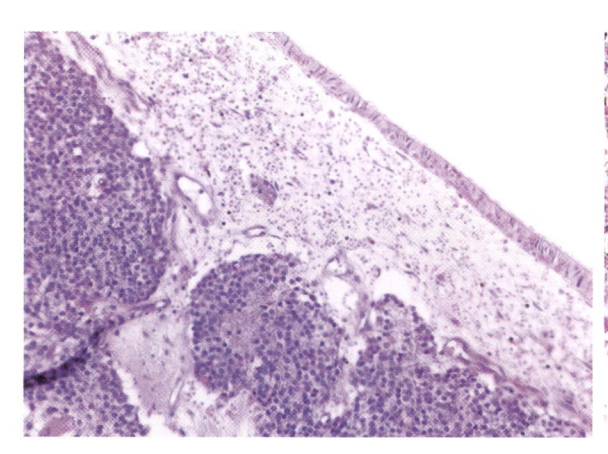

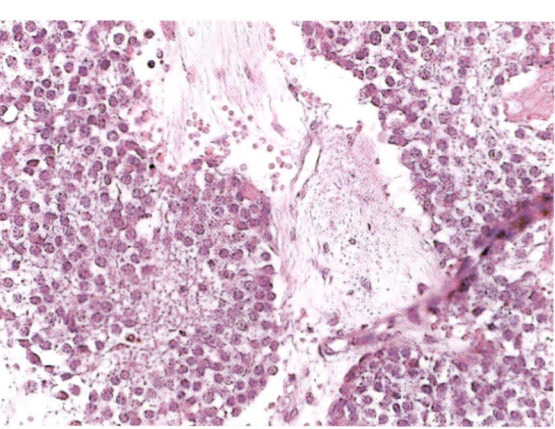

Immunostaining shows that the tumour is positive for: synaptophysin, chromogranin, CD56, neuron specific enolase, NFP. S-100 protein-positive cells are found at the periphery of the tumour lobules. In situ EBER is negative. Neurosecretory granules are present on electron microscopy.
What is the diagnosis?

    a) Sinonasal undifferentiated carcinoma.
    b) Melanoma.
    c) Extranodal NK/T cell lymphoma, nasal type.
    d) Rhabomyosarcoma.
    e) Olfactory neuroblastoma.
    f) Ewing sarcoma/PNET.

24-27) Extended matching questions:

    A) Odontogenic keratocyst
    B) Calcifying odontogenic cyst
    C) Glandular odontogenic cyst
    D) Calcifying epithelial odontogenic tumour
    E) Squamous odontogenic tumour
    F) Dentigerous cyst.
    G) Radicular cyst

For each of the following explanations and descriptions, select the most appropriate answer from the list of options above.

24) Lined partly or completely by non-keratinising epithelium of varying thickness. The connective tissue shows inflammatory infiltrate and cholesterol clefts. Hyaline / Rushton bodies are also found.

25) Lining derived from reduced dental epithelium; consists of 2 – 4 cell layers of non-keratinising epithelium without rete ridges. Wall composed of thin connective tissue appearing immature as it is delivered from the dental papilla.

26) The tumour appears as islands of bland squamous epithelium (no cellular atypia or mitotic figures) without an inflammatory infiltrate. Peripheral palisades are not seen. The epithelial islands are occasionally closely associated with bone spicules.

27) Cyst showing thin parakeratinized lining with basal palisading. Separation of epithelium from connective tissue wall is often seen in surgical specimens. The cyst shows granular layer and in undistinguished basal layer.

Correct answers and explanations:

1) Correct answer: b) Positive serology for herpes simples virus type 1 (HSV-1).

This condition, also known as "cold cores" or "fever blisters," is related to HSV-1. A high percentage of the population have HSV-1; they flourish during periods of stress, particularly in the form of oral and perianal lesions.
Oral candidiasis develops in persons who are immunocompromised.
Mastocytosis is a systemic condition in which most patients have urticaria pigmentosa.
Fibrous histiocytoma (dermatofibroma) is an ill-defined lesion composed of interlacing fascicles of slender spindle cells within a loose collagenous stroma.
A predominantly lymphocytic inflammatory infiltrate on biopsy is non-specific but suggest chronicity.

2) Correct answer: E) Peripheral giant cell tumour (epulis).

This is an unusual inflammatory reaction (not neoplastic), characteristically protruding from the gingiva close to the teeth.

3) Correct answer: D) Palatal papillomatosis (inflammatory papillary hyperplasia).

This is usually associated with poorly fitting dentures and is not a premalignant condition.

4) Correct answer: C) Pyogenic granuloma (lobular capillary haemangioma).

This occurs most commonly in the maxillary labial gingiva.

5) Correct answer: B) Aphthous ulcers (canker sores).

Aetiology unknown.

6) Correct answer: A) Gingivitis/Periodontitis.

Acute forms can become necrotising ("trench mouth"), particularly in the setting of poor oral hygiene, smoking, and stress.

7) Correct answer: G) Mucoepidermoid carcinoma.

This is the most common malignant tumour of the parotid gland and most common malignant salivary gland tumour in children. It can range from being very indolent to overtly malignant. There are four cell types: mucinous, squamous, intermediate, and clear. The tumour is considered high grade when > 20% of the tumour is solid and infiltrative; is predominantly squamous, intermediate, and clear cells. Marked nuclear atypia is generally not seen. Mucin or keratin may escape, causing inflammatory reaction. Prognosis: 5 year survival: 98% for low grade, 60% for high grade.

8) Correct answer: F) Pleomorphic adenoma [benign mixed tumour (BMT)].

This is the most common neoplasm of the salivary glands (60-80%). It accounts for 75% of all parotid neoplasms and <50% of all palatal neoplasms. Parotid: submaxillary: sublingual = 10:1:<< 1. Most frequently seen in women in 30s, it presents as painless and persistent swelling. Grossly it can be: rubbery, well-circumscribed, and bosselated mass with small extensions; cartilage may be present. There are a wide variety of histologic appearances, including glandular, spindled, pseudovascular patterns, and focal squamous metaplasia. It can be very cellular, or be predominantly stroma. Stroma can be myxoid or hyalinised; usually with some cartilage. Focal areas of penetration of surrounding tissue can result in a high recurrence rate, multifocally, if not widely excised. Recurrences may occur over 50 years later.

9) Correct answer: C) Warthin's tumour (Adenolymphoma, papillary cystadenoma lymphomatosum).

This tumour occurs more commonly in men, usually ~60 years old; often multicentric, bilateral in 10-15%. It accounts for 70% of bilateral tumours and is almost exclusive to the parotid. It may arise in ectopic salivary gland tissue in lymph nodes, and is often cystic with papillary growth of two layers of oncocytic epithelial cells associated with numerous inflammatory cells with germinal centres (predominantly B cells). The oncocytic cells are epithelial cells, which are granular, and loaded with mitochondria.

10) Correct answer: b) Leukoplakia.

This is a premalignant lesion which is sharply demarcated; can occur anywhere in oral mucosa. Some common causes include pipe smoking, tobacco chewing and irritation from misaligned teeth or dentures. It may show mucosal thickening with epidermal hyperplasia and hyperkeratosis, or disorderly hyperplasia with varying degrees of dysplasia up to carcinoma in situ.
Erythroplasia (red plaque) is more likely to be dysplastic.
Oral candidiasis (oral thrush) presents in persons who are immunocompromised and the lesion can be removed by rubbing.
Granular cell tumour typically presents as a painless submucosal nodule; it is characterized by sheets, nests or cords of large polyhedral cells with granular cytoplasm and small hyperchromatic nuclei.
Ameloblastoma is a tumour located in the mandible and maxilla (ratio 4:1). It may occur in association with impacted third molars.

11) Correct answer: c) Nasopharynx.

Epstein Barr virus (EBV) is closely associated with almost all carcinomas of the nasopharynx with higher levels of expression in undifferentiated carcinomas than in keratinising carcinomas. While identification of EBV-associated proteins or RNAs is not in itself of prognostic value, evidence of EBV in nodal metastases may point to a nasopharyngeal primary.

12) Correct answer: c) Thyroglossal Cyst.

This is a developmental abnormality that arises from elements of the embryonic thyroglossal duct extending from the foramen cecum of the tongue down to the thyroid gland. Midline is the classic location.
The patient is 16-years old which makes squamous cell carcinoma unlikely.
Thymoma is most commonly found in adults and peak incidence occurs in the fifth to seventh decades. It is the most common solid primary neoplasm of the mediastinum. Thymomas exhibit a range of histologic features.
Branchial cleft cyst is located at the lateral aspect of the neck.
Warthin's tumour was discussed in the answer to question no. 9.

13) Correct answer: a) Allergic polyps.

These are unusual before the age of 20; if it occurs in children, it most likely indicates cystic fibrosis. Histology shows loose oedematous stroma with neutrophils, eosinophils, plasma cells, and lymphocytes, and they can become quite large and may erode bone. Stroma frequently contains isolated bizarre cells.

Angiofibroma occurs almost exclusively in young males between the ages of 10 and 25 years. Histologically, it shows a mixture of numerous haphazardly arranged small thin-walled blood vessels which are rich in mast cells.

Rhinosporidiosis is a hyperplastic polypoid lesion that shows numerous globular cysts containing numerous spores. It is a granulomatous disease affecting the mucous membrane of nasopharynx, oropharynx, conjunctiva, rectum and external genitalia, though the floor of the nose and inferior turbinate are the most common sites.

Sinonasal papilloma is a benign neoplasia of the respiratory mucosa most common in adult men. It typically affects patients ages 30 to 50 and is unilateral in most cases.

14) Correct answer: b) The depth of invasion refers to the depth of greatest spread in presumed continuity below the top of the adjacent mucosa; only in nodular tumours.

b) is an incorrect statement. The maximum depth of invasion (in millimetres) below the luminal aspect of surface should be recorded; if the tumour has ulcerated then the reconstructed surface should be used. The aim should be to provide a best estimate of tumour depth; for large carcinomas this may be an approximation. A more detailed comment on the nature of the tissues invaded (mucosa, muscle, etc.) should occur in the 'Comments' sections.

Note that depth of invasion, defined in this way, is not the same as tumour thickness which will be larger than depth of invasion in exophytic tumours and smaller in ulcerated tumours.

Depth of invasion is significantly related to nodal metastasis for oral T1/T2 carcinomas, although the optimal cut-off point for prognostic purposes is uncertain with 3 mm, 4 mm or 5 mm being suggested by different authors.

Reviews and a meta-analysis suggest that 4 mm is the optimal threshold for prediction of cervical node metastasis. (Royal College of Pathologists Dataset for histopathology reporting of mucosal malignancies of the oral cavity November 2013).

15) Correct answer: b) Distance from invasive carcinoma to surgical margins. From a surgical point of view, <10mm for both mucosal and deep margins is considered involved.

Option d) is an incorrect statement. Measure the distance histologically in millimetres for both mucosal and deep margins. From a surgical point of view, >5 mm is clear, 1–5 mm is close and <1 mm is involved.

Incomplete resection or the presence of dysplasia at the margin is associated with a significantly increased risk of local recurrence. The presence or absence of vascular invasion should be mentioned if it is an obvious feature on medium magnification examination of the tumour.

The presence of carcinoma cells within an endothelial-lined space is the essential criterion and should be distinguished from retraction artefact.

It is not necessary to distinguish between small lymphatics and venous channels. Vascular invasion is a relatively weak predictor of nodal metastasis (Royal College of Pathologists Dataset for histopathology reporting of mucosal malignancies of pharynx. November 2013).

16) Correct answer: a) Human papilloma virus.

There is substantial evidence to link high-risk human papillomaviruses (particularly HPV16) to a subset of oropharyngeal carcinomas. HPV-associated carcinomas are usually HHV8 infection is related to Kaposi sarcoma in HIV patients.

Allergic polyps are not a premalignant lesion.

Oral leukoplakia is related to pipe smoking and tobacco chewing.
Asbestos exposure induces mesothelioma.

17) Correct answer: c) Extended neck dissection.

The main types of neck dissection that may be received are:
Comprehensive neck dissection: this includes both radical and modified radical (functional) dissections. A radical neck dissection includes removal of cervical lymph nodes (levels I–V), sternocleidomastoid muscle, internal jugular vein, spinal accessory nerve and the submandibular salivary gland, while in a functional dissection, the sternocleidomastoid muscle, internal jugular vein, or the spinal accessory nerve may not be removed.

- Selective neck dissection: this involves removal of the nodal group(s) considered to be the most likely site for metastasis, preserving one or more nodal groups that are routinely removed in a radical dissection.
- Extended neck dissection: when additional lymph node groups or non-lymphatic structures are removed.

18) Correct answer: a) For practical purposes, the critical factor influencing the use of adjuvant therapy is involvement of levels IV or V.

In head & neck cancer, the involvement of nodes in level IV or V (together with Stage II or IV disease, perineurial or lymphovascular invasion, is associated with a risk of disease recurrence and is an indication for adjuvant therapy.
ECS is a manifestation of the biological aggression of a carcinoma and is associated with a poor prognosis. ECS should be recorded as present or not identified. If present, the node level(s) showing this feature are recorded. Any spread through the full thickness of the node capsule is regarded as ECS. Involvement of adjacent anatomical structures should also be recorded. If histological evidence of ECS is equivocal, it should be recorded as 'present'; this should prompt the use of adjuvant radiotherapy (Royal College Dataset for nodal excisions & neck dissections, Nov 2013).
The prognostic significance of micrometastases is not certain. Their presence should be included in the number of involved nodes and TNM coded as pN1(mi) or pN2(mi), unless larger metastases are present when the suffix is unnecessary.

19) Correct answer: b) Carcinoma ex-pleomorphic adenoma.

Carcinomas arising in pleomorphic adenomas may be of any histological type, but are thought to be particularly aggressive and the prognosis of the carcinomatous component is poorer than that of comparable carcinomas developing de novo. Evidence for a pre-existing adenoma (remnants of myxochondroid stroma, focal scarring, and hyalinised nodular 'ghost') should be sought in all carcinomas, particularly those showing multiple histological types and a varied histological appearance. The extent of invasion should be measured in these tumours as it is prognostically useful, although precise criteria are not defined. Invasion more than 5–6 mm from the capsule of the residual adenoma is associated with a high risk of local recurrence and distant metastasis.

20) Correct answer: c) T3.

Royal College of Pathologists Dataset for histopathology.

TNM classification of malignant tumours of major salivary glands.

Tx Primary tumour cannot be assessed.
T0 No evidence of primary tumour.
T1 Tumour 20 mm or less in greatest dimension without extraparenchymal extension.non-keratinising and arise in the tonsils or base of tongue, and tend to have better overall and disease-free survivals than non-HPV associated carcinomas.
T2 Tumour more than 20 mm but not more than 40 mm in greatest dimension without extraparenchymal extension.
T3 Tumour more than 40 mm and/or tumour with extraparenchymal extension.
T4a Tumour invades skin, mandible, ear canal or facial nerve.
T4b Tumour invades base of skull, pterygoid plates or encases carotid artery.

21) Correct answer: c) Ameloblastoma.

This is the most common odontogenic neoplasm. Males and female equally affected and the average age at diagnosis is 33 years. The mandible is affected more than the maxilla. It typically grows slowly, is asymptomatic and does not metastasize. It has a typical "soap – bubble" radiological appearance. Histologically shows most commonly a follicular and plexiform pattern. The main differential diagnosis is SCC, however, in this case the clinical history and age of patient does not support this diagnosis.

22) Correct answer: c) Marginal zone lymphoma, MALT type.

This is associated with Sjogren's syndrome and possibly other chronic inflammatory diseases of salivary gland like lymphoepithelial sialadenitis. It is the most common primary extranodal lymphoma and is slightly more common in women. CLL/SLL should be CD5+, CD23+. Follicular lymphomas express CD10+ and Mantle cell lymphomas are CD5+.

23) Correct answer: e) Olfactory neuroblastoma.

Olfactory neuroblastoma (ONB) is an uncommon malignant neuroectodermal nasal tumour & comprises about 2% of all sinonasal tract tumours. ONB may occur at any age (2–94 years), but a bimodal age distribution in the 2nd and 6th decades of life are most common without a gender predilection. The tumour most commonly cause unilateral nasal obstruction, and epistaxis. Less common signs and symptoms include headaches, pain, excessive lacrimation, rhinorrhea, anosmia, and visual disturbances. One of the most important histologic features is a lobular architecture comprised of small round blue cells arranged in nests of tumour below an intact mucosa separated by a vascularised fibrous stroma. The nuclei are small and uniform with hyperchromatic with a "salt-and-pepper" nuclear chromatin. Nucleoli are inconspicuous.

A table for sinonasal tract small blue round cell tumour differentials on the next page:

Sinonasal tract small blue round cell tumour differentials:

Feature	Olfactory neuroblast-oma	Sinonasal undifferen-tiated carcinoma	Melanoma	Extranodal NK/T cell lymphoma	Rhabdomyo sarcoma	Ewing sarcoma/ PNET	Neuroendocrine carcinoma
Age	40–45 y	55–60 y	40–70 y	50–60 y	<20 years	<30 years	50 years
Pattern	Lobular	Sheets and nests	Protean	Diffuse	Sheets, alveolar	Sheets, nests	Ribbons, islands
Neurofib-rillary stroma	Common	Absent	Absent	Absent	Absent	Absent	Absent
Keratin	Focal, weak	>90%	-ve	-ve	-ve	Rare	Positive
CK 5/6	-ve	-ve	-ve	-ve	-ve	n/a	n/a
EMA	-ve	50%	Rare	-ve	-ve	n/a	n/a
NSE	>90%	50%	-ve	-ve	-ve	Positive	Positive
S-100 protein	+ sustentacular	<15%	Positive	-ve	-ve	Rare	Positive
Synaptop hysin	>90% (can be weak)	<15%	-ve	-ve	-ve	Positive	Positive
In situ EBER	Absent	Absent	Absent	Positive (nearly 100%)	Absent	Absent	Absent
Neurose-cretory granules (EM)	Numerous	Rare	Absent	Absent	Absent	Absent	Present

24) Correct answer: G) Radicular cyst.

Radicular cysts are the most common odontogenic cystic lesions of inflammatory origin. Radicular cysts are also known as periapical cysts, apical periodontal cysts, root end cysts or dental cysts. They arise from epithelial residues in the periodontal ligament as a result of inflammation. The inflammation usually follows death of dental pulp. Radicular cysts are found at root apices of involved teeth. These cysts may persist even after extraction of the offending tooth; such cysts are called residual cysts.

25) Correct answer: F) Dentigerous cyst.

The second most common odontogenic cyst is the dentigerous cyst, which develops within the normal dental follicle that surrounds an unerupted tooth. The dentigerous cyst is not thought to be neoplastic. It most frequently is found in areas where unerupted teeth are found: mandibular third molars, maxillary third molars, and maxillary canines, in decreasing order of frequency. Most dentigerous cysts are asymptomatic, and their discovery is usually an incidental finding on radiography.

26) Correct answer: E) Squamous odontogenic tumour.

A rare, benign, locally infiltrative neoplasm of the jaws that appears to originate from the rests of Malassez, gingival surface epithelium or from remnants of the dental lamina. The tumour is often asymptomatic, although it can present with symptoms of pain and tooth mobility. The characteristic radiographic appearance is that of a triangular-shaped unilocular radiolucency associated with the roots of erupted, vital teeth and has a predilection for the anterior maxilla and the posterior mandible.

27) Correct answer: A) Odontogenic keratocyst.

The odontogenic keratocyst (OKC) is the most important of the odontogenic cysts. These lesions are aggressive and can be difficult to remove. OKCs can grow quite rapidly, and recurrences are frequent. This is the third most common odontogenic cyst. These cysts also are found as part of the basal cell nevus syndrome, also known as Gorlin syndrome (Basal cell nevus syndrome).

Histologically, these cysts are formed with a stratified squamous epithelium that produces orthokeratin (10%), parakeratin (83%), or both types of keratin (7%). The epithelial lining appears corrugated when viewed under a microscope. A well-polarized hyperchromatic basal layer is observed, and the cells remain basaloid almost to the surface. No rete ridges are present; therefore, the epithelium often sloughs from the connective tissue. The epithelium is thin, and mitotic activity is frequent. The lumen frequently is filled with a foul-smelling cheese like material that is not pus but rather collected degenerating keratin.

CHAPTER 9

THORACIC PATHOLOGY
Dr Brendan Tinwell

1) A 36 year old Asian woman has a CT-guided core biopsy of a 3cm lung mass. She has mediastinal lymph node and bone metastases. She is reportedly a 'never-smoker' and has no known history of previous malignancy. The lung biopsy shows non-mucinous papillary pattern adenocarcinoma which is TTF-1 positive. The treating oncologist requests molecular investigations on the tumour sample. Which single gene mutation is most likely to be identified?

   a) APC mutation.
   b) BRAF mutation.
   c) EGFR mutation.
   d) KRAS mutation.
   e) Rb gene mutation.

2) A 37 year old woman with right middle lobe bronchiectasis and a 5 pack year smoking history is found to have a polypoid endobronchial tumour in the right middle lobe bronchus on bronchoscopy. Which is the single most likely malignant diagnosis?

   a) Adenoid cystic carcinoma.
   b) Carcinosarcoma.
   c) Pulmonary blastoma.
   d) Squamous cell carcinoma.
   e) Typical carcinoid tumour.

3) A 42 year old woman who has never smoked has a lobectomy for a 33mm peripheral lung mass. The macroscopic appearance and a representative high power field of this tumour are shown below.

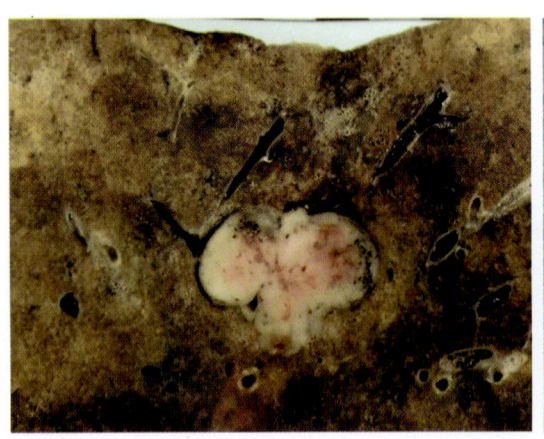

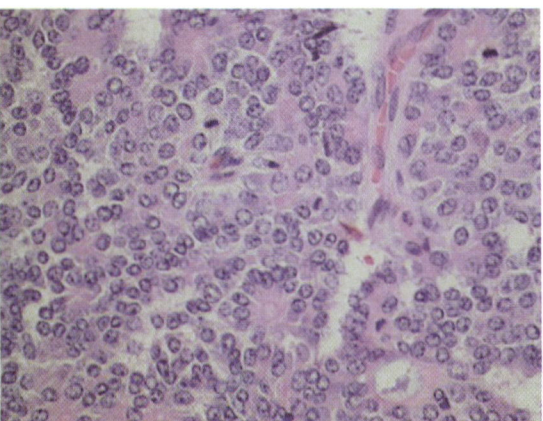

Punctate foci of necrosis were present (not shown). The tumour cells express CK8/18, CD56, Synaptophysin and TTF-1. The average mitotic count is 4 MF/2mm^2 and the Ki67 proliferation index is about 10%. The TNM stage classification is T2a N1 M0 and resection margins are clear.
What is this patient's approximate expected 5 year survival rate?

a) 5%.
b) 15%.
c) 30%
d) 60%.
e) 90%.

4) A 34 year old woman with recently diagnosed hypertension has an oesophago-gastroscopy for unexplained iron-deficiency anaemia. A 2.5cm submucosal mass is biopsied and this shows a neoplasm composed of interlacing fascicles of uniform spindle cells which show immunoreactivity for CD34 and DOG1 (they are negative for CD117, cytokeratin, S100 and SMA). There is little to no mitotic activity or necrosis. A gastric resection is planned, but a CT chest, abdomen and pelvis reveals multiple lung nodules and a bladder mass. A VATS wedge biopsy of lung and intra-operative frozen section (IOFS) is undertaken prior to laparoscopic resection of the gastric nodule. Which is the single most likely histological finding on a wedge excision biopsy (IOFS) of a lung nodule?

a) Metastatic GIST.
b) Metastatic paraganglioma.
c) Neuroendocrine tumorlet.
d) Pulmonary chondroma.
e) Pulmonary hamartoma.

5) A 57 year old man with 6 months of slowly progressive dyspnoea and non-productive cough is referred to a respiratory physician. He has never smoked. He is noted to have finger clubbing and fine end inspiratory crackles on chest auscultation. A high resolution CT (HRCT) chest shows basal predominant peripheral reticular opacity associated with traction bronchiectasis and lesser degrees of ground-glass attenuation and some volume loss. Bronchoalveolar lavage (BAL) fluid contains neutrophil polymorphs with no evidence of eosinophilia, lymphocytosis or malignancy. The clinicoradiological presentation is deemed sufficiently distinctive that a lung biopsy is not requested. Which single combination of histological findings would be expected on a lung excision biopsy?

a) Dense bronchocentric chronic inflammation with frequent loose, poorly formed granulomas and variable interstitial fibrosis.
b) Diffuse interstitial chronic inflammation and fibrosis without any fibroblastic foci or features of lung remodelling.
c) Subpleural and paraseptal fibrosis with intervening areas of relatively normal lung, and scattered fibroblastic foci seen in a background of established fibrosis.
d) Uniform intra-alveolar accumulation of lightly pigmented macrophages and mild interstitial fibrosis.
e) Well-formed epithelioid granulomas distributed along the pleura, bronchovascular bundles and interlobular septa associated with lamellar fibrosis and scanty Schaumann bodies.

6) A 62 year old woman has a VATS pleural biopsy for a new left pleural effusion. She is a non-smoker and has no known asbestos exposure (but there are pleural plaques on chest imaging). She had a mastectomy 15 years previously (pathology not known). Additional investigations reveal likely bone metastases and diffuse peritoneal disease. The photomicrographs below show medium and high power views of the pleural histology.

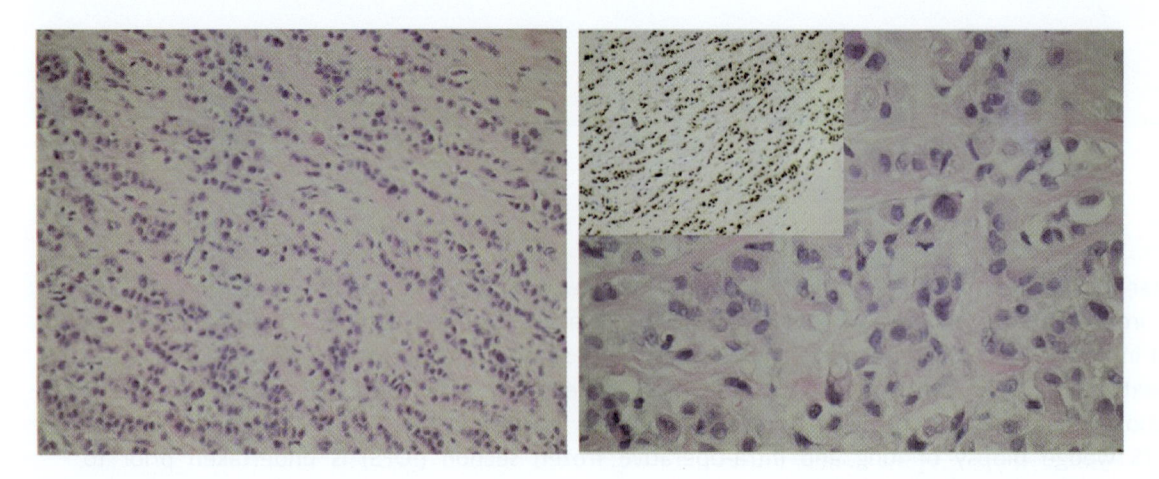

Which is the single most likely immunostain demonstrated in the inset?

   a) ER.
   b) PAX8.
   c) S100.
   d) TTF-1.
   e) WT1.

7) A 72 year old man with a 50 pack year history of cigarette smoking has bronchoscopy and biopsy for a right hilar lung mass.

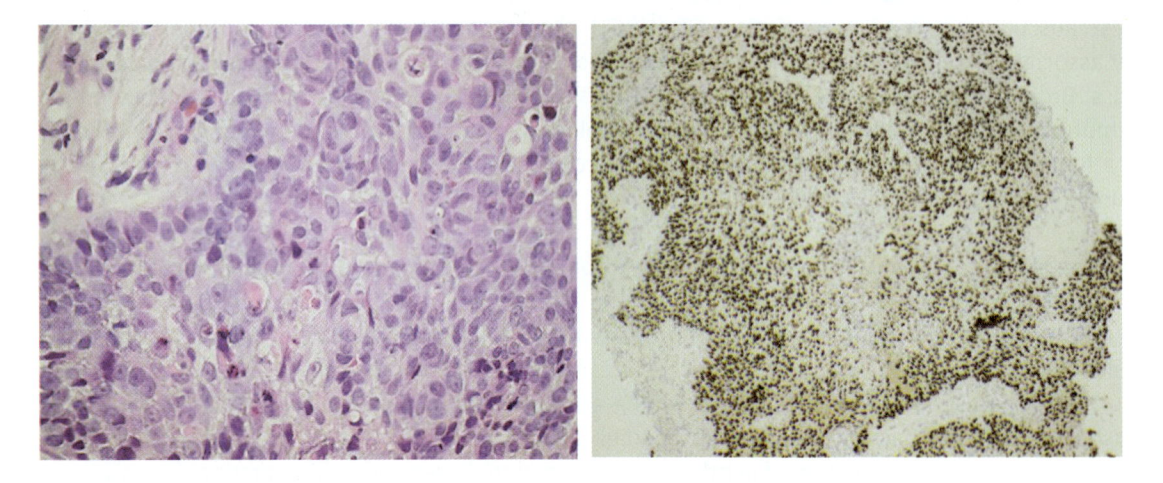

The histology and p63 immunostain is shown. The TTF-1 stain (not shown) is negative.
Which is the single most appropriate histological diagnosis?

   a) Non-small cell carcinoma, favour adenocarcinoma.
   b) Non-small cell carcinoma, favour basaloid carcinoma.
   c) Non-small cell carcinoma, possible large cell neuroendocrine carcinoma.
   d) Non-small cell carcinoma, favour squamous cell carcinoma.
   e) Non-small cell carcinoma, not otherwise specified.

8) A 65 year old man has a wedge excision and intra-operative frozen section for a 25mm peripheral nodule in the right upper lobe. The pleural surface appears thickened and retracted. If malignancy is confirmed the

patient will have a completion lobectomy. What single staging parameter will potentially affect how this fresh specimen is dealt with?

   a) Distance from the carina.
   b) Multiple tumour nodules.
   c) Tumour size.
   d) Vascular invasion.
   e) Visceral pleural invasion.

9) A 59 year old man with a history of smoking-related lung disease has a left lower lobectomy for a 25mm cavitating squamous cell carcinoma. Careful examination of the background lung shows a 4mm, somewhat stellate, solid nodule away from the main tumour. Histology of this and a bronchiole elsewhere is shown below (the histological features are related).

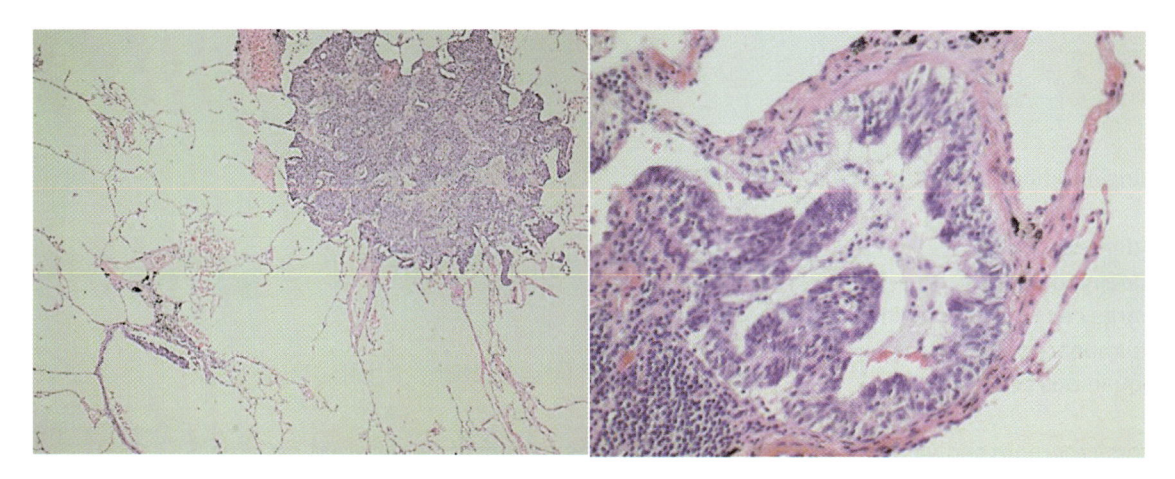

What is the single most important clinical implication of this?

   a) He may have widespread lymphangiitic spread of squamous cell carcinoma.
   b) Multiple neuroendocrine tumourlets may be mistaken for miliary pattern of metastasis.
   c) Smoking-related Langhan's cell histiocytosis may be associated with lung fibrosis.
   d) This indicates coincidental metastatic carcinoid tumour.
   e) This is an incidental minute meningothelial nodule which has no implications.

10) A 19 year old man attends the A&E department with 2 days of fever, non-productive cough and difficulty breathing. He reported having started smoking cigarettes 2-3 weeks previously and had smoked approximately 20 cigarettes in a short period one Saturday evening before the symptoms had started. There was no recent travel history and he was not on any medication. He denied being significantly intoxicated the night of the 'binge' cigarette smoking. He deteriorated rapidly despite antibiotic therapy, requiring intubation and ventilation for hypoxic respiratory failure. A chest radiograph showed bilateral infiltrates. Aggressive management for 3 days resulted in no significant improvement and a VATS lung biopsy was performed. The initial histology result was communicated urgently with the clinical team and rapid clinical improvement followed institution of steroid therapy. Resolution was complete, and there was no relapse. What was the single most likely histological diagnosis?

   a) Acute fibrinous organising pneumonia.
   b) Acute eosinophilic pneumonia.
   c) Acute interstitial pneumonia.
   d) Desquamative interstitial pneumonia.

e) Lymphoid interstitial pneumonia.

11) A 72 year old man has a lobectomy for a lung mass. A section of background, grossly uninvolved lung shows the changes below focally (lesion <5mm size).

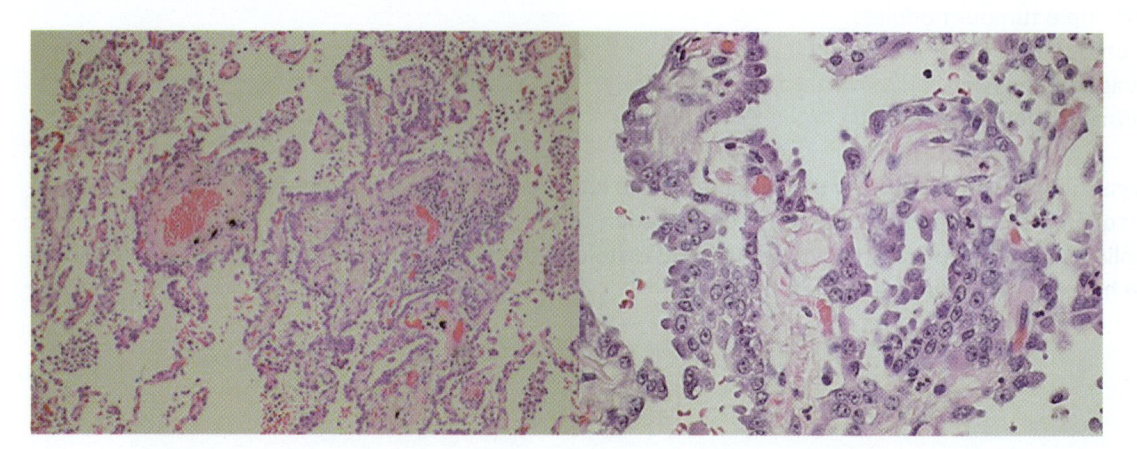

For which single tumour type is this the precursor lesion?

   a) Adenocarcinoma.
   b) Carcinoid tumour.
   c) Large cell neuroendocrine carcinoma.
   d) Pulmonary meningioma.
   e) Squamous cell carcinoma.

12) The photomicrographs shown below are from two different lung cancers in different patients.

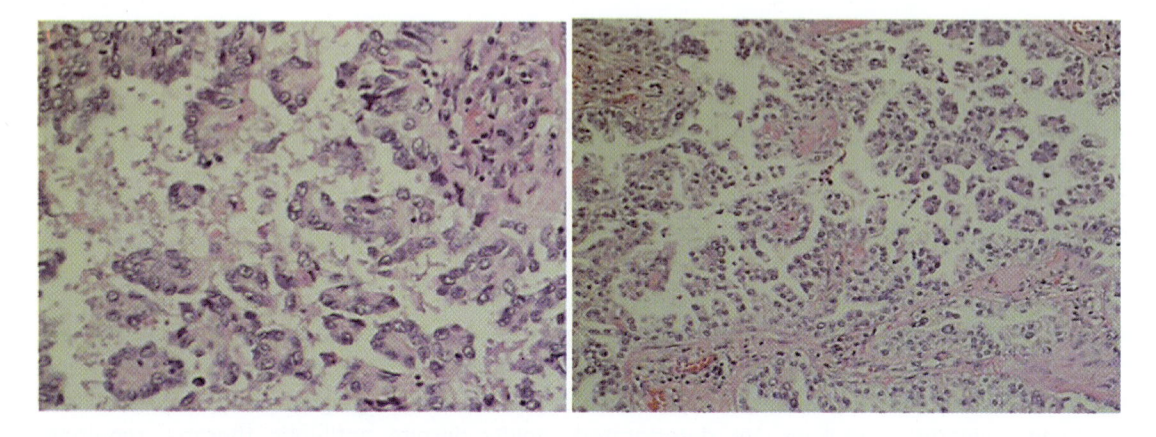

Which is the single most important clinical implication of this appearance?

   a) Confers a better clinical prognosis.
   b) Feature of well differentiated tumours.
   c) Higher risk of lymph node metastases.
   d) Pattern usually associated with ALK chromosomal rearrangement.
   e) Pattern usually associated with metastases to the lung (eg. from breast or kidney).

13) A 75 year old man has a VATS pleural biopsy as part of workup for weight loss, breathlessness and recurrent lymphohistiocytic effusion (cytology negative for malignant cells). The pleura appeared nodular. Images of the H&E and D240 stained sections are presented below. What is the single most likely diagnosis?

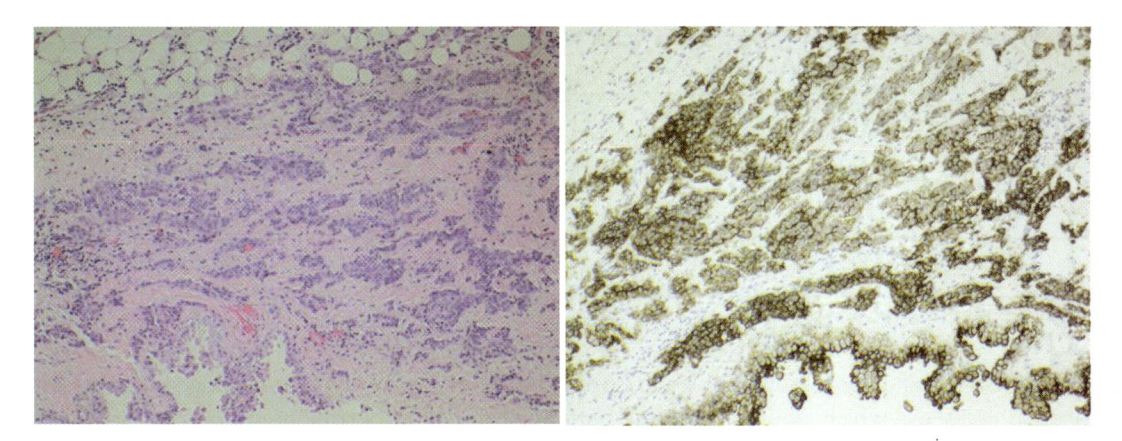

   a) Epithelioid angiosarcoma.
   b) Epithelioid haemangioendothelioma.
   c) Malignant mesothelioma.
   d) Metastatic adenocarcinoma.
   e) Metastatic seminoma.

14) A 69 year old man with a history of asbestos exposure and heavy cigarette smoking has a left pleural effusion and pleural nodularity on a CT scan. He was being followed up for a previous left upper lobectomy for squamous cell carcinoma 4 years previously. A VATS pleural biopsy shows nests and sheets of moderately pleomorphic, eosinophilic tumour cells with prominent nucleoli. Which single immunostain (positive or negative) is most useful in making a diagnosis of metastatic squamous cell carcinoma rather than malignant mesothelioma?

   a) Calretinin.
   b) CK5/6.
   c) Mesothelin.
   d) p16.
   e) p63.

15-19) Match the immunohistochemical profiles below with the most likely correct histological diagnosis.

   A) Malignant fibrous tumour.
   B) Metastatic gastrointestinal stromal tumour.
   C) Metastatic melanoma (spindle cell type).
   D) Metastatic rhabdomyosarcoma.
   E) Pleuropulmonary blastoma.
   F) Pleural thymoma.
   G) Sarcomatoid carcinoma.
   H) Sarcomatoid mesothelioma.
   I) Schwannoma.
   J) Synovial sarcoma.

	CK8/18	EMA	Calret	D240	CD99	CD34	SMA	Desmin	MyoD1	TTF-1	S100	MelanA	CD56	CD117
15	+	+	-	-	-	-	-	-	-	+	-	-	-	-
16	-	-	-	-	-	-	+	+	+	-	-	-	-	-
17	-	-	-	-	+	+	-	-	-	-	-	-	-	-
18	+ focal	+ focal	-	-	+	-	-	-	-	-	-	-	+	-
19	-	-	-	-	-	-	-	-	-	-	+	+	-	-

Calret = Calretinin

20) A 55 year old man with haematemesis has endoscopic treatment for bleeding oesophageal varices. He also has advanced lung disease which would be defined pathologically by 'abnormal permanent enlargement of the airspaces distal to the terminal bronchiole accompanied by destruction of their walls without obvious fibrosis'. What is the single most likely diagnosis?

   a) Alpha 1-antitrypsin deficiency.
   b) Cystic fibrosis.
   c) Primary biliary cirrhosis.
   d) Sarcoidosis.
   e) Wilson disease.

21) A 57 year old man with a fractured femoral neck following a trip and fall is diagnosed with multiple myeloma and receives steroid therapy. He also has tight aortic valve stenosis for which an endovascular valve replacement procedure is planned. While in hospital awaiting surgical repair of the fracture, he is noted to have some consolidation in the right lung and has an episode of coughing up blood-tinged sputum. During the surgery and in the immediate post-operative period he has severe cardiovascular instability with right heart strain and he dies before a CT-angiogram could be performed. His death is referred to Her Majesty's Coroner and he undergoes a post-mortem examination.
Gross examination of the lungs shows large areas of haemorrhagic infarction of the right lung and histology reveals lung necrosis associated with branching fungal hyphae characterised by a broad (15μm), ribbon-like, non-septate appearance. The hyphae invade large, thrombosed branches of the pulmonary artery. Which is the single most likely fungal organism responsible?

   a) *Aspergillus fumigatus*
   b) *Blastomyces dermatitidis*
   c) *Cladosporium carrionii*
   d) *Rhizopus oryzae*
   e) *Sporothrix schenckii*

22) Which is the single most appropriate cause of death sequence (ONS format) for the scenario described in question 21?

   a) 1a Invasive fungal lung infection and fracture neck of femur (operated October 2015)
      1b Multiple myeloma
      2 Aortic valve stenosis.

b) 1a Invasive fungal lung infection and right heart failure
   1b Multiple myeloma (treated with steroid therapy)
   2 Fractured neck of femur (operated October 2015) and aortic valve stenosis.

c) 1a Pulmonary embolus and invasive fungal lung infection
   1b Steroid therapy
   2 Multiple myeloma.

d) 1a Right heart failure and aortic valve stenosis
   1b Invasive fungal lung infection
   2 Multiple myeloma and fractured neck of femur (operated October 2015).

e) 1a Right heart failure, aortic valve stenosis and invasive fungal lung infection
   1b Multiple myeloma and fracture neck of femur (operated October 2015).

23) A 44 year old man with chest pain and facial swelling has a CT scan of his chest which shows an anterior mediastinal mass associated with SVC compression. A transthoracic CT-guided needle core biopsy shows granulomatous inflammation and scanty large, atypical cells with well-defined pale eosinophilic to clear cytoplasm and prominent nucleoli. The atypical cells show immunoreactivity for pancytokeratin, OCT3/4 and D240, and are negative for EMA, CD30, TTF-1, S100 and PAX8. Which is the single most likely serum abnormality?

   a) Elevated alpha fetoprotein (AFP).
   b) Elevated angiotensin converting enzyme (ACE).
   c) Elevated carcinoembryonic antigen (CEA).
   d) Elevated erythrocyte sedimentation rate (ESR).
   e) Elevated lactate dehydrogenase (LDH).

24) A 63 year old woman with Sjögren syndrome has a 37mm peripheral lung mass detected on a CT scan performed as part of investigation into cervical lymphadenopathy. The overlying pleura appears thickened. A needle core biopsy shows a polymorphous lymphoid infiltrate consisting of small, mature centrocytoid cells, monocytoid lymphocytes and plasmacytoid cells. A small bronchiole shows more than 5 lymphocytes infiltrating into the lining epithelium (lymphoepithelial lesion). Infrequent epithelioid granulomas are also observed. Which single statement is correct?

   a) Immunoglobulin heavy chain gene PCR confirms monoclonality in about 60% of cases of MALT lymphoma.
   b) Lymphoepithelial lesions are pathognomonic of MALT lymphoma.
   c) The biopsy should be repeated as this is likely an epiphenomenon related to an epithelial neoplasm.
   d) The heterogenous appearance of the lymphoid cells supports a diagnosis of a reactive proliferation.
   e) The presence of granulomas indicates a likely infective aetiology.

Correct answers and explanations:

1) Correct answer: c) EGFR mutation.

EGFR gene mutations (which are generally mutually exclusive to ALK and KRAS mutations in non-small cell lung cancer) are more commonly observed in lung adenocarcinoma (especially papillary and lepidic subtypes) arising in non-smokers, females or those of Asian descent (up to 75%).
APC (adenomatous polyposis coli) germline mutation is associated with familial adenomatous polyposis (FAP), Gardner syndrome and some families with Turcot syndrome, the principal complication of which is development of colorectal cancer.
BRAF mutations are found in about 50-70% of melanomas and about 90% of sporadic colorectal carcinomas, and only 1-3% of lung adenocarcinoma, usually in smokers.
KRAS mutations may be associated with *mucinous* lung adenocarcinoma, but are usually seen in smokers and are less common in those of Asian descent.
Rb gene mutations were first identified in retinoblastoma, but the gene is functionally inactivated in many human neoplasms including small cell carcinoma.

2) Correct answer: e) Typical carcinoid tumour.

Typical carcinoid tumours are low grade malignant tumours which are peripherally located in only about 10-20% of cases (often spindle cell variant) with an average age of 40-50 years (much lower than lung carcinoma) and not directly related to smoking.
Adenoid cystic carcinoma of the lung usually involves the lower trachea or large bronchi in a circumferential, infiltrative manner forming poorly defined nodular growths that narrow the airway.
Carcinosarcoma is not infrequently at least partly endobronchial, but usually affects elderly patients with a history of heavy smoking.
Pulmonary blastoma is a rare tumour affecting males three times more often than females that usually arises peripherally but may be endobronchial in about 25%.
Squamous cell carcinoma was traditionally regarded as a tumour of central bronchi, but peripheral examples are now just as common as central tumours which tend to be very infiltrative and cavitary, and are more often associated with heavy smoking.

3) Correct answer: d) 60%.

The tumour should be recognisable as showing neuroendocrine morphology (nests and rosette-like structures comprising polygonal cells with granular eosinophilic cytoplasm and fairly regular nuclei exhibiting granular chromatin and small nucleoli) and is positive for neuroendocrine markers. The photomicrograph shows 4 mitotic figures and together with the presence of punctate necrosis should suggest a likely diagnosis of atypical carcinoid tumour. Typical carcinoids *(TC)* (≥5mm in size, mitotic count <2 MF/2mm^2 and no necrosis) NB: at least 3 sets of 2mm^2 should be counted and the mean mitotic count used rather than the single highest rate) and atypical carcinoids *(AC)* (mitotic count 2-10 MF/2mm^2 and/or necrosis, usually punctate but may be in larger zones) occur more often in those <60 years of age and are not necessarily associated with smoking (in contrast to large cell neuroendocrine carcinoma, LCNEC, and small cell carcinoma, SmCC). The 5 year survival rate for TC and AC is about 90% and 60% respectively (higher for resectable tumours), while for LCNEC (stage I disease) and SmCC (without metastases) it is about 33% and 25% respectively. Lymph node metastases are present in 4-14% of TCs and 35-64% of ACs.

4) Correct answer: d) Pulmonary chondroma.

Young individuals with hypertension should raise concern in the first place for secondary causes, some of which include catecholamine-secreting neoplasms. The clinical scenario is very suggestive of Carney triad which is a rare disorder mainly affecting young women, and typically comprising gastric GIST, pulmonary chondroma and extra-adrenal paraganglioma.

Metastatic GIST would be a consideration, but the gastric tumour is small and probably low risk for this.

Metastatic paraganglioma (from a bladder primary) is again a possibility but probably unlikely in the absence of significant pelvic disease and locoregional lymph node spread. Lung carcinoid is an occasional and minor component of multiple endocrine neoplasia (MEN), but the presence of multiple tumourlets would be more in keeping with diffuse idiopathic neuroendocrine hyperplasia (DIPNECH), a very rare condition affecting slightly older women and presenting with respiratory rather than MEN symptoms.

Pulmonary hamartomas have a higher proportion of fibromyxoid stroma, fat, smooth muscle and entrapped epithelium and therefore are reasonably distinct from the chondromas seen in Carney triad which have a thin fibrous pseudocapsule and consist almost exclusively of cartilage ± calcification/ossification.

5) Correct answer: c) Subpleural and paraseptal fibrosis with intervening areas of relatively normal lung, and scattered fibroblastic foci seen in a background of established fibrosis.

The correct answer describes the typical features of usual type interstitial pneumonia (UIP) which is the commonest pattern of idiopathic interstitial pneumonia, correlated clinically with idiopathic pulmonary fibrosis (IPF)/cryptogenic fibrosing alveolitis (CFA), which affects men slightly more often more than women, usually in the 6th to 7th decade. A lung biopsy is usually undertaken only if the presentation is atypical.

Option (A) describes findings that are more suggestive of hypersensitivity pneumonitis which should be considered if poorly defined micronodules are seen on CT (with sparing of lung bases) and may be associated with a specific exposure history eg. bird keeping.

Non-specific interstitial pneumonia (NSIP) usually presents in patients a decade or more younger than those with IPF and finger clubbing is less common. NSIP is frequently associated with lymphocytosis on BAL and ground glass attenuation is a prominent finding on HRCT (in contrast to UIP). Histologically, there is more diffuse, and temporally uniform, interstitial fibrosis and chronic inflammation without fibroblastic foci.

Desquamative interstitial pneumonia (DIP) is only rarely seen in non-smokers and is characterised by basal ground glass attenuation on CT and diffuse accumulation of 'smokers' macrophages within alveoli on histology (as opposed to respiratory bronchiolitis interstitial lung disease (RB-ILD) where the macrophages are concentrated around bronchioles.

Well-formed granulomas distributed along routes of lymphatic drainage typify sarcoidosis (option E), but as with many granulomatous lung conditions, the differential diagnosis might include infective causes amongst others. Patients with pulmonary sarcoidosis tend to be young (20-40 years) and don't usually have inspiratory crackles or clubbing. The radiological features of hilar and mediastinal lymphadenopathy with/without bilateral lung nodules in a lymphatic distribution are quite characteristic.

6) Correct answer: a) ER.

Secondary malignancies of the pleura are probably more frequent considerations in the context of recurrent pleural effusion and abnormal pleural thickening than primary malignancy (mesothelioma). The histology above shows an infiltrative malignant neoplasm composed of a relatively monotonous population of small to medium-sized epithelioid cells which are arranged in discohesive single files and exhibit cytoplasmic vacuoles. The immunohistochemical stain shown in the inset shows nuclear positivity within tumour cells. Although the differential diagnosis might conceivably include mesothelioma (which would normally express WT-1), on balance of probability this is likely to represent metastatic lobular carcinoma from a breast primary (many of which are strongly ER positive).

PAX8 is useful when tumours of renal epithelial or Müllerian origin (amongst others) are considered, the cytoplasmic vacuolation is probably against metastatic melanoma (S100 positive) and although some lung adenocarcinomas have signet ring features (and are sometimes TTF-1 positive), these are rare and tend to exhibit more nuclear pleomorphism. In practice a pathologist might include some of these markers in their panel, but a good history should make it possible to be more directed.

7) Correct answer: d) Non-small cell carcinoma favour squamous cell carcinoma.

About 70% of lung cancers are not resectable at presentation and therefore small biopsies and cytology specimens are the primary method of diagnosis for most patients with lung cancer. Therapeutic advances have driven a need for more accurate terminology and diagnostic criteria for all major histological types of lung cancer encountered in these small specimens. Four important therapeutic advances are recognised: (1) Use of tyrosine kinase inhibitors in advanced lung adenocarcinoma with EGFR mutations. (2) Lung adenocarcinoma with ALK rearrangements responding to ALK inhibitors such as Crizotinib and Ceritinib. (3) Patients with adenocarcinoma or non-small cell lung cancer (NSCLC-NOS) are more responsive to Permetrexed than squamous cell carcinoma (SqCC) and (4) SqCC associated with life-threatening haemorrhage when treated with Bevacizumab. In poorly differentiated NSCLC ie. without glands/papillae, mucin, keratin or intercellular bridges, immunohistochemistry is utilised to refine the diagnosis and avoid a diagnosis of NSCLC-NOS as far as possible (ideally <5% of biopsies with NSCLC): a simple panel of TTF-1 (stains 75-85% of lung adenocarcinoma) and p63 (or the more specific squamous marker p40) may be sufficient in most instances) if both are negative, a cytokeratin stain should be employed to confirm the tumour is carcinoma, and if negative, other stains for epithelioid 'mimickers' eg. melanoma, lymphoma, mesothelioma, angiosarcoma etc should be applied as appropriate.

If neuroendocrine (NE) morphology is present (organoid nesting, rosettes, abundant cytoplasm and prominent nucleoli) then NE markers (CD56, Synaptophysin and/or Chromogranin) might be able to suggest a diagnosis of large cell neuroendocrine carcinoma (LCNEC)) do not use if NE morphology is not suspected.

8) Correct answer: e) Visceral pleural invasion.

Visceral pleural invasion becomes important in small (sub 3cm) tumours (difference between pT stage 1a/b and 2a for example) and therefore at frozen section it is best to sample the tumour from the parenchymal aspect, leaving the overlying visceral pleura intact where possible. This may help to avoid retraction artefacts that might hamper assessment of visceral pleura invasion.

Distance of the tumour from the carina is generally not relevant for a peripheral tumour abutting the visceral pleura and although multiple tumour nodules (same lobe metastases) are an important staging consideration, it would not affect examination of this fresh specimen (but careful examination of the fixed specimen is still required).

Similarly, although best assessed in the fresh state, tumour size can be obtained by careful slicing after fixation (if necessary correlating with the radiological staging).

Vascular invasion is not a staging parameter.

9) Correct answer: b) Multiple neuroendocrine tumourlets may be mistaken for miliary pattern of metastasis.

Diffuse neuroendocrine (NE) cell hyperplasia and multiple tumourlets (as shown in the above photomicrographs) are uncommon (in contrast to incidental tumourlets), and are usually seen in the following 3 settings: (1) Chronic lung injury such as bronchiectasis and fibrosis (commonest, do not usually progress to carcinoid tumours), (2) Diffuse idiopathic pulmonary NE cell hyperplasia (DIPNECH) which is a pre-neoplastic condition associated with carcinoid tumours and (3) That seen in lungs resected for carcinoid tumours both adjacent to and away from the tumour.

In lungs resected for localised lesions (NSCLC or carcinoid) a diagnosis of DIPNECH may not be appropriate, but its presence should be noted in the report as it may have important implications including: (1)

Progression to carcinoid tumours may occur, (2) patients may rarely develop respiratory symptoms and (3) follow-up CT scans may show multiple nodules which could mimic a miliary pattern of metastasis.
Carcinoid tumours by definition are >5mm in size, and minute meningothelial nodules are usually perivenular consisting of whorls of meningothelioid cells with intranuclear inclusions lacking NE marker positivity.

10) Correct answer: b) Acute eosinophilic pneumonia.

Eosinophilic pneumonia (EP) represents a heterogenous collection of inflammatory or infectious diseases of the lung, causes of which include drug/inhalant hypersensitivity reactions, infections (especially parasitic), vasculitides (including Churg-Strauss syndrome and Wegener granulomatosis), allergic bronchopulmonary aspergillosis (ABPA), connective tissue disorders and idiopathic processes. Acute eosinophilic pneumonia is an acute febrile illness which may present with acute respiratory distress (ARDS) like features which responds to corticosteroid therapy. Peripheral eosinophilia may or may not be present, but bronchoalveolar lavage (BAL) should show >20% eosinophils (lung biopsy is therefore usually not necessary in this setting). The condition resolves on steroid therapy without relapse.
Acute fibrinous organising pneumonia (AFOP) differs from diffuse alveolar damage (DAD) by the absence of hyaline membranes, and is associated with a higher mortality rate (53%) than cryptogenic organising pneumonia (COP)(<15%). Acute interstitial pneumonia (AIP) and DAD have similar clinical and histological features and are also associated with high mortality rates.
Desquamative interstitial pneumonia (DIP) classically presents with insidious symptoms (cough and breathlessness on exertion), but may rarely present with severe, acute disease, and is associated with smoking. Patients who do not respond to smoking cessation may require steroid therapy.
Lymphoid interstitial pneumonia (LIP) is more common in women, usually in their 5th decade, and typically presents with gradual onset of cough and dyspnoea over 3 years or more. It is often associated with collagen vascular disease or immunodeficiency. Most patients with LIP improve with steroid therapy.

11) Correct answer: a) Adenocarcinoma.

The following precursor lesions are recognised in lung cancer: atypical adenomatous hyperplasia (AAH, as shown in the photomicrographs above) the first pre-invasive lesion for adenocarcinoma), squamous dysplasia (for squamous cell carcinoma) and diffuse idiopathic neuroendocrine cell hyperplasia (DIPNECH), which predisposes to the development of carcinoid tumours. AAH is usually smaller than 5mm (but this is not an absolute criterion) and tends to show less cytological atypia and cellular crowding when compared to adenocarcinoma in situ (AIS) the second pre-invasive lesion for lung adenocarcinoma), but the distinction may be difficult histologically as there is a continuum of changes between AAH and AIS. Gaps along the basement membrane are seen between the mild to moderately atypical epithelial cells. AAH is seen in the background lung of 5-23% of resection specimens for lung adenocarcinoma and shows similar molecular genetic abnormalities. The AAH lesion is usually not targetable by CT-guided biopsy and therefore is not usually a consideration on needle core biopsy.

12) Correct answer: c) Higher risk of lymph node metastases.

The histological images show lung adenocarcinoma with a micropapillary pattern. The papillary tufts lack fibrovascular cores and appear to 'float' in air-spaces or sometimes within stroma. This pattern is well recognised as indicating a poorer prognosis in ovarian, breast and bladder cancer. In lung adenocarcinoma, the main patterns recognised are lepidic, acinar, papillary, micropapillary and solid and studies have shown that the micropapillary pattern correlates with lymphovascular invasion (LVI) and lymph node metastases resulting in reduced survival.
Tumours with this pattern are more likely to be associated with EGFR mutations rather than ALK chromosomal rearrangements. Although a pattern sometimes seen in metastases from other organs, careful

use of immunohistochemistry and correlation with clinical and radiological findings should not result in misclassification.

13) Correct answer: c) Malignant mesothelioma.

The photomicrographs show an epithelioid tumour exhibiting short, solid cords and small nests of eosinophilic cells, focally involving fat (upper right of picture). Papillaroid projections are present on the surface and there is strong membranous expression of D240.
D240 is often used as a marker for lymphatic endothelium (to confirm tumoral lymphovascular invasion for example), but also stains tumour cells of, amongst others, Kaposi sarcoma, epithelioid mesothelioma and seminoma.
It does not typically stain adenocarcinoma and the morphological features shown (albeit at low magnification) should exclude angiosarcoma, haemangioendothelioma and seminoma (the latter also being unusual in this age group).

14) Correct answer: e) p63.

p63 strongly stains a high proportion of squamous cell carcinoma (SCC) (nuclear staining) and is generally not expressed in mesothelioma (some examples may show a few positive cells).
Calretinin may stain a significant proportion of lung SCC, as well as some thymic carcinomas (which also express squamous markers) and CK5/6 does not allow distinction between mesothelioma and squamous cell carcinoma.
Mesothelin stains mesothelioma, but also stains just over 25% of squamous cell carcinomas.
p16 is usually used as a surrogate marker for high risk HPV in squamous carcinoma of the cervix, penis and oral cavity for example, but is not necessarily specific for this (eg. it is expressed by some small cell carcinomas).

15) Correct answer: G) Sarcomatoid carcinoma
16) Correct answer: D) Metastatic rhabdomyosarcoma
17) Correct answer: A) Malignant fibrous tumour
18) Correct answer: J) Synovial sarcoma
19) Correct answer: C) Metastatic melanoma (spindle cell type)

Explanations for 15-19)
These are the typical immunoprofiles for the above entities, as shown in the table on page 108.

20) Correct answer: a) Alpha1-antitrypsin deficiency.

Emphysema is defined pathologically as abnormal, permanent enlargement of the airspaces distal to the terminal bronchiole accompanied by destruction of their walls without obvious fibrosis. This is in contrast to the honeycomb lung associated with end stage interstitial lung disease. The commonest pattern of emphysema, centrilobular (or centriacinar) emphysema, is seen in heavy smokers who will typically show most pronounced changes at the apices of the lobes (better ventilated and less well perfused, therefore relatively lower levels of protective anti-proteases, especially alpha1-antitrypsin). The panlobular (or panacinar) emphysema seen in alpha1-antitrypsin deficiency accounts for a small number of emphysema patients and usually affects the basal portions of the lungs to a greater degree (better perfusion means more neutrophils and therefore more destructive proteases without the benefit of normal serum alpha-1 antitrypsin). This condition is also associated with liver cirrhosis (the most likely explanation for this man's varices) due to damage caused by accumulation of the alpha-1 antitrypsin in hepatocytes.

21) Correct answer : d) *Rhizopus oryzae*.

Pulmonary mucormycosis is the second most important cause of fungal pneumonia (after Aspergillosis) in immunosuppressed patients and is associated with significant mortality (40-76%), especially in those with haematological malignancy. The older term of zygomycosis is no longer used. Agents of mucormycosis include *Rhizopus, Mucor, Cunninghamella, Lichtheimia* (formerly *Absidia*) and *Rhizomucor* species, which are ubiquitous spore-forming fungi that are frequently harmless for the immunocompetent individual. The fungal hyphae of mucorales organisms are typically described as broad (6-16μm), ribbon-like, non-septate (or sparsely septate) and irregularly shaped with non-dichotomous, right angle branching.

22) Correct answer: b) 1a Invasive fungal lung infection and right heart failure, 1b Multiple myeloma (treated with steroid therapy), 2 Fracture neck of femur (operated October 2015) and aortic valve stenosis.

The clinical scenario refers to cardiovascular instability and right heart strain and that he dies before a CT-angiogram could be performed (the clinical team suspected pulmonary embolism). Thus the mode of death is probably one of right heart failure. The presence of angio-invasive fungal disease (mucormycosis) and pulmonary infarction suggests that this is the main cause for the right heart failure. For simplicity, these could be incorporated as '1a Invasive fungal lung infection and right heart failure.' His underlying multiple myeloma would most likely have predisposed him to this invasive fungal infection (but steroid therapy may not have helped) and therefore is listed as (1b). The fracture neck of femur and aortic valve stenosis are probably only indirectly contributory.
Option (A) is incorrect because he did not die as a direct result of the femoral fracture (this just 'revealed' the myeloma), option (C) is not correct because he did not have a pulmonary embolus and myeloma is listed under part 2, and option (D) may have been considered if myeloma was listed as (1c). Option (E) is probably the closest distractor, but reads a little clumsily, as multiple different entities are lumped together in (1a) and should not have the femoral fracture as contributing *directly* to his death.

23) Correct answer: e) Elevated lactate dehydrogenase (LDH).

The description and provided immunoprofile supports a diagnosis of mediastinal seminoma (provided metastasis from a testicular primary has been excluded) which is often associated with elevations of serum LDH, but not usually AFP. The latter is more likely to be seen with non-seminomatous germ cell tumours such as embryonal carcinoma and yolk sac tumour. The presence of granulomatous inflammation is a potential diagnostic pitfall, especially in small biopsies or FNA cytology samples, as the malignant infiltrate may be obscured by the inflammatory response.

24) Correct answer: a) Immunoglobulin heavy chain gene PCR confirms monoclonality in about 60% of cases of MALT lymphoma.

Primary pulmonary lymphoma is rare, comprising <0.5% of primary lung tumours, <1% of all lymphomas and about 3-4% of extranodal lymphomas. Marginal zone lymphoma (MZL) of mucosa-associated lymphoid tissue (MALT) is the commonest type of primary lung lymphoma (70-80% of cases), usually presenting in older patients (but sometimes in younger patients with immunosuppression or HIV infection). There is an association with autoimmune disease (eg. Sjögren syndrome, rheumatoid arthritis or SLE) and about a third of patients are asymptomatic at presentation. Radiological patterns include ground-glass infiltrates, peripheral mass with pleural thickening or pulmonary consolidation with air bronchograms and apparent cavitation. MALT lymphoma of the lung may also occasionally present as multiple and bilateral nodules. Lymphoepithelial lesions (LEL) are not pathognomonic and granulomas and amyloid deposition may be observed. The polymorphous appearance of the lymphocytes favours a diagnosis of MALT lymphoma over other low grade B cell lymphomas such as small lymphocytic lymphoma (SLL/CLL), follicular lymphoma and mantle cell lymphoma, but immunohistochemistry and molecular investigations are also required to make the distinction between MALT lymphoma and reactive or other neoplastic proliferations. Monoclonality is demonstrated in about 60% of cases (Ig heavy chain gene PCR) and about 40-50% of MALT lymphomas harbour MALT1 gene rearrangement, t(11;18)(q21;q21).

CHAPTER 10

DERMATOPATHOLOGY
Dr Limci Gupta

1) A middle aged male presents with well-defined erythematous plaques with silvery white scales on the scalp and elbows associated with onycholysis of the nails. What is the likely diagnosis?

   a) Lichen planus
   b) Lichen sclerosus
   c) Psoriasis
   d) Interface dermatitis

2-4) Extended matching questions : Please match the following histological features with the listed diagnosis.

   A) Lichen planus
   B) Lichen sclerosus
   C) Lichen simplex chronicus
   D) Pemphigus vulgaris
   E) Bullous pemphigoid

2) A middle aged male presents with scaly, erythematous and nodular lesions. He gives a history of persistent rubbing and scratching. Histology shows psoriasiform epidermal hyperplasia, orthokeratosis and parakeratosis. There is vertically orientated fibrosis in the papillary dermis. There are no other significant findings.

3) A patient presents with a skin condition manifesting as hypo-pigmented plaques on the trunk. Histologically, there is hyperkeratosis and epidermal atrophy with loss of rete ridges. There is collagenisation of superficial dermis with band like chronic inflammatory infiltrate beneath it.

4) An elderly lady presents with pruritic, purplish plaques on the flexor aspect of her arms. Microscopically, there is hypergranulosis and saw-tooth like rete ridges. A lichenoid band like chronic inflammatory infiltrate is present beneath the basement membrane. Colloid bodies are present. What is this skin condition?

Scenario for questions 5 and 6: A patient presents with a skin condition characterised by flaccid bullae on the scalp and trunk. Similar lesions were present within the oral cavity. Histologically, the lesions show suprabasal acantholysis with the basal layer attached to the basement membrane. There is focal involvement of follicular epithelium. There are scattered acantholytic cells within the blister and adjacent epidermis.

5) What is the diagnosis?

   a) Bullous pemphigoid
   b) Pemphigus vulgaris
   c) Toxic epidermal necrolysis
   d) Dermatitis herpetiformis

6) Direct immunoflorescence would show the following pattern:

a) Intercellular deposition of IgG
   b) IgG deposits around the blood vessels in the dermis
   c) Ig A granular deposits in the papillary dermis
   d) Linear IgG in the basement membrane.

7) A 5 year old girl presents with flesh coloured papulo-nodular lesions on the arms and back, associated with central umbilication. Microscopically, the lesion shows profound papillomatous epidermal hyperplasia. The epithelial cells are enlarged and show eosinophilic cytoplasmic inclusions. What is the likely diagnosis?

   a) Cytomegalovirus
   b) Molluscum contagiosum
   c) Herpes simplex infection
   d) Fungal infection

Extended matching questions choices for questions 8-13: Please match the following clinical picture and/or histological features with the listed diagnosis.

   A) Bullous pemphigoid
   B) Pemphigus vulgaris
   C) Dermatitis herpetiformis
   D) Impetigo
   E) Porphyria
   F) Epidermolysis bullosa
   G) Herpes simplex
   H) Eczema
   I) Erythema multiforme

8) A 15 year old girl presents with variable sized blisters at the corner of mouth and ulcers within the oral mucosa. There is a history of recent pneumonia. Histology shows necrotic keratinocytes and multinucleate cells with basophilic intranuclear inclusions.

9) An elderly gentleman presents with large tense bullae on the flexor aspect of forearms, abdomen and thighs.

10) A young male presents with multiple itchy vesicles on the extensor aspects of the elbow.

11) A newborn develops multiple blisters soon after birth. Histology shows pauci-cellular sub-epidermal blisters.

12) A 5 year old child develops multiple pustules on the face, some of which rupture and are covered by honey coloured crust. Histology shows sub-corneal blisters with aggregates of neutrophil polymorphs. Culture grows staphylococcus.

13) A middle aged female presents with multiple oozing crusted lesions on both hands. Histology shows intra-epidermal spongiosis and occasional lympho-histiocytic collections within the epidermis. Mild chronic inflammatory infiltrate is present within the dermis.

14) A patient presents with a skin condition characterised by recurrent targetoid lesions mainly involving distal extremities. On microscopy, the lesion shows basal vacuolar degeneration with subepidermal bullae and intraepidermal dyskeratotic keratinocytes. What is the diagnosis?

a) Erythema nodosum.
b) Erythema multiforme.
c) Granuloma annulare
d) Scabies

15-17) Extended matching questions: Please match the following with the listed diagnosis.

A) Xanthelasma
B) Necrobiosis lipoidica
C) Erythema nodosum
D) Xantho-granuloma
E) Granuloma annulare

15) A 45 year old diabetic female presents with multiple yellow coloured plaques on the shin. Histology shows sandwich-like horizontal layers of altered collagen alternating with inflammatory cells consisting of lymphocytes, plasma cells, histiocytes and a few multinucleated giant cells within the deep dermis.

16) An elderly female presents with multiple ring shaped lesions on her forearms. Histology shows bundles of necrotic collagen surrounded by palisades of histiocytes. The immunostaining is positive for CD68 and vimentin.

17) A young girl presents with bilateral tender nodules on both legs which gradually decrease in size over 3-4 weeks. Histologically, the dermis contains chronic peri-vascular and peri-adnexal inflammation. The connective tissue septa are oedematous and contain acute and chronic inflammation sparing fat lobules.

18) A dermal tumour, unattached to the overlying epidermis, is composed of islands of uniform basaloid cells with some peripheral nuclear palisading. Scattered abortive hair follicle-like structures are present. There are no retraction artefacts.
What is the diagnosis?

a) Trichoepithelioma.
b) Squamous cell carcinoma.
c) Basal cell carcinoma.
d) Poroma.

19-22) Extended matching questions: Please match the following with the appropriate diagnosis.

A) Spitz naevus.
B) Dysplastic naevus.
C) Blue naevus
D) Naevus sebaceous
E) Congenital naevus

19) A young male presents with a 1cm dark pigmented lesion on the upper arm. Histology shows architectural atypia with fusion of rete ridges. There is nested and lentiginous proliferation of melanocytes with lamellar fibroplasia and random cytological atypia.

20) This is histology of a 10 mm reddish lesion on the forearm of a 30 year old male.

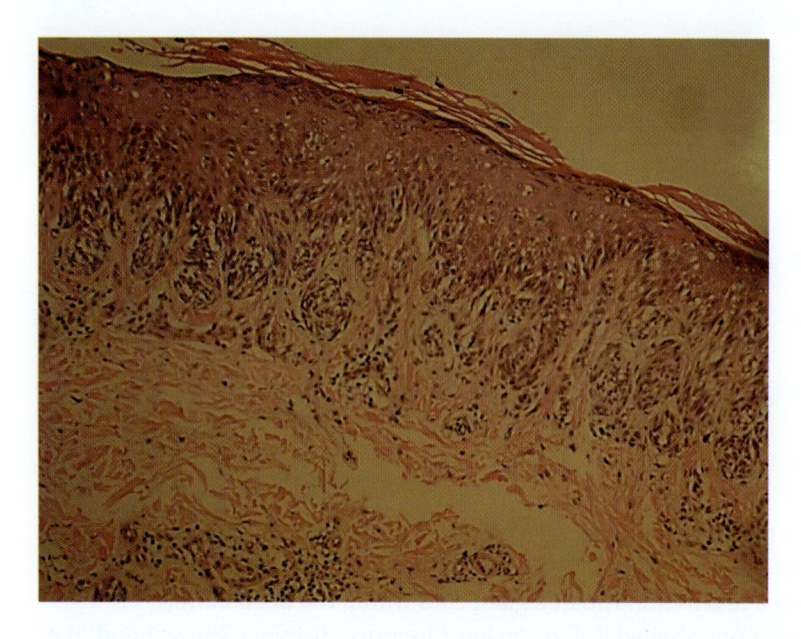

It shows acanthosis of the epidermis and fascicles of spindle cells perpendicular to the dermis (as seen in the figure). There is clefting around the nests with Kamino bodies in the basal layer. Central pagetoid spread is also present.
What is the diagnosis?

21) A patient presents with a well circumscribed nodule on the dorsum of the right hand. Microscopically, the lesion is well circumscribed and is composed of a two-cell population with polygonal cells and pigmented dendritic melanocytes.

22) A young male presents with a hairless patch on the scalp. Histology shows prominent sebaceous lobules with malformed hair follicles and apocrine glands within the dermis.

23) In malignant melanoma, Breslow thickness constitutes a vital prognostic factor and T stage parameter. The Breslow thickness is measured from:

    a) Stratum corneum to the deepest point of invasion of the tumour.
    b) Granular layer to the deepest point of invasion of the tumour.
    c) Dermo-epidermal junction to the deepest point of invasion of the tumour.
    d) Stratum spinosum to the deepest point of invasion of the tumour.

24) Which one of the following is a feature of a radial growth phase melanoma?

    a) Always invasive.
    b) Invading Clark level 3.
    c) Tumour nests compressing the surrounding tissue.
    d) Absence of dermal mitoses.

25) An ulcerated nodule on the forehead shows dermal proliferation of hyperchromatic pleomorphic spindle shaped cells surrounded by desmoplastic stroma. There are lymphoid collections at the advancing edge of the lesion. The tumour cells are positive for S100 and negative for HMB45. What is the diagnosis?

    a) Benign intradermal naevus
    b) Congenital naevus

c) Desmoplastic melanoma
d) Deep penetrating naevus

26) Which subtype(s) of basal cell carcinoma is/are low risk according to the growth pattern :

a) Superficial
b) Infiltrative
c) Morphoeic
d) Nodular

27) Which one of the following is a high risk pathological factor for basal cell carcinoma:

a) Basi-squamous type
b) Tumour < 10 mm from margin
c) Clark level 4
d) Nodular pattern

28) According to the Royal College of Pathologists minimum dataset, which of the following is NOT a risk factor for the recurrence of a squamous cell carcinoma of the skin:

a) Perineural invasion.
b) Lymphovascular invasion.
c) Grade of tumour.
d) Tumour invading Clark level 3.

29) The following is a high risk feature for the primary tumour (T) staging of squamous cell carcinoma of the skin.

a) Tumour on forehead.
b) Tumour on the hair bearing lip.
c) Tumour on nose.
d) 1mm thickness.

30) Which of the following is NOT a staging determinant(s) for squamous cell carcinoma of the skin?

a) Depth of invasion.
b) Lymphovascular invasion.
c) Poorly differentiated carcinoma.
d) Perineural invasion.

31) A 65 year old female presents with extensively ulcerative and locally invasive squamous cell carcinoma involving her lower right cheek. A CT scan shows tumour invading the maxilla focally. What is the T stage of this tumour?

a) pT2.
b) pT3.
c) pT4.
d) pT1.

32) An ulcerated nodule on the scalp of an elderly man shows a dermal proliferation of spindle shaped and polygonal pleomorphic cells some of which have vacuolated cytoplasm. The tumour cells are negative for MNF 116, Desmin, S100, Melan A and HMB45. What is the likely diagnosis?

   a) Spindle cell carcinoma.
   b) Spindle cell melanoma.
   c) Atypical fibroxanthoma.
   d) Leiomyosarcoma.

33) A violaceous nodule on the face of an elderly female is microscopically composed of a sheet-like proliferation of round cells with hyperchromatic and dusty nuclei, inconspicuous nucleoli and increased mitoses. The tumour cells are negative for TTF1 and CD45. They show perinuclear dot like positivity with CK20. The cells are positive for chromogranin and CD56. What is the diagnosis?

   a) Merkel cell carcinoma.
   b) Lymphoma.
   c) Small cell carcinoma of the lung.
   d) Melanoma.

34) A 40 yr old lady presented with an abdominal wall mass which was locally excised. Histology under high magnification shows:

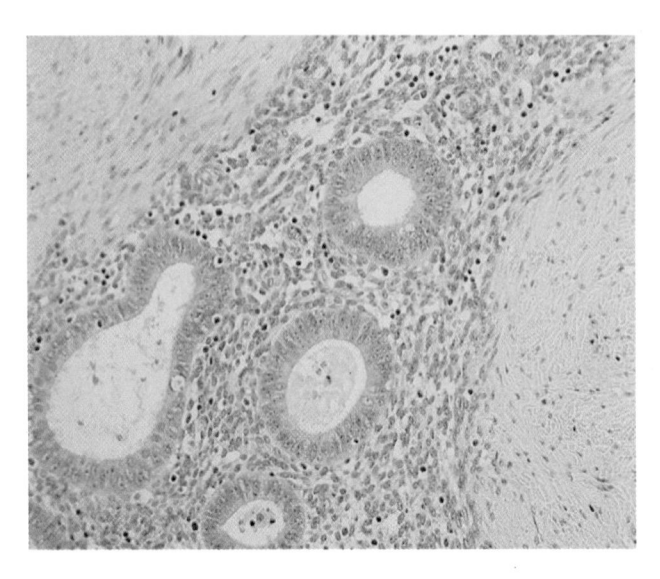

What is this condition associated with?
a) Skin adnexal tumour
b) Abdominal wall scar from caesarean section
c) Syringocystadenoma
d) GIST

Correct answers and explanations:

1) Correct answer : c) Psoriasis.

Psoriasis is an auto-immune disorder characterised by erythematous plaques with a silvery white scale on the extensor aspects of elbows, scalp and extremities. They can be associated with nail changes including onycholysis and nail pitting.
Histology shows a parakeratotic and acanthotic epidermis with thickened and elongated rete ridges. The granular layer ranges from thin to absent. There is a collection of neutrophils within the stratum corneum called Munro microabscesses. There is supra-papillary thinning with dilated blood vessels in the dermal papillae with a perivascular chronic inflammatory infiltrate. The presence of these blood vessels are responsible for minor skin bleeds when the scale is lifted (Auspitz sign).

2) Correct answer : C) Lichen simplex chronicus (also called prurigo nodularis).

Lichen simplex chronicus presents with erythematous, scaly and thickened lesions. Prurigo nodularis refers to a raised pruritic nodule. Microscopically, there is an orthokeratotic, parakeratotic epidermis with psoriasiform hyperplasia. The dermis shows lamellar fibrosis within dermal papillae.

3) Correct answer :B) Lichen sclerosus et atrophicus.

This manifests mostly on the genitalia and trunk. Microscopically, the epidermis is thin and atrophic with basal cell vacuolation. The superficial dermis shows hyalinisation with a band of chronic inflammation beneath it.

4) Correct answer : A) Lichen planus.

This is the classical microscopic picture of lichen planus. This is an auto-immune lichenoid dermatosis which manifests as purple and pruritic plaques, mostly on the flexor aspects of the extremities. There may be involvement of the lips, oral mucosa and scalp.

5) Correct answer : b) Pemphigus vulgaris .

6) Correct answer : a) Intercellular deposition of IgG.

Explanations for answers 5 & 6:
This is a classical histological picture of this entity.
Pemphigus vulgaris is an auto-immune acantholytic disorder characterised by painful flaccid blisters on the skin with a predilection for the trunk, groin, axillae, scalp, face, pressure points and on mucous membrane lining the mouth, nose, throat and genitals.
The immunofluorescence shows a 'fish-net' appearance in the epidermis with IgG (with IgG deposition in intercellular spaces). In this disorder, auto-antibodies are formed against desmoglein 3, which is present in intercellular bridges and acts like a glue for epidermal cells to stick together, thereby causing acantholysis and blistering.

Bullous pemphigoid and dermatitis herpetiformis show sub-epidermal blistering.
On immunofluorescence, dermatitis herpetiformis shows IgA granular deposits in the papillary dermis and bullous pemphigoid shows linear IgG pattern at the basement membrane.

7) Correct answer : b) Molluscum contagiosum.

This presents as small, firm, raised papules on the face, neck, trunk and limbs but rarely affects palms and soles. It is caused by the molluscum contagiosum virus (pox virus).
Microscopic features consist of molluscum bodies which present as intra-cytoplasmic inclusions.

8) Correct answer : G) Herpes simplex.

This is a classic picture seen with this infection.

9) Correct answer : A) Bullous pemphigoid.

There is a vesiculo-bullous rash typically on the flexor aspects of the limbs and trunk.
Histology shows subepidermal blisters. There is patchy basal layer vacuolation with abundant degranulated eosinophils. There are linear deposits of IgG and C3 at dermo-epidermal junction.

10) Correct answer : C) Dermatitis herpetiformis .
This presents with itchy blisters on the extensor aspects of the upper and lower limbs. It is associated with gluten sensitive enteropathy. Histology shows sub-epidermal blisters with neutrophil aggregates at the dermal papillae. The direct immunofluorescence shows granular IgA deposits at the dermal papillae.

11) Correct answer : F) Epidermolysis bullosa.

This is a condition caused by defects in structural proteins that normally tether the skin to underlying structures. They are characterised by pauci-cellular blisters at the site of trauma and pressure.

12) Correct answer : D) Impetigo.

This is a communicable infection that often involves hands and face. Characteristic histological features are subcorneal accumulation of neutrophils with pustule formation. Gram stain shows Gram positive cocci.

13) Correct answer : H) Eczema.

This is also called spongiotic dermatitis. It presents as an itchy rash or papulo-vesicular lesions and sometimes oozing and crusted lesions. With time, these lesions can develop scaly plaques. Microscopically, there is intra-epidermal spongiosis causing separation of keratinocytes followed by blister formation. The dermis usually shows a perivascular lymphocytic inflammatory infiltrate.

14) Correct answer : I) Erythema multiforme.

This is a classical presentation of this vesiculo-bullous disease with erythematous plaques resembling targetoid lesions on the distal extremities especially palms, soles and also the mucosal surfaces. It can also present as papules, macules and bullae.
Microscopically, there is a variable vacuolar degeneration with intra-epidermal dyskeratotic keratinocytes, and interface inflammation. Severe forms include Steven-Johnson syndrome and toxic epidermal necrolysis.

15) Correct answer : B) Necrobiosis lipoidica.

This skin condition is usually associated with diabetes. The histology shows large irregular areas of granulomatous inflammation with areas of collagen degeneration centering around deep dermis and subcutis. The macrophages, admixed with lymphocytes and prominent plasma cells, dissect between collagen bundles.

16) Correct answer : E) Granuloma annulare.

This is a classical presentation and histology of this dermatological auto-immune condition. This is a necrobiotic lesion associated with other autoimmune conditions such as SLE, rheumatoid arthritis & autoimmune thyroiditis.

17) Correct answer : C) Erythema nodosum .
This condition is associated with bacterial, mycoplasma, fungal infections, tuberculosis, drugs, sarcoidosis and certain malignancies. It presents as multiple tender nodules on the shins, forearms or trunk. It is a subtype of septal panniculitis associated with septal acute inflammation followed by chronic inflammation. There may be some involvement of fat lobules. Occasional giant cells and eosinophils are present. Radial granulomas can also be present (histiocytic collections with giant cells surrounding cholesterol clefts).

18) Correct answer : a) Trichoepithelioma.

Trichoepithelioma is a benign skin adnexal tumour. Basal cell carcinoma is often mistakenly diagnosed due to the resemblance with this tumour but identification of hair follicle like structures, absence of retraction artefacts and absence of atypical mitoses are characteristic of trichoepithelioma. Most often, there is no attachment of the basaloid nests to the overlying epidermis in trichoepithelioma whereas attachment is usually seen in BCCs.
BCL 2 and CD34 are good markers for confirmation. BCL2 diffusely stains the basaloid nests in BCC whereas staining is focal in trichoepithelioma.
CD34 diffusely stains basaloid cells in trichoepithelioma whereas it stains only around the nests only in BCC.

19) Correct answer : B) Dysplastic naevus.

Dysplastic naevi develop as larger lesions (usually > 1cm) which are irregularly shaped and variably pigmented. Microscopically, there is architectural atypia with bridging of rete ridges. There is a lentiginous and nested proliferation of melanocytes. This proliferation extends beyond the dermal component and is called 'shouldering effect'. There is lamellar fibroplasia of the dermis and random cytological atypia.

20) Correct answer : A) Spitz naevus.

This is a benign melanocytic lesion commonly seen in children and young adults and more frequently involves the head and neck regions. It usually presents as a pink-red coloured nodule. The overlying epidermis is acanthotic with fascicles of spindle shaped and epithelioid cells perpendicular to the basement membrane arranged in a 'rain-drop' pattern. Kamino bodies are present in the basal layer of epidermis. Central pagetoid spread can be seen in this lesion.
It can be confused with melanoma but melanoma is generally seen in older people. Cytological atypia will be marked and malignant melanocytes will be arranged parallel rather than perpendicular to the basement membrane. In the dermal component of melanoma, deeper maturation will be absent and mitoses (including atypical mitoses) can be present in the deeper aspect in melanomas.

21) Correct answer : C) Blue naevus .

The blue naevus is a slightly elevated or sometimes dome shaped lesion on the dorsum of the hands or feet. It is a well circumscribed pigmented lesion within the dermis mainly with the two cell population of pigmented dentritic melanocytes and polygonal cells. This lesion may extend to the subcutaneous fat.

22) Correct answer : D) Naevus sebaceous.

This usually occurs as a congenital, hairless yellowish plaque on the scalp or face.
Histology shows malformed hair follicles with heterotopic apocrine glands. It can be associated with skin adnexal tumours and skin carcinomas.

23) Correct answer : b) Granular layer to the deepest point of invasion of the tumour.

An increase in Breslow thickness is associated with increased risk of metastasis and decreased survival. Breslow thickness is measured from the granular layer to the deepest point of invasion.

24) Correct answer : d) Absence of dermal mitoses.

Radial growth phase melanomas can be invasive or non – invasive (also called in situ malignant melanoma). The cells in the invasive growth phase are either solitary or in the form of small clusters. Dermal mitoses are absent. The tumour nests do not compress or distort the surrounding tissue. This type is usually less than 1 mm and restricted to Clark level 2.

25) Correct answer : c) Desmoplastic melanoma.

This is a classic histological picture of this entity.
- Commonly encountered in elderly patients.
- Deeply infiltrative, sometimes paucicellular with marked interstitial fibrosis and collagenisation.
- Look for mitosis (scanty, abnormal) and lymphocytic aggregates at the infiltrative edges.
- In case of difficulty in differentiating from neural elements of scar tissue, do neurofilament markers and look for adnexal structures (absent in scar tissue)
- Look carefully at the epidermis for any atypical melanocytic proliferation.
- S100 + ; HMB45- (or positive only in the superficial papillary dermis); NF – ; Melan A + (can be positive in about 45% cases)
-

Benign congenital naevus has uniform benign melanocytes tracking around skin adnexae and extending into the subcutaneous tissue along the septa and in single cells. It does not have deep mitoses or pleomorphism. Congenital naevi have a higher risk of developing into melanoma compared to other benign naevi.
Deep penetrating naevus mostly occur on the head and neck. It has a characteristic microscopic architecture with a dermal and subcutaneous proliferation of spindled and epithelioid cells in a wedge shaped manner. Scattered melanophages are typically seen. There are no deep mitoses. The naevus cells are positive for S100 and HMB45. These lesions rarely recur and never metastasize.

26) Correct answers : a) Superficial and d) Nodular.

Two groups are described according to growth pattern which correlate with the biological risk of the tumour.
Low risk: superficial, nodular, fibroepithelial.
High risk: micronodular, infiltrating/morphoeic/sclerosing

27) Correct answer : a) Basi-squamous type.

Risk assessment is required for skin cancer MDT discussion. It is associated with risk of recurrence or persistent disease.

Pathological high risk factors include: basi-squamous, vascular invasion in basi-squamous, perineural invasion, Clark level 5 or deeper, involved margin of less than 1mm margin. The presence of moderate to severe squamous atypia or a malignant squamous component is associated with high risk of recurrence and metastasis. These lesions are usually categorised as basi-squamous type.

28) Correct answer : d) Tumour invading Clark level 3.

Perineural invasion is associated with an increased risk of local recurrence and is also a staging determinant. Lymphovascular invasion is associated with increased risk of local recurrence and distant metastasis.
A poorly differentiated or an undifferentiated SCC is both a risk of recurrence and a staging determinant.
Tumour invading Clark level 4 is a risk of recurrence and a staging determinant (Please also see staging in explanation for answer 30).

29) Correct answer : b) Tumour on the hair bearing lip.

The high risk features for SCC are : >2mm thickness, Clark level 4 or more, perineural invasion, tumour on the ear or hair bearing lip and poorly differentiated or undifferentiated.
(Please also see staging in explanation for answer 30)

30) Correct answer : b) Lymphovascular invasion.

The factors for staging of the tumour are size, thickness of tumour, level of invasion, perineural invasion and grade of the tumour.
Lymphovascular invasion is associated with increased risk of local recurrence and distant metastasis. Margin positivity is associated with increased risk of local recurrence.

pT staging of skin SCC:

T1 : Tumour 20 mm or less in greatest dimension and with less than two high-risk features
T2 : Tumour greater than 20 mm in greatest dimension or any size and with two or more high-risk features
T3 : Tumour with invasion of maxilla, mandible, orbit or temporal bone
T4 : Tumour with invasion of skeleton (axial or appendicular) or perineural invasion of skull base

High-risk features for the primary tumour (T) staging for skin SCC:

Depth/invasion >2 mm thickness
Clark level 4 or more
Perineural invasion
Anatomic location : Primary site ear
                           Primary site hair-bearing lip
Differentiation :   Poorly differentiated or undifferentiated

31) Correct answer : b) pT3.

Invasion of the facial or cranial bone (maxilla, mandible, orbit or temporal bone) is pT3 (Please see staging in explanation for answer 30).

32) Correct answer : c) Atypical fibroxanthoma (AFX).

This is a low grade tumour, usually presenting as an ulcerative nodular lesion. It occurs on the head and neck regions usually in old age and usually in sun damaged and /or radiation damaged skin. It rarely recurs or metastasizes.

Microscopically, it shows polygonal/spindled pleomorphic cells, some of which have vacuolated cytoplasm within the dermis. Numerous multi-nucleated giant cells are present. **This is a diagnosis of exclusion and the differentials include SCC, melanoma and leiomyosarcoma.** The tumour cells are negative for epithelial, melanocytic and muscle markers.

33) Correct answer : a) Merkel cell carcinoma.

This is an aggressive neuroendocrine tumour of the skin which can present in the skin, oral cavity and other sites. It presents as a violaceous nodule usually in the head and neck region. Microscopically, the tumour is composed of sheets of uniform round cells with hyperchromatic and dusty nuclei, inconspicuous nucleoli and scanty cytoplasm. The tumour has increased mitotic activity. There is dot like positivity for CK 20. The cells are positive for NSE, chromogranin and synaptophysin. The differentials include metastatic neuroendocrine carcinoma (Merkel cell Ca is negative for TTF1) and lymphoma (CD45+).

34) Correct answer : b) Abdominal wall scar from caesarean section.

This is a classical picture of endometriosis. Abdominal wall endometriosis usually occurs in the surgical scar of Caesarean sections or abdominal hysterectomies.
It should contain at least two of the following features:
- Endometrial stroma
- Endometrial glands
- Haemosiderin laden macrophages

CHAPTER 11

SOFT TISSUE & OSTEOARTICULAR PATHOLOGY
Dr Rukma Doshi & Dr Jayson Wang

For questions 1 till 6 : Extended matching item questions:

A) Fibromatosis
B) Inflammatory myofibroblastic tumour
C) Low grade fibromyxoid sarcoma
D) Myxoinflammatory fibroblastic sarcoma
E) Nodular fasciitis
F) Solitary fibrous tumour

For each of the following descriptions of *soft tissue lesions*, select the **most appropriate** diagnosis from the list of options above. Each option may be used once, more than once, or not at all.

1) A 50 year old woman presents with a 4cm mass in the jaw. Excision shows a variably cellular lesion with areas of storiform, fascicular and patternless architecture, composed of fusiform cells in myxoid or collagenous stroma with branching blood vessels. The cells are positive for STAT6.

2) A 68 year old man with a 5cm mass found in the mesenteric fat. Histology shows wavy tapered spindled cells arranged in sweeping fascicles. There are scattered blood vessels with sclerosis and mast cells in the vessel walls. The nuclei of the cells are positive for beta-catenin.

3) A 35 year old man with 3cm mass in the left ankle. Biopsy shows an infiltrative spindle cell tumour within myxoid stroma, containing stellate cells with some highly pleomorphic Hodgkin-like cells and vacuolated cells. There is a background of lymphocytes, eosinophils and neutrophils.

4) A 25 year old man presents with a rapidly growing 3cm mass in the deltoid muscle. Histology shows a circumscribed lesion with bland spindled cells arranged in a loose storiform pattern, with extravasated red blood cells and keloid-like collagen.

5) An 18 year old female presents with a 5cm mass in the thigh. Histology shows a lobulated lesion composed of short spindled cells, some with oval or oblong morphology, arranged in a fascicular pattern in collagenous stroma with focal whorled arrangement into rosettes. The cells are positive for MUC4.

6) A 35 year old woman found to have a 2cm mass in the mediastinum. Histology shows spindled and stellate cells arranged in tight storiform pattern with mixed inflammatory cells including numerous plasma cells.

7) A 34 year old woman presents an 8cm mass in the gluteus region. Excision of the mass show a lobulated cellular tumour composed of small oval and epithelioid cells with vacuolated cytoplasm suggestive of lipoblasts. The scant stroma is myxoid with fine branching capillary sized blood vessels. Which of the following is the diagnosis?

   a) Angiolipoma
   b) Chondroid lipoma
   c) Hibernoma

d) Round cell liposarcoma
e) Well-differentiated liposarcoma

8) A 65 year old woman presents with constipation. CT scan shows a 20cm mass arising from the retroperitoneum encasing the kidney and colon. Resection of the mass shows a tumour with areas of lipomatous differentiation with adipocytes showing hyperchromatic nuclei, as well as separate areas of atypical spindled cells arranged in fascicles. Which of the following molecular tests will support the diagnosis?

a) 11q13 rearrangements
b) Amplification of MDM2 gene
c) Fusion of FUS and DDIT3 genes
d) HMGA2-LPP fusion gene
e) PDGFRA gene mutations

9) A 45 year old woman presents with a 1cm nodule in the proximal calf. Excision shows a poorly demarcated spindle cell lesion in the dermis arranged in fascicles and storiform patterns. Which of the following features favour a dermatofibrosarcoma protruberans over a dermatofibroma?

a) Confined within the dermis with occasional vertical extension into fat
b) Entrapment of collagen bundles at the periphery
c) Grenz zone over the lesion with overlying hyperplastic epidermis
d) Positivity for CD34
e) Positivity for CD68

10) A 50 year old woman presents with a nodule in the forearm. Excision shows nests of epithelioid cells in the dermis, with fine granular eosinophilic cytoplasm. The cells are positive for D-PAS, CEA and S100. Which of the following features is NOT indicative of malignancy in this tumour?

a) Infiltrative borders
b) Prominent nucleoli
c) Spindled morphology of the cells
d) Necrosis
e) Mitoses >2/10hpf

11) A 75 year old man presents with a mass in the proximal forearm. Excision shows an infiltrative tumour with pleomorphic spindled cells with sclerotic stroma in the subcutaneous tissues. Which of the following will favour a malignant peripheral nerve sheath tumour over a spindle cell melanoma?

a) B-RAF mutation detected
b) Focal S100 positivity
c) Known NF1 germline mutation
d) Positive HMB45 and MelanA staining
e) Pigmentation within the cells

12) A 10 year old male was found to have a large retroperitoneal mass above the kidney. Core biopsies show a cellular tumour composed of medium sized round and polygonal cells with frequent mitoses in fibrotic stroma. Occasional glomeruloid bodies are seen. The tumour is positive for desmin and WT1 (both C-terminus and N-terminus directed antibodies), but negative for myogenin, CD99 and NB84. Which is the most likely diagnosis?

a) Desmoplastic small round cell tumour
   b) Embryonal rhabdomyosarcoma
   c) Ewing sarcoma
   d) Neuroblastoma
   e) Wilms tumour

13) A 79 year old female presents with per vaginal bleeding. A transvaginal ultrasound scan showed a large 17cm submucosal fibroid in the uterus. The patient underwent a hysterectomy, which showed a spindle cell lesion which is positive for desmin and h-caldesmon, with weak positivity for ER. Which of the following features are diagnostic for malignancy in this tumour?

   a) No atypia, mitotic count <1/10 hpf, necrosis
   b) Widespread atypia, mitotic count 1/10hpf, no necrosis
   c) Widespread atypia, mitotic count 2/10hpf, necrosis
   d) Focal atypia, mitotic count of 5/10hpf, no necrosis
   e) No atypia, mitotic count of 12/10hpf, no necrosis

14) A 12 year old female presents with a mass in the post nasal space. The biopsy shows nests and sheets of small polygonal cells with scant eosinophilic cytoplasm and frequent mitoses. The cells are positive for desmin, myogenin and myoD1 immunostaining. Molecular testing showed the cells harbour a t(2:13) translocation with fusion of the PAX3 and FOXO1 genes. Which is the correct diagnosis?

   a) Alveolar rhabdomyosarcoma
   b) Embryonal rhabdomyosarcoma
   c) Epithelioid leiomyosarcoma
   d) Pleomorphic rhabdomyosarcoma
   e) Rhabdomyoma

15) A 52 year old woman had a previous wide local excision and radiotherapy of the left breast for invasive ductal carcinoma. She now presents with small erythematous papules in the left breast near the scar. An excision biopsy of one of the lesions shows numerous irregular dilated vascular channels in the superficial dermis, lined by mildly pleomorphic endothelial cells. On immunohistochemistry, the vessels are partly lined by SMA positive cells. Which is the most likely diagnosis?

   a) Angiosarcoma
   b) Arteriovenous malformation
   c) Atypical vascular lesion
   d) Capillary haemangioma
   e) Retiform haemangioendothelioma

16) A 28 year old male presents with a mass above the knee. Excision of the mass shows an infiltrative tumour, comprising uniform short spindled and ovoid cells arranged in tight fascicles with areas of calcification. There are frequent mitoses. The cells are positive for EMA, bcl2 and CD99. Which of the following translocations is this tumour associated with?

   a) t(X:17) ASPL-TFE3
   b) t(X:18) SS18-SSX1
   c) t(1:13) PAX7-FOXO1
   d) t(7:16) FUS-CREB3L2
   e) t(12:22) EWSR1-ATF1

17) A 40 year old man present with a nodule in the proximal thumb. Excision of the nodule shows a tumour as below.

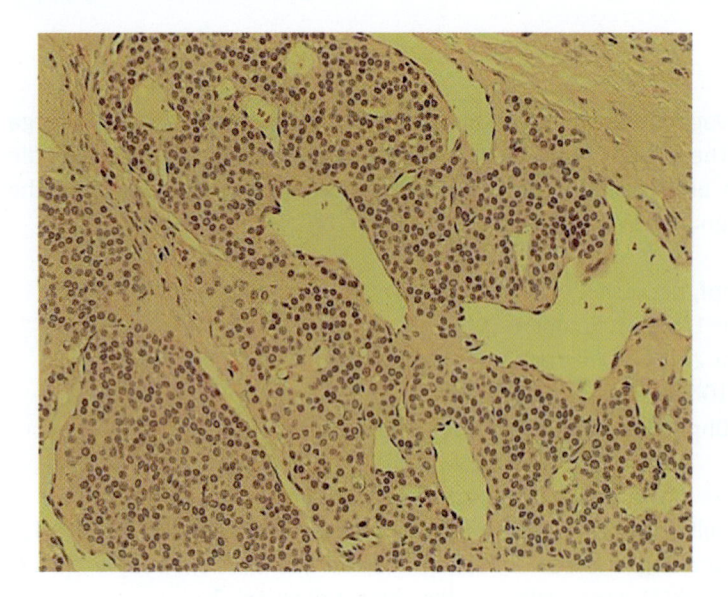

Which is the correct diagnosis?

   a) Glomangioma
   b) Granular cell tumour
   c) Haemangiopericytoma
   d) Lobular capillary haemangioma
   e) Myopericytoma

18) A 31 year old woman presents with the mass in the dorsum of the foot. Excision shows a tumour as shown below. The tumour is positive for D-PAS, S100, HMB45 and MelanA. Molecular testing showed the tumour is positive for t(12:22) EWSR1-ATF1 translocation. Which is the most appropriate diagnosis?

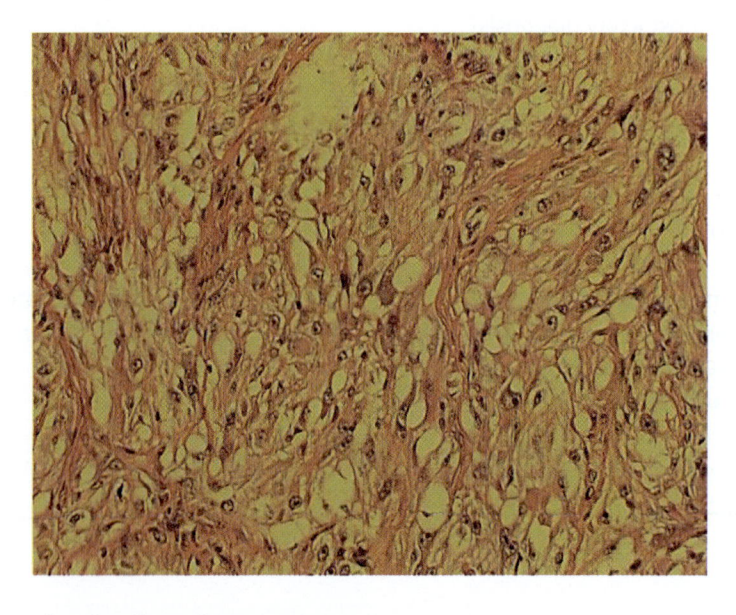

   a) Alveolar soft part sarcoma

b) Clear cell sarcoma
   c) Epithelioid schwannoma
   d) Melanoma
   e) Renal clear cell carcinoma

For questions 19 till 23 : Extended matching item questions:

   A) Avascular necrosis
   B) Gout
   C) Osteoarthritis
   D) Rheumatoid Arthritis
   E) Septic arthritis

For each of the following descriptions of *diseases of the joints,* select the **most appropriate** diagnosis from the list of options above. Each option may be used once, more than once, or not at all.

19) A 75 year old man presents with pyrexia and a hot swollen right knee. Aspiration of the knee effusion shows numerous neutrophils. No crystals are identified on polarised light. Cultures of the joint fluid revealed Gram positive cocci.

20) A 68 year old man presents with a hot swollen left knee. Aspiration of the knee effusion shows numerous neutrophils. Elongated birefringent crystals are identified on polarised light. He also has a nodule in the foot, excision of which showed granulomatous inflammation with central refractile material.

21) A 45 year old woman presents with bilateral wrist pain. She also has subcutaneous nodules in the elbow, the excision of which showed granulomatous inflammation with central necrotic collagenous material.

22) A 64 year old obese man presents with bilateral knee pain. X-ray of the knees revealed thinning of the joint spaces with subcortical sclerosis and cyst spaces.

23) A 72 year old man on long term steroids present with right hip pain. X-ray of the hip shows sclerotic distorted acetabulum. A total hip replacement was performed, and histology of acetabulum shows empty lacunae in the cortical bone with sclerosis in the marrow spaces.

24) A 47 year old woman presents with a pain in the right knee. On examination, there was an effusion as well as a mass in the lateral border of the joint. Excision of the mass showed a cellular lesion composed of epithelioid cells with interspersed multinucleated giant cells. There is admixed chronic inflammatory cells and haemosiderin deposition confirmed with Perl's stain. Which is the correct diagnosis?

   a) Juvenile xanthogranuloma
   b) Langerhans cell histiocytosis
   c) Pigmented villonodular synovitis
   d) Tuberculosis
   e) Undifferentiated pleomorphic sarcoma with giant cells

25) A 25 year old woman presents with a 1cm nodule on the lateral border of the left wrist. Excision of this shows a cyst lined by loose fibrous wall with no discernible epithelial lining. What is the correct diagnosis?

   a) Bursitis
   b) Ganglion
   c) Juxta-articular myxoma

d) Superficial fibromatosis
e) Synovial cyst

26) An 82 year old woman presents with a fractured neck of femur following a fall. She underwent a hemiarthroplasty, and histology of the femoral head shows thin bony trabecula with microfractures. Which of the following is NOT a cause of the underlying condition?

    a) Age
    b) Hormone replacement therapy
    c) Immobility
    d) Lack of sunlight exposure
    e) Type 1 diabetes mellitus

27) A 51 year old man presents with collapse of the T12 vertebra. Decompression and laminectomy of the vertebral body shows necrotic bone with foci of granulomatous inflammation with central necrosis. Which of the following special stains would help confirm the most likely diagnosis?

    a) AE1/3 immunostain
    b) Elastic van Gieson
    c) Periodic Acid Schiff with Diastase
    d) Perl's
    e) Ziehl-Neelsen

28) A 10 year old boy presents unwell with pyrexia and a swollen right lower leg. X-ray of the right tibia shows a well-circumscribed area of radiolucency within the shaft of the bone. Debridement of the bone mass shows mixed acute and chronic inflammation with a central area of necrotic bone. Which is the most important investigation recommended to perform?

    a) Blood cultures
    b) Bone mineral density analysis
    c) Genetic testing for collagen gene mutations
    d) Isotope bone scan
    e) Serum calcium levels

29) A 32 year old man present with a hard bony nodule on his forehead. During surgery, the nodule was found adherent to the skull. Excision showed a lesion with surface cartilage and perichondrium with underlying woven bone. No atypia is seen. Which is the correct diagnosis?

    a) Chondrosarcoma
    b) Chondroblastoma
    c) Osteochondroma
    d) Osteoid osteoma
    e) Osteosarcoma

30) A 79 year old woman presents with weight loss, anaemia and pain in the left knee. X-ray of the distal femur shows a destructive sclerotic lesion. Bone biopsies were performed and showed nests of pleomorphic epithelioid cells infiltrating bone. On immunohistochemistry, the cells are positive for AE1/3, GCDFP-15 and ER. Which of the following would you recommend for further investigation to determine the primary site?

a) Bone scan
   b) Bronchoscopy
   c) Colonoscopy
   d) Mammogram

31) A 12 year old girl as a mass in the left iliac crest. A bone biopsy shows sheets of uniform polygonal cells with scant cytoplasm and frequent mitoses. There is focal rosette formation. The cells are positive for CD99, and negative for CD45, desmin, WT1, NB84, S100, TTF1 and AE1/3. Which translocation is the most likely tumour associated with?

   a) t (11:22) EWSR1-FLI1
   b) t (11:22) EWSR1-WT1
   c) t (12:22) EWSR1-ATF
   d) t (12:22) EWSR1-CREB1
   e) t (12:22) EWSR1-DDIT3

32) A 22 year old man presents with lymphadenopathy, hepatosplenomegaly and multiple osteolytic bone lesions. A bone biopsy shows sheets of cells with convoluted nuclei and eosinophilic cytoplasm, as well as scattered multinucleated giant cells. There are also admixed plasma cells, eosinophils and neutrophils. Which of the following immunostains would support the most likely diagnosis?

   a) AE1/3 and CAM5.2
   b) CD1a and S100
   c) CD4 and CD8
   d) CD45 and CD20
   e) Kappa and lambda

33) An 18 year old woman has multiple radiolucent lobulated lesions in the phalanges of both hands. Biopsies show a cartilaginous tumour composed of mildly pleomorphic chondrocytes in loose hyalinised and myxoid cartilage. With which of the following syndromes is this underlying condition associated?

   a) Gardner syndrome
   b) McCune-Albright syndrome
   c) Multiple Endocrine Neoplasia-2 (MEN-2)
   d) Ollier-Maffucci syndrome
   e) Paget's disease of the bone

34) A 25 year old man presents with a 3cm hard lump deep in the forearm. Excision revealed a mass within the muscle plane close to the bone. There is a peripheral rim of woven trabecular bone with osteoclasts rimming. In the centre are plump enlarged spindle shaped cells arranged haphazardly, with extravasated red blood cells. Which is correct diagnosis?

   a) Exostosis
   b) Extraskeletal osteosarcoma
   c) Myositis ossificans
   d) Ossifying fibromyxoid tumour
   e) Periosteal osteosarcoma

35) A 35 year old man presents with a mass in the proximal tibia. Biopsy of the mass is shown below. Which is the correct diagnosis?

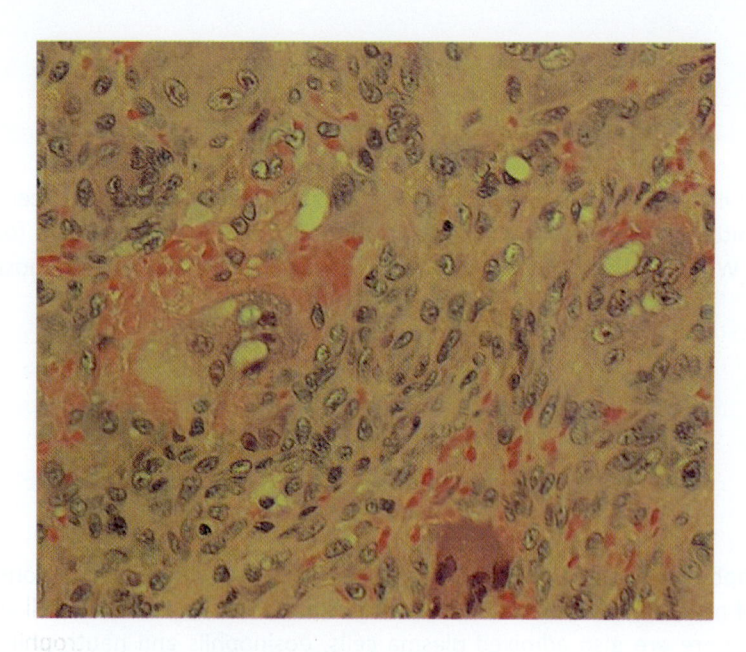

a) Bone remodelling following fracture
b) Giant cell tumour of bone
c) Granulomatous inflammation
d) Osteosarcoma
e) Undifferentiated pleomorphic sarcoma

Correct Answers and Explanations:

1) Correct answer: F) Solitary fibrous tumour

Solitary fibrous tumour (SFT) is composed of spindle cells arranged in what is described as a patternless pattern. They usually contain prominent stromal collagen and branching staghorn or hemangiopericytoma-like vessels. They were first recognised in the pleura. However, over the years, these tumours have been described in many organs. Most SFTs are benign but about 15% behave aggressively. They demonstrate diffusely positive immunohistochemical staining for CD34. STAT6 is a newly described highly sensitive and specific immunohistochemical marker for SFT and can be helpful to distinguish this tumour type from histological mimics.

2) Correct answer: A) Fibromatosis

Fibromatosis is subdivided into superficial and deep fibromatoses. Superficial fibromatoses arise in the palmar (Dupuytren's contracture) or plantar (Ledderhose disease) soft tissues. Deep fibromatoses include desmoid-type and aggressive fibromatoses, and comprises a clonal proliferation of myofibroblasts. Both tumours show an infiltrative growth of myofibroblasts and have a tendency toward local recurrence but an inability to metastasize. Most deep fibromatoses have diffuse nuclear staining for beta catenin. Deep fibromatoses may also be associated with familial adenomatous polyposis.

3) Correct answer: D) Myxoinflammatory fibroblastic sarcoma

Myxoinflammatory fibroblastic sarcoma (MIFS) is a low grade sarcoma with myxoid stroma, inflammatory infiltrate and virocyte-like or Reed-Sternberg-like cells. The tumour commonly involves the hands and feet. The most striking feature at low magnification is a prominent dense, mixed acute and chronic inflammatory infiltrate associated with alternating hyaline and myxoid zones in variable proportions. Remember other tumours such as proliferative myositis and proliferating fasciitis can show large ganglion like cells.

4) Correct answer: E) Nodular fasciitis

Nodular fasciitis is a rapidly growing myofibroblastic proliferation that usually occurs in the subcutaneous tissue. It is composed of plump but uniform fibroblastic/myofibroblastic cells and typically displays a loose or tissue culture-like growth pattern. Mitotic figures may be plentiful, but atypical mitoses are usually not seen. Extravasated red cells, chronic inflammatory cells, and multinucleated osteoclast like giant cells are other frequently identified features. The lesional border is typically infiltrative, at least focally, although it may be well delineated. USP6 gene rearrangement is a recurrent genetic event in nodular fasciitis, which makes it a valuable ancillary tool for the pathological diagnosis of these lesions.

5) Correct answer: C) Low grade fibromyxoid sarcoma

Low grade fibromyxoid sarcoma is a rare low grade malignant spindle cell tumour. The tumour is composed of heavily collagenised and myxoid zones and very bland spindled cells showing a whorled growth pattern. A characteristic arrangement of blood vessels which is described as arcades of curvilinear blood vessels helps in identifying these lesions. Most tumours occur in young to middle aged adults. These tumours are mainly deep seated and occur in the extremities and trunk. They show a balanced translocation of t(7; 16) or t(7; 11).

6) Correct answer: B) Inflammatory myofibroblastic tumour

Inflammatory myofibroblastic tumour (IMT) is a lesion composed of myofibroblastic spindle cells admixed with a prominent inflammatory infiltrate of plasma cells, lymphocytes, and eosinophils. It occurs primarily in soft tissue and viscera of children and young adults (most frequently in the first two decades of life). The most common sites are the lung, mesentery, and omentum. A subset of these tumours show positive staining for ALK (about 2/3 of paediatric tumours and 1/3 of adult tumours).

7) Correct answer: d) Round cell liposarcoma

Round cell liposarcoma is a malignant tumour which is part of the spectrum of myxoid and round cell liposarcoma. It usually presents as a large painless mass within the deep soft tissues of the limbs in young adults, with the age at presentation on average a decade younger than with other histological subtypes of liposarcoma. The tumour is composed of uniform round to oval shaped primitive mesenchymal cells and a variable number of small signet-ring lipoblasts in a prominent myxoid stroma with a characteristic branching vascular pattern. The branching ("crow's feet"-like) and anastomosing network of vessels, are pathognomonic in the myxoid end of the spectrum. The round cell areas are characterized by solid sheets of round cells with a high nuclear/cytoplasmic ratio and conspicuous nucleoli, with no intervening myxoid stroma. In the majority of cases this shows a diffuse staining for S100 protein. The karyotypic hallmark of myxoid and round cell liposarcoma is the t(12; 16) translocation.

8) Correct answer: b) Amplification of MDM2 gene

Atypical lipomatous tumour or well differentiated liposarcoma is a low grade adipocytic neoplasm, which can dedifferentiate to a sarcoma of variable histological grade. Histologically dedifferentiated liposarcoma shows a transition from well differentiated liposarcoma of any type to a non-lipogenic sarcoma which, in most cases, is high grade. The dedifferentiated component usually has a pleomorphic spindle cell morphology, although heterologous components (such as cartilaginous or osteoid differentiation) may be seen. The tumours (both well differentiated and dedifferentiated) are typically positive for MDM2 and CDK4 by immunohistochemistry and can be demonstrated by In Situ Hybridisation (ISH) to test for MDM2 amplification.

9) Correct answer: d) Positivity for CD34

Dermatofibrosarcoma protuberans (DFSP) are malignant tumours with a tendency to recur locally and a low risk for metastasis. They present in mid adult life in the chest or proximal extremities. The tumour is composed of a storiform proliferation of monotonous, short spindled cells with dark nuclei and pale cytoplasm. Characteristically the tumour infiltrates the subcutaneous adipose tissue in honeycomb pattern architecture. Dermatofibrosarcoma protuberans harbours a supernumerary ring chromosome derived from portions of chromosomes 17 and 22 resulting in a gene fusion between the platelet-derived growth factor β (PDGFβ) gene and the collagen type 1, alpha 1 (COL1A1) gene.

DFSPs can be distinguished from dermatofibromas by their more cellular nature, storiform architecture, and immunoreactivity for CD34. Dermatofibromas have a more mixed population of tumour cells, a storiform growth pattern, collagen trapping, and usually demonstrate immunoreactivity for Factor XIIIa.

10) Correct answer: a) Infiltrative borders

Granular cell tumour is a neural tumour comprising round or spindle cells with abundant pink, granular cytoplasm due to abundant intracytoplasmic lysosomes. The granules are also S100 protein and CD68 positive, and ultrastructurally contain lysosomes. Criteria for malignancy include: 1) necrosis, 2) spindling of cells, 3) vesicular nuclei with large nucleoli, 4) greater than 2 mitoses per 10 high power fields at 200X magnification, 5) high nuclear to cytoplasmic ratio, and 6) pleomorphism. Neoplasms that meet 3 or more of these criteria are classified as malignant. Those that meet only one or two criteria are regarded as atypical while those that show only focal pleomorphism but none of the other features are classified as benign.

11) Correct answer: c) Known NF1 germline mutation

Malignant peripheral nerve sheath tumours (MPNST) are sarcomas arising from the nerve sheath. They frequently but not invariably arise in the setting of neurofibromatosis type I (about 50%). The tumours show spindled morphology consisting of sweeping fascicles of hyperchromatic spindled cells that often have wavy nuclear contours with varying cellularity. Heterologous differentiation is common. This most typically takes the form of rhabdomyoblastic differentiation (Triton tumour), but angiosarcomatous, osteosarcomatous, and chondrosarcomatous areas may be present as well. MPNST are often S100 negative (50%) or only focally positive. Melanomas are usually diffusely positive for S100 and show positive staining with HMB-45 and Melan-A, although some melanomas may show only focal positivity.

12) Correct answer: e) Wilms tumour

Wilms tumour is the commonest renal tumour of childhood typically presenting around the age of three years with the majority occurring before 6 years of age. Most cases are sporadic. However a small minority are associated with syndromes such as Beckwith-Weidemann syndrome, Denys-Drash syndrome and WAGR (Wilms tumour, aniridia, genitourinary abnormalities and mental retardation) Syndrome. They can be biphasic or triphasic. Triphasic tumours are composed of fibroblast-like stroma, epithelial elements (abortive tubules or glomeruli) and undifferentiated blastema (diffuse, nodular, cordlike or basaloid pattern of densely packed primitive small blue cells with scanty cytoplasm, overlapping nuclei with finely dispersed chromatin). Anaplastic features are a marker for poor prognosis. A common differential diagnosis of Wilms tumour is desmoplastic small round cell tumour (DSRCT), but although both can be desmin positive, DSBRT is positive with only antibodies directed at the C-terminus of WT1, while Wilms tumour is positive with both C-terminus and N-terminus antibodies.

13) Correct answer: c) Widespread atypia, mitotic count 2/10HPF, necrosis

Leiomyosarcoma is a malignant tumour composed of cells showing distinct smooth muscle features which usually occurs in middle-aged or older persons. They are the commonest sarcoma of the uterus, and also make up a significant proportion of retroperitoneal (including pelvic) sarcomas. They are composed of intersecting groups of spindle cells bearing elongated cigar or blunt ended nuclei. The typical features are (1) diffuse moderate to marked nuclear atypia, appreciable at low-power magnification; (2) a high mitotic rate (≥10 mf/10 hpf, and can often be much higher); and (3) tumour cell necrosis. A diagnosis of uterine leiomyosarcoma usually requires two of the three above criteria (different criteria apply to myxoid and epithelioid leiomyosarcomas).

14) Correct answer: a) Alveolar rhabdomyosarcoma

Alveolar rhabdomyosarcoma is a primitive, malignant, round cell neoplasm that cytologically resembles lymphoma and which shows partial skeletal muscle differentiation. They are common in younger children and more often occur in adolescents and young adults arising in the extremities. A translocation was found in the majority of alveolar rhabdomyosarcoma, juxtaposing the PAX3 or PAX7 genes on chromosomes 2 and 1, respectively, with the FOXO1/FKHR gene on chromosome 13, to generate chimeric genes which encode PAX3/FOXO1 and PAX7/FOXO1 fusion proteins. Alveolar rhabdomyosarcomas are high grade neoplasms that are inherently more aggressive than embryonal rhabdomyosarcomas. All the other forms of tumours of skeletal muscle lineage do not have a consistently reproducible genetic abnormality.

15) Correct answer: c) Atypical vascular lesion

Atypical vascular proliferations are vascular proliferations which arise after radiotherapy, predominantly in the breast. They usually develop on an average 3 years after radiotherapy. The clinical appearance is that of small 5mm red to brown papules. Histologically they are well circumscribed, superficial dermal lesions, usually with no extension into the subcutaneous tissue composed of dilated and or narrow vascular structures lined by a single layer of slightly enlarged endothelial cells. There is no endothelial multilayering, prominent cytological atypia, increased mitoses, confluent haemorrhage or necrosis. The main differential diagnosis is a well differentiated angiosarcoma which tends to be deeper and more infiltrative with significant atypia, multilayering and mitoses. They show an elevated proliferation index (Ki-67) and express p63. Angiosarcomas do not have a SMA-positive myopericyte layer around the endothelial cells, unlike atypical vascular lesions which show an incomplete myopericyte layer.

16) Correct answer: b) t(X; 18) SS18-SSX1

Synovial sarcoma is a mesenchymal spindle cell tumour which has a specific (more than 90%) balanced t(X: 18) (p11; q11)SS18-SSX1 translocation. They occur mainly before 50 years of age (90%). Over 80% arise in deep soft tissue of extremities especially around the knee, and the tumour frequently arises adjacent to joints or tendon sheaths. Histologically, synovial sarcoma is biphasic or monophasic. Biphasic synovial sarcomas have epithelial and spindle cell components in varying proportions. TLE1 is a newly described marker which has been found to be an excellent discriminator of synovial sarcoma from other sarcomas such as malignant peripheral nerve sheath tumours.

17) Correct answer: a) Glomangioma

Glomus tumours are mesenchymal neoplasms composed of cells that closely resemble the modified smooth muscle cells of the normal glomus body. They typically occur in young adults predominantly in the distal extremities, especially the subungual region, the hand, the wrist and the foot. They are composed of nests of glomus cells surrounding capillary sized vessels. Glomangiomas are characterised by dilated veins surrounded by small clusters of glomus cells. The diagnosis of "malignant glomus tumour" should be made in tumours showing: 1) size >2 cm and subfascial or visceral location; 2) atypical mitotic figures; or 3) marked nuclear atypia and any level of mitotic activity. Glomus tumours of all types express smooth muscle actin.

18) Correct answer: b) Clear cell sarcoma

Clear cell sarcoma is also called malignant melanoma of soft parts. They are more common in young adults (third and fourth decades). The foot is the commonest site. The gastrointestinal tract (ileum) is the commonest visceral site. Histologically the tumour demonstrates monotonous clear to spindled cells arranged in nests/packets which usually infiltrate tendon sheaths. The cells contain vesicular nuclei with prominent nucleoli and contain abundant PAS positive glycogen. They are usually positive for S100, Melan-A and HMB-45. The main differential diagnosis of this lesion is melanoma, but unlike melanoma, clear cell sarcomas have characteristic translocations: t(12;22)(q13;q12) and t(2;22)(q13;q12).

19) Correct answer: E) Septic arthritis

Microorganisms of all types can seed in joints during haematogenous dissemination, either by direct inoculation or from contiguous spread from an adjacent abscess. Infectious arthritis is a serious rapidly destructive lesion of the joint which can lead to permanent deformities. Bacterial infections usually cause an acute suppurative arthritis. *H. influenzae* arthritis predominates in children less than 2 years of age, *S. aureus* is seen in older children and adults, and gonococcus is more common in adolescence and young adulthood. Patients with sickle cell disease are prone to infection with *Salmonella*. Other aetiologies include Mycobacterium tuberculosis and viral organisms.

20) Correct answer: B) Gout

Crystal deposits in joints are associated with a variety of acute and chronic joint disorders. Endogenous crystals include monosodium urate (MSU, gout), calcium pyrophosphate dihydrate, and basic calcium phosphate (hydroxyapatite). Exogenous crystals, such as corticosteroid ester crystals and talcum, and the biomaterials polyethylene and methyl methacrylate, may also induce joint disease. Gout is marked by transient attacks of acute arthritis which eventually lead to chronic gouty arthritis and the appearance of tophi (large aggregates of urate crystals surrounded by inflammation). The crystals are long and slender needle shaped crystals which are negatively birefringent.

21) Correct answer: D) Rheumatoid arthritis

Rheumatoid arthritis is a chronic systemic inflammatory disorder which affects the joints as well as skin, blood vessels, heart, lungs, and muscle. In the joints it presents as a non-suppurative inflammatory synovitis which if untreated results in destruction of the articular cartilage and ankylosis of the joints. It affects women more often than men in those aged 40 to 70 years old. The pattern of joint is involvement is pathognomonic. It is classically symmetrical, starting in smaller joints (metacarpophalangeal, proximal phalangeal and feet) followed by the larger ones (wrists, ankles, elbows, and knees). The lumbosacral region and hips are usually spared. The diagnosis is based primarily on the clinical features and includes the presence of four of the following criteria: (1) morning stiffness, (2) arthritis in three or more joint areas, (3) arthritis of hand joints, (4) symmetric arthritis, (5) rheumatoid nodules, (6) serum rheumatoid factor, and (7) typical radiographic changes.

22) Correct answer: C) Osteoarthritis

Osteoarthritis is a degenerative joint disease which usually occurs in patients older than 55 years. It affects the interphalangeal joints of the hands, metacarpophalangeal joint of the thumb, hips and knees, cervical and lumbar vertebral joints. Secondary osteoarthritis may result from Legg-Calvé-Perthes disease, previous history of gouty arthritis, rheumatoid arthritis, infectious arthritis, pseudogout, and trauma. Histologically the articular cartilaginous surface is fibrillated, frayed, and thinned or denuded. Chondrocytic hyperplasia is seen represented by aggregates of chondrocytes surrounded by basophilic staining matrix. Cartilage may ultimately form loose bodies or "joint mice" by proliferation of chondrocytes, with subsequent fragmentation into the joint space.

23) Correct answer: A) Avascular necrosis

Avascular necrosis, also known as osteonecrosis, is infarction of bone and marrow that can occur in the medullary cavity of the metaphysis or diaphysis as well as the subchondral region of the epiphysis. Avascular necrosis can be idiopathic or secondary. The common causes are trauma, corticosteroid administration, radiation or sickle cell disease. Histologically medullary infarcts are geographic and involve the cancellous bone and marrow. The dead bone, recognised by its empty lacunae, is surrounded by necrotic adipocytes.

24) Correct answer: c) Pigmented villonodular synovitis

Pigmented villonodular synovitis commonly affects women in their third and fourth decades. The knee is the most common joint involved (80%). The other joints which can be involved include hip, shoulder, and ankle. They presents with a long history of pain, swelling, limitation of motion, and joint stiffness. Grossly the synovium is brown and thickened and contains papillary villous projections. Under the microscope cellular infiltrates of mononuclear cells are seen within the subsynovial connective tissue. The mononuclear cells have oval nuclei with vesicular or clumped chromatin and abundant cytoplasm. Hemosiderin-laden mononuclear cells, multinucleated giant cells, and foam cells are present.

25) Correct answer: b) Ganglion
A ganglion is a small (1–1.5 cm) cyst that is almost always located near a joint capsule or tendon sheath. A common location is around the joints of the wrist, where it appears as a firm, fluctuant, pea sized translucent nodule. It arises as a result of cystic or myxoid degeneration of connective tissue; hence the cysts characteristically have a fibrous wall showing myxoid change and no true cyst lining. The lesion may be multilocular. The fluid that fills the cyst is similar to synovial fluid; however, there is no communication with the joint space.

26) Correct answer: b) Hormone replacement therapy
Osteoporosis, or thinning bones, can result in painful fractures. Risk factors for osteoporosis include aging, being female, low body weight, low sex hormones or menopause, smoking, and some medications. The diagnosis of osteoporosis relies on the quantitative assessment of bone mineral density (BMD), usually by central dual energy X-ray absorptiometry (DEXA). BMD at the femoral neck provides the reference site. General management includes assessment of falls risk. Maintenance of mobility and correction of nutritional

deficiencies, particularly of calcium, vitamin D and protein, should be advised to prevent or reduce the risk of falls.

27) Correct answer: e) Ziehl-Neelsen

Tuberculous arthritis and osteomyelitis are chronic diseases that can occur in all age groups, especially adults. They usually develop as complications of haematogenous dissemination from a visceral (usually pulmonary) site of infection. The affected synovium may grow as a pannus over the articular cartilage and erode the bone along the joint margins. Chronic disease results in severe destruction with fibrous ankylosis and obliteration of the joint space. The weight-bearing joints are usually affected, especially the spine, hips, knees, and ankles. The histological hallmark is confluent granulomas which comprise a focus of chronic inflammation consisting of a microscopic aggregation of macrophages that are transformed into epithelium-like cells, surrounded by a collar of lymphocytes and occasionally plasma cells. The granulomas usually have central caseous necrosis.

28) Correct answer: a) Blood cultures

The history given is suggestive of an infectious aetiology, with a bone focus (or nidus) suggestive of osteomyelitis. The most appropriate initial investigation is therefore of blood cultures for disseminated bacteraemia.

29) Correct answer: c) Osteochondroma

Osteochondroma is a cartilage capped bony projection arising on the external surface of bone which contains a marrow cavity that is continuous with that of the underlying bone. The most common sites of involvement are the metaphyseal region of long bones, which includes the distal femur, upper humerus, upper tibia and fibula. Patients often present with a hard mass which has been present for years.

30) Correct answer: d) Mammogram

The skeletal system is the third most common site of metastatic tumours, after the lungs and liver. In adults, metastases commonly originate from cancers of the breast, lung, prostate, kidney and thyroid gland. Metastases are rare in children. However, the commonly metastasising paediatric tumours are neuroblastoma, rhabdomyosarcoma and clear cell sarcoma of kidney. Tumours metastasise to bones which contain red bone marrow such as in the vertebra, femur, ribs, sternum, pelvis and skull. X-rays of the involved bones can show either lytic, blastic or mixed lesions. Lung, breast, thyroid and kidney metastases are lytic. Prostate tumour metastases are osteoblastic. In some cases, immunohistochemistry may be required to identify the primary site of the tumour. For example, breast cancers are usually positive for GCDFP-15, ER and PR.

31) Correct answer: a) t (11:22) EWSR1-FLI1

Ewing's sarcoma/Peripheral Neuroectodermal Tumours (PNET) are small round cell tumours showing a varying degree of neuroectodermal differentiation. They are one of the commonest tumours of childhood and occur in bone and within soft tissues.

Many soft tissue tumours are associated with translocations involving the EWSR1 gene. PNET is associated with the EWSR1-FLI1 fusion gene; desmoplastic small round cell tumour is associated with EWSR1-WT1 fusion; clear cell sarcoma and angiomatoid fibrous histiocytoma are associated with EWSR1-CREB1 or EWSR1-ATF fusion; and myxoid/round cell liposarcoma is associated with EWSR1-DDIT3 fusion.

32) Correct answer: b) CD1a and S100

Langerhans cell histiocytosis usually affects infants or young adults. There are three classic clinical syndromes:
a. Eosinophilic granulomas (localised form commonly in the lung or bone)
b. Letterer-Siwe (systemic form with multiorgan involvement)
c. Hand-Schüller-Christian disease (overlaps with the systemic and localised from)

Histology shows Langerhans cells which are generally mononuclear with pale, eosinophilic cytoplasm and bland, bean-shaped nuclei accompanied by numerous eosinophils. Characteristically Langerhans cells are positive for S-100 protein and CD1a. Electron microscopy shows tennis racket shaped Birbeck granules.

33) Correct answer: d) Ollier-Maffucci syndrome

Ollier disease is a development disorder caused by failure of enchondral ossification with production of multiple enchondromas in the metaphysis of long bones. Maffucci Syndrome has the features of Ollier disease along with angioma of the soft tissue. The hand is the most common site. Patients are in the first two decades and the risk of progression to chondrosarcomas is higher in patients with the syndrome. Radiologically the tumours can be radiolucent or mineralised. Chondromas are hypocellular, with abundant hyaline cartilage matrix containing regularly dispersed chondrocytes situated within lacunar spaces. The chondrocytes have pale eosinophilic cytoplasm with regular nuclei.

34) Correct answer: c) Myositis ossificans

This is seen mostly in young, active adults and affects the extremities. Clinically the patients are usually athletes who present with a solitary painful or tender mass. A history of trauma can be elicited in a majority of the cases. Histologically a triphasic pattern with distinct zonation is characteristic. This comprises a central cellular area with a reactive tissue culture like appearances with regular nuclei. The central zone is bordered by a rim of immature osteoid. This is in turn covered by a zone of mature lamellar bone.

The main differential diagnosis includes extraskeletal osteosarcoma which is characterised by a more disorderly growth of pleomorphic cells admixed with a characteristic delicate lace-like osteoid. There is no zonation. Myositis ossificans is a benign, self-limited process with an excellent prognosis. Spontaneous regression can occur.

35) Correct answer: b) Giant cell tumour of bone

These tumours occur in skeletally mature patients in their second, third, and fourth decades. The epiphyses of the distal femur, proximal tibia, distal radius, and sacrum are the common sites of occurrence. Radiologically the tumour presents as a lytic epiphyseal mass with no reactive sclerosis or periosteal reaction. Histological sections show many almost evenly distributed multinucleated giant cells admixed with sheets of mononuclear histiocytic cells. Multinucleated giant cells have features of osteoclasts and contain numerous nuclei, sometimes greater than 100. Remember that other tumours and lesions with osteoclast type giant cells include giant cell reparative granulomas, aneurysmal bone cysts, and giant cell rich osteosarcoma.

CHAPTER 12

UROPATHOLOGY
Dr Jayson Wang & Dr Brendan Tinwell

For questions 1 till 5 : Extended matching item questions:

- A) Haemorrhagic cystitis
- B) Interstitial cystitis
- C) Malakoplakia
- D) Polypoid cystitis
- E) Inverted papilloma

Match each of the following descriptions with the appropriate *diseases of the bladder* from the list of the options above.

1) A 40-year-old woman with dysuria. Bladder biopsy shows chronic inflammation, with numerous histiocytes in the lamina propria containing round deposits positive for von Kossa stain.

2) A 68-year-old man receiving Cyclophosphamide chemotherapy presents with heavy haematuria. Bladder biopsy shows mixed acute and chronic inflammation with hyperaemia and extensive extravasation of red blood cells in the lamina propria, as well as extensive ulceration.

3) An 82-year-old man with a long-term catheter in situ develops haematuria. Biopsy shows broad based oedematous papillary lesions with mixed acute and chronic inflammation.

4) A 25-year-old woman presents with haematuria and urgency. Bladder biopsy shows mixed acute and chronic inflammation with scattered cells staining with toluidine blue.

5) A 50-year-old woman with haematuria found to have a large pedunculated polypoid lesion in the bladder. Excision show irregular invaginating cords and nests of non-hyperplastic transitional epithelium in the lamina propria.

6) A 65-year-old woman with a previous papillary urothelial carcinoma underwent a check cystoscopy. A small polypoid lesion was found at the scar of the previous tumour. A cold cup biopsy was performed. The histology showed small papillary fronds lined by cuboidal cells with clear cytoplasm and uniform nuclei, as well as underlying similar cells arranged as small tubules in the underlying lamina propria. Which is the most likely diagnosis?

- a) Cystitis glandularis
- b) Nephrogenic adenoma
- c) Polypoid cystitis
- d) Papillary urothelial carcinoma
- e) Urothelial carcinoma with glandular differentiation

7) A 55-year-old man presents with haematuria and dysuria. On cystoscopy, the bladder was congested. A bladder biopsy showed squamous metaplasia of the transitional epithelium. In addition, below is a photomicrograph of the underlying lamina propria. What is the underlying aetiology?

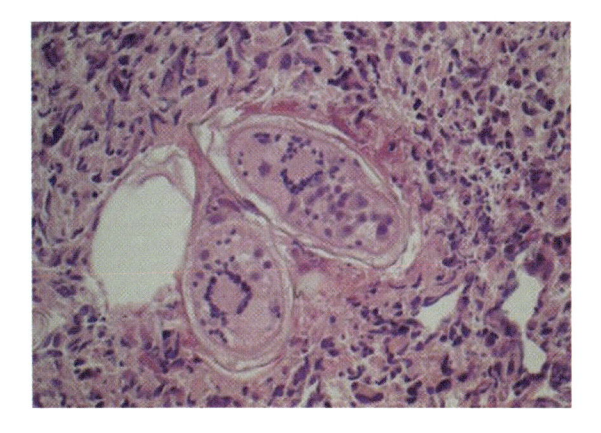

a) Calcium deposition
b) Inflammatory myofibroblastic tumour
c) Interstitial cystitis
d) Malakoplakia
e) Schistosomiasis infection

8) A bladder was found to contain foci of adenocarcinoma with no urothelial component. Which one of the following immunostains is most helpful in distinguishing a primary bladder adenocarcinoma from a metastasis from a bowel primary?

   a) Beta-catenin
   b) CDX2
   c) CK7
   d) CD20
   e) PSA

9) A cystectomy was performed for a bladder tumour. Histologically, it shows a high-grade urothelial carcinoma with a papillary component, as well as invasion as nests into the lamina propria, and to the outer half of the muscularis propria. What is the correct T stage of the tumour?

   a) pT1
   b) pT2a
   c) pT2b
   d) pT3a
   e) pT3b

10) At the same operation, bilateral pelvic lymph nodes were resected. Histology showed one out of eight lymph nodes on the left were involved by metastatic urothelial carcinoma, while two out of ten lymph nodes were involved on the right. What is the correct N stage of the tumour?

   a) pNx
   b) pN1
   c) pN2
   d) pN3
   e) pN4

11) A 29-year-old man presents with painless haematuria. Light microscopy of the renal biopsy appeared unremarkable. Below is an electron micrograph of the glomerulus. Which is the correct diagnosis?

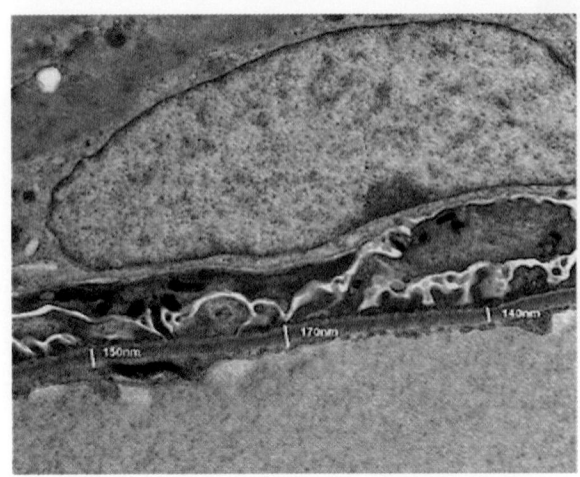

(Photomicrograph courtesy of Dr R Rajab)

   a) Alport syndrome
   b) Thin basement membrane disease
   c) Minimal change glomerulonephritis
   d) IgA nephropathy
   e) Membranous glomerulonephritis

12) A 45-year-old man was admitted with severe burns and shock. Within a few days, he develops acute renal failure. The renal biopsy show preserved glomeruli, with vacuolisation and loss of brush borders of the cells in the tubules, as well as sloughing of the cells into the lumen. There is interstitial oedema. Which is the correct diagnosis?

   a) Acute tubular necrosis
   b) Acute tubulo-interstitial nephritis
   c) Papillary necrosis
   d) Obstructive nephropathy
   e) Malakoplakia

13) A 21-year-old male develops acute renal failure; the renal biopsy shows cortical necrosis. The capillaries within the glomeruli as well as the small arterioles show deposition of fibrinoid material, associated with swelling of the endothelial cells and splitting of the capillary walls. Which of the following organisms may be a cause of this condition?

   a) Group B Streptococcus
   b) Syphilis
   c) E coli type O157
   d) Mycobacterium Tuberculosis
   e) Malaria

14) All of the following conditions may be associated with focal segmental glomerulosclerosis, except:

   a) IgA nephropathy
   b) HIV

c) Sickle cell disease
d) Family history of renal problems
e) All of the above can be associated

15) A 42-year-old woman presents with acute renal failure and pulmonary haemorrhage. A renal biopsy showed crescents in the glomeruli. Immunofluorescence shows linear deposition of IgG and C3 in the glomerular basement membrane. Which is the correct diagnosis?

   a) Systemic Lupus Erythematosus
   b) Churg-Strauss syndrome
   c) Henoch Schonlein purpura
   d) Goodpasture syndrome
   e) HIV associated nephropathy

For questions 16 till 20 : Extended matching item questions: Match each of the following descriptions with the appropriate *diseases of the kidney*:

   A) Diabetic glomerulosclerosis
   B) Membranous glomerulonephritis
   C) Minimal change glomerulonephritis
   D) Membranoproliferative glomerulonephritis
   E) Acute proliferative glomerulonephritis

16) A 28-year-old woman presenting with nephrotic syndrome. The glomeruli show expansion of the mesangium with endocapillary cell proliferation. Silver staining shows double layers of staining.

17) A 34-year-old man presents with proteinuria (4g/day) with hypoalbulinaemia and hyperlipidaemia. The glomeruli show diffuse thickening of the glomerular capillary wall. Silver staining shows spikes extending from the glomerular aspect of the wall.

18) A 5-year-old boy presents with proteinuria (5g/day) and generalised oedema. The glomeruli appear unremarkable.

19) An 81-year-old woman presents with chronic renal failure. The glomeruli show nodular areas of hyalinisation in the mesangium.

20) An 8-year-old boy presents with fever, sore throat, followed by dark urine and periorbital oedema. The renal biopsy shows glomeruli filled with neutrophils and histiocytes. Electron microscopy showed several large dome-shaped dense subepithelial deposits in the glomeruli.

21) A 59-year-old woman presents with a painful 1cm polypoid lesion at the urethral meatus. Histologically, it is composed of cellular fibrous stroma with dense mixed acute and chronic inflammation. The lining is extensively ulcerated with focal residual transitional epithelium. No koilocytosis or atypia is seen. Which is the most likely diagnosis?

   a) Bartholin gland cyst
   b) Condyloma acuminatum
   c) Squamous papilloma
   d) Urethral caruncle
   e) Urothelial papilloma

22) A 65 year old man with a 2.4cm endophytic left renal tumour undergoes CT guided biopsy of this (photomicrograph images shown below). He has previously had an anterior resection for a tumour of the lower sigmoid colon. Immunohistochemistry of the renal tumour cells shows the following profile: CK8/18+, EMA+ (cytoplasmic), CD117+, CK7-, Vimentin- and CD10-. Which is the single most appropriate diagnostic conclusion?

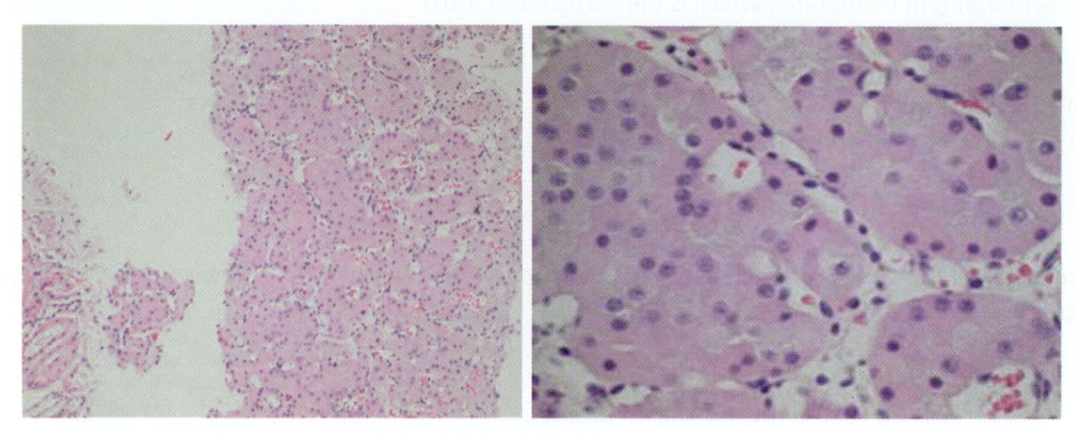

a) Oncocytic neoplasm favour chromophobe RCC.
b) Oncocytic neoplasm favour eosinophilic variant clear cell RCC.
c) Oncocytic neoplasm favour metastatic colonic GIST (epithelioid variant).
d) Oncocytic neoplasm favour renal carcinoid tumour.
e) Oncocytic neoplasm favour renal oncocytoma.

23) A 48 year old woman with an incidental 2.5cm multicystic renal mass on a CT scan undergoes a partial nephrectomy. The report describes the tumour as grossly well circumscribed with a solid and cystic cut surface. Histology shows a tumour composed of microcysts and tubules lined by bland cuboidal epithelial cells, set within variably cellular stroma containing spindle-shaped cells which show immunoreactivity for ER, SMA and Desmin. What is the single most likely diagnosis based on the above description?

a) Cystic nephroma.
b) Mixed epithelial stromal tumour.
c) Multilocular cystic clear cell RCC.
d) Synovial sarcoma.
e) Tubulocystic carcinoma.

24) A 69 year old woman has a partial nephrectomy for a 35mm left kidney mass. The histology shows a well-circumscribed, encapsulated tumour composed of cells with clear cytoplasm and small, dark nuclei. The tumour cells are arranged in a compact tubulopapillary pattern. Immunohistochemistry shows tumour positivity for CK7 and CA-IX (membranous staining sparing the luminal border). The cells do not express CD10 or AMACR. Which is the single most likely additional histological feature that is characteristic of this tumour?

a) Foamy macrophages within fibrovascular cores.
b) Hobnail nuclei with prominent nucleoli.
c) Intratumoral oxalate crystals.
d) Separate population of spindle-shaped cells within mucinous stroma.
e) Tumour nuclei oriented away from the basal aspect of the cells.

25) An 82 year old woman on long term steroid therapy complicated by recurrent episodes of pyelonephritis, flank pain and haematuria has a radical nephrectomy for a 12cm kidney mass. Grossly the kidney showed a necrotic, yellowish mass with ill-defined borders. Representative photomicrographs of the mass are shown.
Which is the single most likely histological diagnosis?

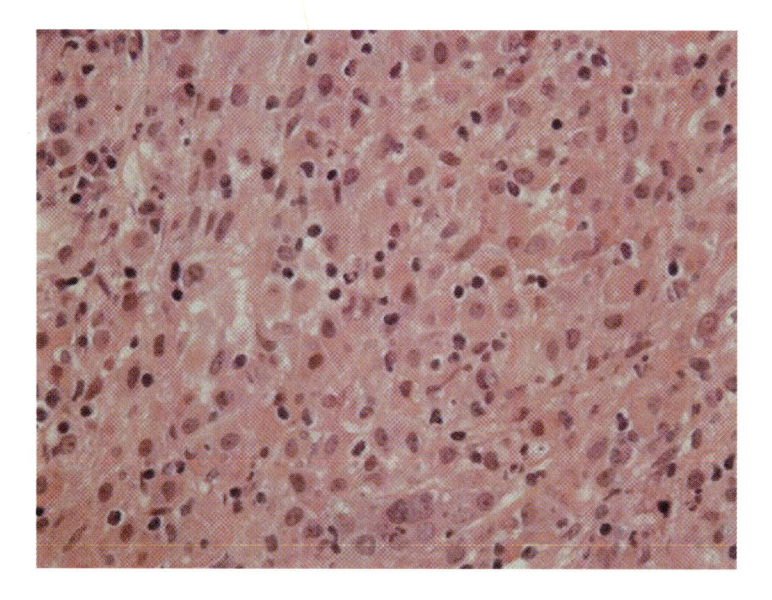

a) Chromophobe renal cell carcinoma.
b) Epithelioid angiomyolipoma.
c) Renal malakoplakia.
d) Renal oncocytoma.
e) Renal tuberculosis.

26) Which single example of cystic renal disease has the most significant risk of development of renal cell carcinoma?

a) Acquired cystic kidney disease
b) Autosomal dominant polycystic kidney disease
c) Cystic renal dysplasia
d) Tuberous sclerosis
e) Von Hippel Lindau disease

27) A 42 year old man with well controlled systemic hypertension related to autosomal dominant polycystic kidney disease (ADPKD) dies suddenly. His wife described him as being well, without any complaints, prior to collapsing in a restaurant. Which single condition is a well recognised cause of sudden death in an adult with ADPKD and the most likely explanation for this man's demise?

a) Aortic dissection
b) Pulmonary thromboembolism
c) Retroperitoneal haemorrhage
d) Subarachnoid haemorrhage
e) Urosepsis

A scenario for questions 28 to 32: A 52 year old woman has a needle core biopsy of a large tumour which obliterates the upper pole of the left kidney and adrenal gland. The biopsy is reported as showing sheets of rounded, eosinophilic tumour cells with large vesicular nuclei and prominent nucleoli. Different immunohistochemical profiles are outlined below.

Match the immunohistochemical findings with the single most likely histological diagnosis.

	CK8/18	EMA	CK7	CD117	CD10	Vimentin	SMA	MelanA	PAX8
A	+	+	+	-	+	+	-	-	+
B	-	-	-	-	-	+	+	+	-
C	+	+	-	-	+	+	-	-	+
D	+	+	+	+ weak	-	-	-	-	+
E	-	-	-	-	+	+	-	-	-
F	-	-	-	-	+	+	-	+	+
G	+ weak	-	-	-	+	+	-	+	-
H	+	+	-	+	-	-	-	-	+

28) Adrenocortical carcinoma

29) Chromophobe renal cell carcinoma

30) Collecting duct carcinoma

31) Epithelioid angiomyolipoma

32) Clear cell renal cell carcinoma

Correct Answers and Explanations:

1) Correct answer: C) Malakoplakia

Malakoplakia is a chronic inflammatory condition, thought to be due to defects in the cytoplasmic phagocytic activity of macrophages. It is related to bacterial infections and is commoner in immunosuppressed patients. Grossly, the lesions appear as yellowish plaques. Histologically there are sheets of macrophages with eosinophilic cytoplasm (PAS-positive), as well as scattered intracytoplasmic laminated calcium deposits with a targetoid appearance which are termed Michaelis-Gutmann bodies and stain with von Kossa stain.

2) Correct answer: A) Haemorrhagic cystitis

Haemorrhagic cystitis refers to a mixed inflammation of the bladder where there is a major component of haemorrhage. This type of cystitis is commonly seen following chemotherapy or local radiotherapy, and may be accompanied by atypia in the urothelial or stromal cells.

3) Correct answer: D) Polypoid cystitis

Polypoid cystitis is an inflammatory condition of the bladder due to chronic physical irritation. The commonest causes are long-term indwelling catheters or bladder stones, and it is usually seen on the dome or posterior wall. Apart from the inflammatory component, the urothelium is polypoid and oedematous. This broad based oedematous polypoid appearance is distinct from papillary urothelial carcinomas, which are often narrow papillae with a narrow fibrovascular core and exhibit hyperplasia and atypia in the urothelium.

4) Correct answer: B) Interstitial cystitis

This is a lesion of unknown aetiology which usually affects young and middle-aged women, and presents with dysuria and suprapubic pain which can sometimes be severe and debilitating. The inflammation may contain prominent mast cells (stained with toluidine blue). The overlying mucosa may also show chronic ulceration, so-called Hunner ulcers. The diagnosis requires exclusion of other specific causes of inflammation.

5) Correct answer: E) Inverted papilloma

This is a rare lesion which often presents with haematuria. Cystoscopically, it is a smooth polypoid lesion with flat surface urothelium, but with invaginations of the urothelium into the lamina propria as cords and nests. They can be distinguished from inverted type urothelial carcinomas by the bland cytology, uniform size of nests, normal layers of urothelial cells which are non-hyperplastic and lack of stromal reaction.

6) Correct answer: b) Nephrogenic adenoma

Nephrogenic adenoma or nephrogenic metaplasia is a reactive condition, following previous surgery or trauma, in which the urothelium is replaced by a single-layered cuboidal epithelium which may either show surface papillary architecture or small nests and tubules in the lamina propria. It can be distinguished from adenocarcinoma (from the bladder, prostate or other source) by the bland or mild atypia, circumscription and clinical history. If necessary, immunostaining with PSA and PSAP may be used to rule out a prostatic carcinoma.

7) Correct answer: e) Schistosomiasis infection

Schistosoma haematobium is a parasitic infection endemic in certain countries in the Middle East. Infection occurs retrograde along the urethra, and the parasite deposits ova in the bladder wall (as seen in the photomicrograph). This results in a chronic inflammatory reaction with prominent eosinophils and fibrosis, as well as squamous metaplasia of the urothelium. Schistosomiasis infection predisposes to both squamous cell carcinoma and urothelial carcinoma of the bladder.

8) Correct answer: a) Beta-catenin

Primary adenocarcinoma of the bladder is a rare tumour, accounting for 2.5% of bladder malignancies. It should be distinguished from the commoner urothelial carcinoma with focal glandular differentiation. For therapeutic purposes, it is important to determine whether the adenocarcinoma has originated in the bladder or is a metastasis as a result of direct invasion from other organs such as the prostate, uterus or bowel. The presence of a surface component is not a useful discriminant, as adjacent carcinoma in-situ may not be identified due to ulceration. While a prostatic or gynaecological primary may be excluded by PSA or PAX8 immunostaining, urinary and bowel adenocarcinomas share the immunoprofile of positivity for CDX2 and CK20. Beta-catenin is found to have nuclear positivity in the majority of bowel cancers, but will either be negative or show cytoplasmic positivity in bladder adenocarcinomas.

9) Correct answer: c) pT2b
10) Correct answer: c) pN2

For answers 9 & 10) : The following table shows a brief overview of the TNM staging for bladder cancers.

Stage	Level of involvement
Tumour	
pTa	Non-invasive papillary
pT1	Invades subepithelial connective tissue
p T2a	Invades inner half of the muscularis propria
pT2b	Invades outer half of the muscularis propria
pT3a	Invades perivesical tissue microscopically
pT3b	Invades perivesical tissue macroscopically
pT4	Invades adjacent organs/structures
Nodes	
p N0	No lymph nodes involved
pN1	1 pelvic lymph node involved
pN2	2 or more pelvic lymph nodes involved
pN3	Common iliac node(s) involved
Metastasis	
pM0	No metastasis
p M1	Distant metastasis

11) Correct answer: b) Thin basement membrane disease

The normal adult glomerular basement membrane has a thickness of 300-400nm (measured by electron microscopy). Thin basement membrane disease (also known as benign familial haematuria) is a genetic disease due to a defect in genes coding for components in type IV collagen. This results in a diffusely thin glomerular basement membrane of 150-250nm. Patients present with asymptomatic haematuria or proteinuria, but do not develop renal failure.

Thin basement membrane disease is related but distinct from Alport syndrome, which is a X-linked disorder that is also due to defects in genes involved in type IV collagen. In Alport syndrome, the glomerular basement membrane shows splitting, with a basket weave appearance. Clinically, these patients may progress to nephrotic syndrome and renal failure. This condition also causes hearing defects.

12) Correct answer: a) Acute tubular necrosis/injury

Acute tubular necrosis/injury is a condition characterised by damage to the epithelial cells in the proximal and distal convoluted tubules. This is most commonly due to ischaemia (related to vasculitis, thromboembolism or circulatory shock) or various drugs. ATN/I presents acutely as oliguric acute renal failure, followed by a polyuric phase as the tubular epithelium regenerates. The prognosis depends on the underlying cause, and requires supportive measures to prevent secondary damage to the kidney and other organs.

13) Correct answer: c) E coli type O157

The histological description given is that of thrombotic microangiopathy, and is caused by endothelial cell injury and platelet aggregation with thrombi formation in capillaries and small arterioles. There are several causes for this, the most characteristic being haemolytic uraemic syndrome (HUS) and thrombotic thrombocytopaenic purpura (TTP). TTP is caused by a genetic defect in the ADAMTS-13 gene, while HUS may be caused by a variety of aetiologies, including verocytotoxin-releasing bacteria (including Escherichia coli type O157), antiphospholipid antibodies, pregnancy, renal vascular diseases, drugs or radiation.

14) Correct answer: e) All of the above can be associated

Focal segmental glomerulosclerosis (FSGS) is a histological lesion characterised by sclerosis which affects some but not all the glomeruli (focal), and only a part of each glomeruli (segmental). It is not a single condition, but rather a collection of entities giving rise to the histological picture described, and with a clinical presentation of proteinuria, possibly with nephrotic syndrome. FSGS may be characterised as either idiopathic/primary (with no associations identified) or secondary. Secondary forms of FSGS include those associated with HIV infection (HIV nephropathy), heroin abuse, other drugs, sickle cell disease, obesity and other renal problems (eg. scarring from IgA nephropathy, reflux nephropathy, hypertensive renal disease or renal agenesis). There are also familial cases of FSGS due to the inheritance of genes regulating podocyte function, such as NPHS1 and NPHS2.

15) Correct answer: d) Goodpasture syndrome

Goodpasture syndrome, also known as anti-GBM (glomerular basement membrane) antibody-induced disease, is an autoimmune condition caused by the production of anti-GBM antibodies specifically targeting a component of type IV collagen. This results in a linear deposition of the IgG antibody, together with the C3 complement factor in the glomeruli. This in turn results in glomerular injury and secondary proliferation of the parietal epithelium, forming so-called crescents. Clinically, the syndrome may present in a variety of ways, including haematuria, proteinuria +/- nephrotic syndrome or acute/chronic renal failure. In addition, there is cross-reactive antigenicity with the pulmonary alveolar basement membrane. Patients may also develop or present with haemoptysis or pulmonary haemorrhage. Treatment is by plasmapheresis and/or immunosuppression and steroids.

Apart from Goodpasture syndrome, there are other causes of crescentic (rapidly progressive) glomerulonephritis, including immuno-complex mediated diseases, particularly Systemic Lupus Erythematosus (SLE); and pauci-immmune glomerulonephritis, usually associated with cytoplasmic or perinuclear anti-neutrophilic cytoplasmic antibodies (c-ANCA or p-ANCA). Conditions which give rise to pauci-immune glomerulonephritides include granulomatosis with polyangiitis (previously Wegener's granulomatosis), Churg-Strauss disease and microscopic polyangiitis.

16) Correct answer: D) Membranoproliferative glomerulonephritis

Membranoproliferative glomerulonephritis (MPGN) is a collection of glomerular diseases, characterised by mesangial hypercellularity and capillary wall thickening, resulting in a lobular appearance of the glomeruli. They may be classified as either primary (usually affecting children) or secondary (in adults). There are two common types of primary MPGN: Type I – Mesangiocapillary glomerulonephritis (with subendothelial immune-complex deposits), Type II – Dense Deposit Disease (with intramembranous deposits). Both show predominantly deposition of C3. There is also a rarer Type III MPGN.
Causes of secondary forms include infections, malignancies, autoimmune diseases and thrombotic events. MPGN may present in a variety of ways, including nephritic or nephrotic syndromes, haematuria or proteinuria.

17) Correct answer: B) Membranous glomerulonephritis

Membranous glomerulonephritis is characterised by diffuse thickening of the glomerular basement membrane, due to the deposition of electron-dense immunoglobulin complexes along the subepithelial side of the basement membrane. The remaining intervening basement membrane gives rise to the spikes of silver staining. On immunostaining, there is diffuse deposition of IgG and C3 in the glomerular capillary wall. The diffuse deposition is also seen under electron microscopy. It is associated with a variety of causes, including many drugs, underlying malignancy, autoimmune diseases and infections. It is the commonest cause of nephrotic syndrome in adults. The prognosis is variable.

18) Correct answer: C) Minimal change glomerulonephritis

Minimal change glomerulonephritis characterised by the diffuse flattening of the foot processes of the podocytes in glomeruli. Under light microscopy, the glomeruli appear normal, but the changes can be seen using electron microscopy. No immune complex deposition is seen. In contrast to membranous glomerulonephritis, it is the commonest cause of nephrotic syndrome in children. It is associated with a variety of causes, including Hodgkin lymphoma, atopic disorders, infections and immunisations. The prognosis is very good, with 90% of cases treated successfully with steroids.

19) Correct answer: A) Diabetic glomerulosclerosis

Both type I and II diabetes mellitus may affect kidneys in a variety of ways, including arteriolar sclerosis, infections and papillary necrosis. In the glomeruli, glycosylation of proteins results initially in capillary basement membrane thickening followed by diffuse mesangial sclerosis and nodular glomerulosclerosis (termed Kimmelstiel-Wilson disease). Patients may present with proteinuria, nephrotic syndrome, or chronic renal failure. End-stage renal failure occurs in up to 40% of diabetic patients. Occasionally, patients with undiagnosed diabetes mellitus may have renal problems as their initial presentation.

20) Correct answer: E) Acute proliferative glomerulonephritis

This disease is characterised by hypercellular glomeruli due to the aggregates of leukocytes. It usually is caused by a recent infection, classically Group A Streptococcus, although there are other causes such as SLE. This results in deposition of circulating immune complexes and complement factors such as C3 in the subepithelial basement membrane, usually as large humps. Clinically, they frequently affect children, who present with nephritic syndrome (mild proteinuria and haematuria, periorbital oedema and mild hypertension). This disease is usually self-resolving in children.

21) Correct answer: d) Urethral caruncle
Caruncles are inflammatory lesions commonly affecting women, particularly of post-menopausal age. The aetiology is uncertain, but may be due to urethral prolapse followed by chronic irritation. Grossly, they appear as red friable polyps, and the histology is as noted. The differential diagnoses include condyloma acuminatum/genital warts (distinguished by squamous epithelium with koilocytes), urothelial papilloma (papillae of transitional epithelium with thin fibrovascular cores), squamous papilloma (hyperplastic squamous epithelium) and Bartholin gland cyst (cyst in the stroma lined by transitional/squamous/columnar epithelium).

22) Correct answer: e) Oncocytic neoplasm favour renal oncocytoma.

Eosinophilic/oncocytic renal neoplasms constitute a significant proportion of renal tumours and often the main question is whether such a tumour meets the criteria for renal oncocytoma. In the case of a low grade tumour without papillary architecture, lack of CK7 staining may help favour oncocytoma over chromophobe RCC, but the correct cytomorphology is probably the most important consideration as hybrid tumours occur and some oncocytomas may show significant focal expression of CK7 while a subset of chromophobe RCC may not express CK7 at all. It is also important to note that a radiological finding of central scar is not helpful, as about 30% of oncocytomas lack this feature and it may be seen in other low grade renal tumours. Definitive diagnosis of renal oncocytoma on needle core biopsy is controversial, and because oncocytic features may uncommonly be encountered in renal cell carcinoma and hybrid tumours it seems more prudent to render a somewhat descriptive diagnosis and favour oncocytoma or chromophobe RCC as appropriate.

23) Correct answer: b) Mixed epithelial stromal tumour.

Predominantly cystic renal tumours are diagnostically challenging and include cystic nephroma/mixed epithelial stromal tumour, multilocular cystic renal cell carcinoma and synovial sarcoma, but occasionally angiomyolipoma and oncocytoma can have a prominent cystic component. Other tumours with a characteristic cystic architecture include tubulocystic carcinoma and acquired cystic disease associated RCC.
Mixed epithelial stromal tumour (MEST) is a rare tumour exhibiting a striking predilection for perimenopausal females (F:M = 6:1) which is notable for the combination of bland tubules and small cysts (lacking cytological atypia or mitotic activity) and spindle cell, ER/PR positive, stroma.
Cystic nephroma is typically predominantly multicystic without solid elements. The septa are thinner than in MEST and contain similar 'ovarian type' stroma. The term renal epithelial stromal tumour (REST) has been proposed as a unifying term for MEST and cystic nephroma.
The identification of small aggregates of clear cells in the wall of cysts is important for the diagnosis of *multilocular cystic RCC* (but must be distinguished from the expansile clear cell areas seen in clear cell RCC with cystic changes).
The spindle cells of *synovial sarcoma* do not express SMA or Desmin, but are immunoreactive for EMA, CD56 and sometimes CD99.
Tubulocystic carcinoma is composed of tubules and cysts lined by cells with prominent nucleoli and a small percentage metastasise. Ovarian type stroma is absent.

24) Correct answer: e) Tumour nuclei oriented away from the basal aspect of the cells.

The tumour histology and immunohistochemical profile is very suggestive of a *clear cell (tubulo) papillary renal cell carcinoma (CC-PRCC)*. Some distinctive features of this tumour include cells with abundant clear cytoplasm, low grade nuclei lacking prominent nucleoli oriented in a linear fashion away from the basal aspect of the cells and lack of foamy macrophages.
Hobnail nuclei and prominent nucleoli are a feature of *tubulocystic RCC* which usually lacks appreciable papillary architecture and expresses CD10 and AMACR.
Intratumoral oxalate crystals are a feature of *acquired cystic disease-associated RCC* (which may have clear cell features and papillary architecture, but are typically negative or only focally positive for CK7 and positive for AMACR) and *mucinous tubular and spindle cell carcinoma (MTSCC)* is characterised by slit-like tubules and cords of spindle cells set within mucinous stroma (and usually positive for AMACR).

25) Correct answer: c) Renal malakoplakia.

Malakoplakia is a rare, multisystem, chronic granulomatous inflammatory condition more often seen in immunocompromised or debilitated patients and mainly associated with *Escherichia coli* infection. The term malakoplakia is derived from the Greek words *malakos* (soft) and *plakos* (plaque), which describes the gross appearances quite well. The condition is encountered most frequently in the genitourinary system, but other sites affected include the skin, gastrointestinal and respiratory tracts. It is thought to result from defective phagolysosomal activity within macrophages that have ingested bacteria. The deposition of calcium and iron on residual bacterial glycolipid material results in the characteristic rounded, targetoid, intracytoplasmic basophilic inclusion, the Michaelis-Gutmann body, which is considered pathognomonic for malakoplakia. Perls and von Kossa stains can be used to highlight the inclusions. Pseudotumoral renal masses are an unusual presentation, and the diagnosis of malakoplakia may be challenging on small needle core biopsies of such masses. Appropriate use of special stains and/or immunohistochemistry should aid correct diagnosis.

26) Correct answer: e) Von Hippel Lindau disease.

The risk of renal cell carcinoma (clear cell type) is highest in Von Hippel Lindau (VHL) disease (60%) - up to 1/3 of VHL patients die from metastatic RCC. Acquired cystic kidney disease (ACKD) occurring in patients on prolonged renal dialysis, follows with a 5-10% risk, which increases with the length of dialysis. ACKD-associated RCC and clear cell PRCC are the commonest subtypes of RCC occurring in ACKD. Autosomal dominant polycystic kidney disease (ADPKD) has an unknown risk of RCC, but is probably not increased. RCC occurs in about 3% of tuberous sclerosis (TS) patients, but more importantly this condition is associated with angiomyolipoma (including epithelioid variant, which may behave aggressively).

27) Correct answer: d) Subarachnoid haemorrhage.

Autosomal dominant polycystic kidney disease (ADPKD) is associated with sudden death, and this may be caused by sepsis due to pyelonephritis (unlikely in this scenario as the individual was reportedly well), mitral valve prolapse and subarachnoid haemorrhage due to ruptured intracranial aneurysm. Ruptured kidney cysts may bleed (the cystic kidneys are susceptible to blunt trauma) and cause haematuria or anaemia but are unlikely to be associated with catastrophic retroperitoneal haemorrhage (unless from rupture of aortic dissection or aneurysm). Although these patient may develop complications related to hypertension such as cardiac disease and atherosclerotic abdominal aortic aneurysm, it is perhaps less likely in this man who is young and has apparently well controlled hypertension. Very large cysts may be associated with compression of the inferior vena cava (IVC) and therefore potentially with thrombosis thereof and pulmonary embolism, but this is probably an exceedingly rare complication. Electrolyte disturbances, seizures and gastrointestinal haemorrhage may cause sudden death in uraemic patients with end stage renal disease.

28) Correct answer: G) see immuno table

The presence of finely vacuolated, bubbly cytoplasm is an important feature of adrenocortical neoplasms, and when this possibility is raised by the clinical context, the lack of EMA and PAX8 expression together with positivity for MelanA and Inhibin is helpful in confirming a diagnosis of a tumour of adrenocortical origin.

29) Correct answer: D) see immuno table

Chromophobe RCC may be difficult to distinguish from renal oncocytoma and some eosinophilic variants of CCRCC, but attention to the nuclear features and application of immunohistochemistry should classify most cases. Diffuse CK7 staining is helpful, but some examples may not show this.

30) Correct answer: A) see immuno table

The immunoprofile of CDC is not specific, but most show CK7 positivity and express PAX8. The latter is not entirely helpful in distinguishing CDC from high grade urothelial carcinoma (UCa) as about 25% of upper tract UCa express this nuclear marker.

31) Correct answer: B) see immuno table

Epithelioid angiomyolipoma (eAML) may be confused with high grade RCC, but is characterised by a lack of epithelial marker positivity, PAX8 negativity and positive staining with SMA, Desmin, MelanA and HMB45 (S100 is negative).

32) Correct answer: C) see immuno table

High grade CCRCC often shows more noticeable cytoplasmic eosinophilia and careful observation and sampling at specimen dissection may help to identify conventional clear cell areas.

CHAPTER 13

MALE GENITAL TRACT PATHOLOGY
Dr Limci Gupta & Dr Jayson Wang

1) A 38 year old HIV positive male presents with a testicular enlargement with vague nodule formation and pain. Partial orchidectomy demonstrated a poorly defined white grey nodule on cut section. The microscopic sections revealed an inflammatory mass composed of poorly formed granulomas, lymphocytes and occasional giant cells. Which of the following stains would you like to request in the first instance? Choose the most appropriate.

   a) PLAP.
   b) ZN.
   c) CD68.
   d) DPAS.
   e) OCT 3/4.

2) A 32 year old male presented to his GP with a two day history of testicular enlargement and pain. He also reported a history of pain and swelling in both cheeks a week prior to presentation. On examination, both testes were enlarged and red. The patient reported pain on palpation. Given the clinical history, which is the most likely diagnosis?

   a) Non seminomatous germ cell tumour of the testis.
   b) Testicular infarction.
   c) Tuberculous orchitis.
   d) Mumps orchitis.
   e) Un-descended testis.

3) A 19 year old male presents to the accident and emergency department with testicular pain radiating to the abdomen. There was no history of trauma or injury to testis. On examination, his right testis was enlarged and tender. There was no fever. A diagnosis of 'acute scrotum' was made clinically and a right sided orchidectomy was performed. Histological sections showed interstitial oedema, fresh haemorrhage, lymphovascular congestion and infarcted seminiferous tubules. No vasculitis was identified. Which was most likely aetiology for this condition?

   a) Testicular torsion.
   b) Acute viral orchitis.
   c) Testicular haematoma.
   d) Vasculitis.

4) Which of the following is the preferred fixative for good nuclear details on a testicular biopsy?

   a) Formalin.
   b) Bouin's fluid.
   c) Alcohol.
   d) Glutaraldehyde.

5) A 75 year old male had an orchidectomy for a testicular tumour. Grossly, the tumour had a uniform grey-tan surface with no necrosis or haemorrhage. Microscopically, the tumour was composed of sheets of large

atypical cells with vesicular nuclei and prominent nucleoli. There is a relative sparing of seminiferous tubules. Frequent mitosis and apoptosis were present. The tumour cells were CD45+, CD20 + and BCL2+. They were negative for OCT3/4, PLAP, CD30, AFP, CD2, CD3 and CD5. The Ki67 proliferative index was 70%. What is the most likely diagnosis in this case?

    a) Spermatocytic seminoma.
    b) Classical seminoma.
    c) Diffuse large B- cell lymphoma.
    d) Burkitt's lymphoma.

Extending matching questions for 6-10 : Match the characteristic features / immunohistochemistry / clinical presentation with corresponding testicular tumours:

    A) Seminoma.
    B) Teratoma.
    C) Embryonal carcinoma.
    D) Yolk sac tumour.
    E) Sertoli cell tumour.

6) The testicular tumour is composed of round to polyhedral cells with a clear cytoplasm, central nucleus with one or two prominent nucleoli. These cells are arranged in nests separated by vascular stroma containing lymphocytes.

7) The testicular tumour is composed of tubulo-alveolar pattern and sheets of tumour cells with indistinct cell borders, overlapping nuclei, pleomorphic nuclei and prominent nucleoli and abundant cytoplasm. The tumour cells are positive for Oct 3/4 and CD30.

8) The patient comes with gynaecomastia and on examination, is found to have a testicular tumour, composed of tubules of bland cuboidal cells in hyalinised stroma. The cells are positive for CK8/18 and EMA.

9) The testicular tumour comprises cartilage, thyroid tissue and skin appendages in different proportions.

10) Tumour cells are cuboidal and arranged in reticular pattern with glomeruloid structures and clinically raised serum AFP.

11) Testicular tumour is confined to testis with lymphovascular invasion. Which tumour stage is it at?

    a) pT1.
    b) pT2.
    c) pT3.
    d) pT4.

12) Which feature of a spermatocytic seminoma helps differentiation from a classical seminoma?

    a) Absence of intra-tubular germ cell neoplasia.
    b) Oct3/4 positive.
    c) Younger age.
    d) Prominent lymphocytic inflammatory infiltrate.

13) This is a 25yr old male with a tumour in the epididymis. The tumour is composed of tubules of variable sizes lined by flattened and cuboidal bland cells within a fibrous stroma. The tumour cells are positive for calretinin, EMA and cytokeratins. What is the likely diagnosis?

    a) Papillary cystadenoma.
    b) Adenocarcinoma.
    c) Adenomatoid tumour.
    d) Epididymal carcinoma.

14) Which is/are the poor prognostic factor(s) in radical prostatectomy for prostate cancer? (More than one answer is accepted)

    a) Presence of high grade prostatic intra-epithelial neoplasia (PIN).
    b) Bladder neck involvement.
    c) Squamous metaplasia.
    d) Perineurial invasion.

15) Which of the following testicular tumours are **NOT** associated with intratubular germ cell neoplasia?

    a) Seminoma.
    b) Spermatocytic seminoma.
    c) Embryonal carcinoma.
    d) Mixed germ cell tumour.

Extending matching questions for questions 16-18 : Please match the following with their correct descriptions:

    A) Prostatic adenocarcinoma, Gleason grade 2.
    B) Prostatic adenocarcinoma, Gleason grade 5.
    C) Prostatic atrophy.
    D) Prostatic adenocarcinoma, Gleason grade 3.
    E) Prostatic adenocarcinoma, Gleason grade 4.

16) On core needle biopsy of the prostate, a malignant infiltrate of small smooth rounded glands lined by atypical cells showing prominent nucleoli and luminal crystalloids is best regarded as which of the above?

17) On core needle biopsy of the prostate a malignant infiltrate of poorly defined small glands with complex cribriform structures, lined by atypical cells showing prominent nucleoli is seen. This is best regarded as which of the above?

18) On core needle biopsy of the prostate a malignant infiltrate of solid nests of atypical cells showing prominent nucleoli is seen. There are associated single cells and areas of comedo -necrosis. This is best regarded as which of the above?

19) Gleason score on core needle biopsy of the prostate is determined by which of the following?

    a) The sum of the highest Gleason grade pattern and the next highest grade pattern.
    b) The sum of the most prominent and second most prominent Gleason grade pattern.
    c) The sum of the most prominent Gleason grade and the highest grade pattern.
    d) The worst Gleason grade present.

e) The most prominent Gleason grade present.

20) A 65 year old man is diagnosed with prostatic adenocarcinoma on core biopsy. He undergoes radical prostatectomy and the resection confirms prostatic adenocarcinoma, Gleason grade 3+4=7, in both right and left lobes of the prostate. The tumour is confined to the prostate with no extraprostatic extension. The resection margins are clear of tumour. Which pathological T stage is correct?

a) pT1b.
b) pT2.
c) pT4.
d) pT3a.
e) pT1a.

21) A 57 year old man was found to have a firm craggy fixed prostate on digital rectal examination. Transrectal core biopsies of the prostate gland were performed and the histology is shown below. The cells are positive for PSA, PSAP, CD56, synaptophysin and negative for TTF-1. Which is the correct diagnosis?

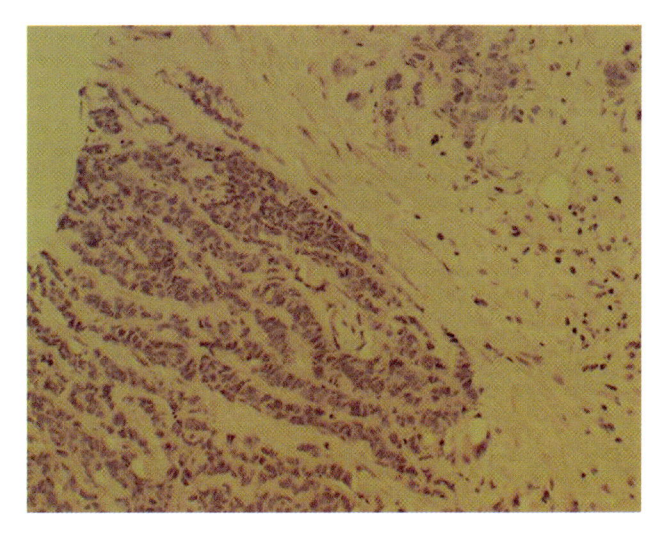

a) Acinar prostate carcinoma, Gleason 4.
b) Acinar prostate carcinoma, Gleason 5.
c) Ductal type prostatic carcinoma, Gleason 4.
d) Small cell carcinoma from the lung.
e) Small cell carcinoma of the prostate.

22) Penile cancer is predominantly associated with which of these viral infections?

a) HPV2.
b) HPV11.
c) HPV16.
d) HSV2.

23) Which subtype of penile cancer has better prognosis?

a) Verrucous.
b) Basaloid.
c) Acantholytic.

d) Sarcomatoid.

24) A 43 year old man presents with raised whitish plaques on the glans of his penis. An incisional biopsy was performed. The histology is shown below. What is the correct diagnosis?

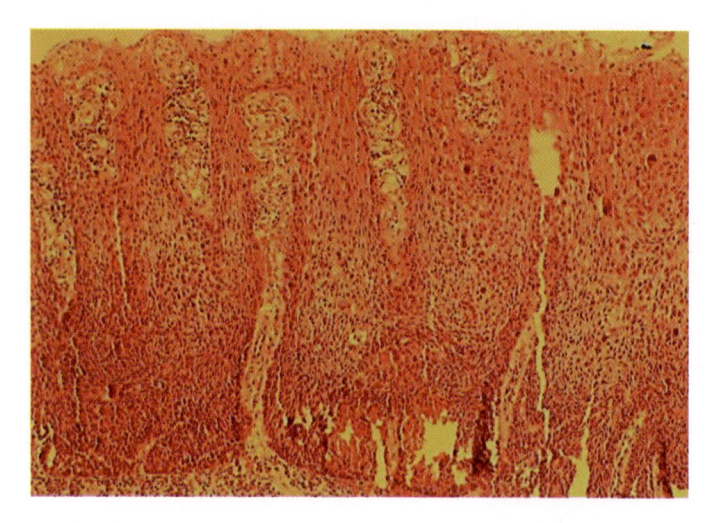

a) Differentiated penile intraepithelial neoplasia.
b) Undifferentiated penile intraepithelial neoplasia.
c) Lichen sclerosus et atrophicus.
d) Squamous cell carcinoma, basaloid type.
e) Zoon's balanitis.

25) A young male presented with multiple reddish brown plaques on the dorsal aspect of penis. Histology showed reactive epithelial changes, oedema and dense plasma cell infiltrate admixed with lymphocytes and siderophages in underlying stroma. What is this condition called?

a) Balanitis xerotica obliterans.
b) Peyronie disease.
c) Zoon's balanitis.
d) Psoriasis.

26) A middle aged male presented with right sided groin pain on exertion. There was a past history of vasectomy 5 years ago. On examination the nodule was palpated above the testis. What is the likely diagnosis?

a) Testicular torsion.
b) Mumps.
c) Spermatic granuloma.
d) Inguinal hernia.

27) An elderly male presented with urethral bleeding after 6 weeks of transurethral resection of prostate. On examination the prostate was nodular. Histology showed a highly cellular lesion with fascicles of relatively bland spindles cells and extravasated red blood cells. There were many mitoses but no atypical mitotic figures were seen. What is the diagnosis?

a) Kaposi sarcoma.

b) Post-operative spindle cell nodule.
   c) Leiomyoma.
   d) Adenocarcinoma.

28) A middle aged male presented with a firm and a slightly cystic 10mm lesion in scrotum. The microscopy showed calcium deposits in the dermis with lymphohistiocytic reaction around them. What is the diagnosis?

   a) Gout.
   b) Scrotal calcinosis.
   c) Rheumatoid arthritis.
   d) Amyloidosis.

29) Which of the following syndromes / diseases is bilateral papillary cystadenoma of epididymis associated with?

   a) Klinefelter's Syndrome.
   b) Cryptorchidism.
   c) Von Hippel-Lindau disease.
   d) Osler's disease.

30) A young man presents with raised reddish-purple plaques on his scrotum. Excisions of the lesions were performed. Below is the histology on one of these lesions. What is the most likely diagnosis?

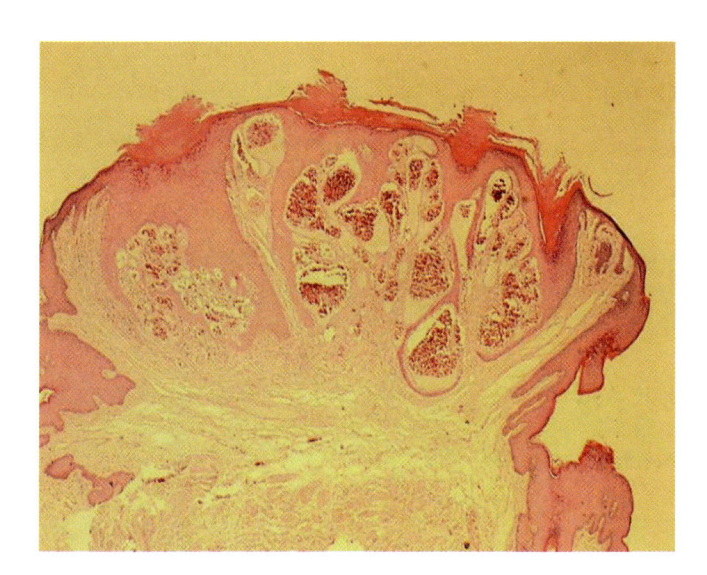

   a) Kaposi's sarcoma.
   b) Capillary haemangioma.
   c) Angiosarcoma.
   d) Angiokeratoma.
   e) Pyogenic granuloma.

Correct answers and explanations:

1) Correct answer : b) ZN.

The histological appearances are of an inflammatory lesion (orchitis) with poorly formed granulomas. Once the diagnosis of granulomatous orchitis is made, tuberculosis must be excluded and as such, a ZN is the first line stain to perform in this case. The patient is immunocompromised as a result of HIV infection, increasing the susceptibility to mycobacterial infections.

2) Correct answer : d) Mumps orchitis.

The history describes a case of acute orchitis. This may occur secondary to viral or bacterial infection. The history of pain and swelling of the cheeks in the week prior to presentation suggests an episode of bilateral parotitis. Mumps infection can present as acute parotitis and orchitis.

3) Correct answer : a) Testicular torsion.

Torsion of the spermatic cord is the most common cause of testicular infarction although infarction may also occur secondary to trauma, incarcerated inguinal hernia, vasculitis and epididymitis. Clinically it presents acutely with scrotal pain, enlargement, congestion and in severe cases, black discolouration.

4) Correct answer : b) Bouin's fluid.

It is a fixative composed of picric acid, acetic acid and formaldehyde. It provides good nuclear detail and is a good preservative for chromosomes, especially to observe meiosis and also cellular morphology.

5) Correct answer : c) Diffuse large B cell lymphoma (DLBCL).

DLBCL is the most common testicular tumour in elderly males. These can be unilateral or bilateral. Histologically, the tumour has interstitial growth pattern with relative sparing of seminiferous tubules. The tumour is composed of large cells with large nuclei and prominent nucleoli. Lymphomas are not associated with intratubular germ cell neoplasia (ITGCN).
Seminomas are germ cell tumours with characteristic features (see answer 6). They are positive with OCT3/4 and would be negative for lymphoid markers.
Burkitt's lymphoma in the testis is rare and has been reported in children. It shows a 'starry sky' pattern with almost 100% Ki67 positivity.

6) Correct answer : A) Classical seminoma.

These occur in 30-40 years of age. The tumour is composed of polygonal cells with clear cytoplasm, uniform nuclei and prominent nucleoli with lymphocytic inflammatory infiltrate. The tumour cells are positive with OCT3/4, PLAP and CD117.

7) Correct answer : C) Embryonal carcinoma.

These usually are a part of mixed germ cell tumour and may present as a testicular mass or metastasis. The histological features are as above. The tumour cells are pleomorphic with necrosis, arranged in the form of papillary, solid or glandular patterns. Tumour cells are positive for cytokeratins & CD30 and are negative for EMA.

8) Correct answer : E) Sertoli cell tumour.

An uncommon tumour which presents as a testicular mass or gynaecomastia due to oestrogen production by tumour cells. It occurs in middle aged men. Histologically, the tumour is composed of trabecula and cords of clear cells resembling seminiferous tubules. The tumour is positive for cytokeratins & vimentin and are negative for OCT 3/4.

9) Correct answer :B) Teratomas.

These are second most common testicular germ cell tumour in children. They are usually benign in children and malignant in post-pubertal males. They are classified into mature, immature and malignant forms. All three germ layer derivatives are seen. A variable mixture of respiratory epithelium, cartilage, bone, skin or neural tissue is present. Malignant transformation in tumours for example squamous cell carcinoma can occur.

10) Correct answer : D) Yolk sac tumour.

These are the most common testicular tumour in children. The tumour cells are polygonal, eosinophilic to cuboidal or columnar. Schiller Duval bodies/glomeruloid structures are characteristic. Tumour cells are usually AFP positive.

11) Correct answer : b) pT2

According to TNM staging:
pTis: Intra-tubular germ cell neoplasia
pT1: Tumour confined to testis or epididymis without lymphovascular invasion; invading tunica albuginea but not vaginalis.
pT2: Tumour confined to testis or epididymis with lymphovascular invasion or invading tunica vaginalis.
pT3: Tumour invading the spermatic cord with or without lymphovascular invasion.
pT4: Tumour involving scrotum with or without lymphovascular invasion.

12) Correct answer : a) Absence of intra-tubular germ cell neoplasia.

Spermatocytic seminomas occur in older people and show heterogenous population of cells. The tumour is composed of small cells resembling lymphocytes with smudged chromatin and scanty cytoplasm, intermediate cells and large cells with prominent nucleoli. There is no lymphocytic infiltrate and no association with ITGCN. Prognosis is excellent. The tumour is negative for OCT3/4, PLAP and EMA.

13) Correct answer : c) Adenomatoid tumour.

It is the second most common tumour of the epididymis after a lipoma. It occurs in young patients with an age range of 20-40 years. These are mesothelial in origin and often painful. The tumour is composed of tubules of variable sizes lined by flattened and cuboidal bland cells within a fibrous stroma. The tumour is positive for calretinin, EMA and cytokeratins.

14) Correct answer : b) Bladder neck involvement

Following radical prostatectomy, features which have been shown to have prognostic significance, and should be reported include: Gleason grade, extraprostatic extension, seminal vesicle involvement, lymphovascular invasion, lymph node metastasis and margin positivity (including at the bladder neck). The presence of high grade PIN and perineurial invasion is very common in prostatectomy specimens, but has no proven prognostic significance.

15) Correct answer : b) Spermatocytic seminoma.

Intratubular germ cell neoplasia IGCN is considered a precursor for germ cell tumours. It can be present in all germ cell tumours in adults except spermatocytic seminoma. In children ITGCN is not seen in yolk sac tumours and teratomas.

16) Correct answer : D) Prostatic adenocarcinoma, Gleason grade 3.

17) Correct answer : E) Prostatic adenocarcinoma, Gleason grade 4.

18) Correct answer : B) Prostatic adenocarcinoma, Gleason grade 5.

Explanations for questions 16-18:
The diagnosis of prostatic adenocarcinoma is dependent on both architectural and cytological criteria. Architecturally, malignant infiltrates tend to demonstrate rigid glands with specific patterns as described below. Cytologically, these malignant glands will be lined by cells showing nuclear and nucleolar enlargement. The nucleoli may be prominent. Absence of a basal layer is diagnostic of adenocarcinoma, and while this may be appreciated on H&E, immunohistochemistry may be required in difficult cases.
Prostatic adenocarcinoma is graded by the **Gleason system**.
Prostatic adenocarcinoma grades 1 and 2 cannot be diagnosed on core biopsy.
A grade 2 diagnosis can be made only on TURP or prostatectomy.
Grade 3 adenocarcinoma shows small smooth rounded glands. No large cribriform glands, solid sheets or single cells should be seen.
Grade 4 adenocarcinoma shows poorly defined small glands, poorly formed glands (they don't have a central lumen) and complex cribriform structures or hypernephroid glands.
Grade 5 adenocarcinoma shows solid nests with comedo necrosis, cribriform aggregates with comedo necrosis, solid sheets or single cells.

19) Correct answer : c) The sum of the most prominent Gleason grade and the highest grade pattern.

E.g. On a core biopsy showing 60% Gleason grade 3 pattern adenocarcinoma, 35% component of Gleason grade 4 pattern adenocarcinoma and 5% component of Gleason grade 5, the Gleason score would be recorded as Gleeson 3+5=8.

20) Correct answer : b) pT2.

Evaluation of the (primary) tumour ('T'):
TX: cannot evaluate the primary tumour
T0: no evidence of tumour
T1: tumour present, but not detectable clinically (by palpation) or with imaging
  T1a: tumour was an incidental histologic finding in less than 5% of prostate tissue resected (for other reasons)
   T1b: tumour was an incidental histologic finding in greater than 5% of prostate tissue resected
    T1c: tumour was found in a needle core biopsy performed due to an elevated serum PSA
T2: tumour palpated on examination, but has not spread outside the prostate
T3: tumour has spread through the prostatic capsule or invaded seminal vesicles

T3a: tumour has spread through the capsule on one or both sides
T3b: tumour has invaded one or both seminal vesicles
T4: the tumour has invaded adjacent structures

21) Correct answer : e) Small cell carcinoma of the prostate.

Small cell carcinoma of the prostate is a rare high grade malignant tumour arising in the prostate gland. It is characterised by small to medium sized polygonal cells with nuclear moulding which express neuroendocrine markers, similar to the more common small cell carcinoma of the lung. They may express PSA although this expression may be lost. They may also express TTF1 in a significant proportion of cases, although negative TTF1 is helpful in distinguishing this from metastasis from a lung primary. Small cell carcinoma is associated with conventional adenocarcinoma of the prostate in up to 50% of cases. They may occur either de novo or develop in patients with adenocarcinoma of the prostate following treatment.

22) Correct answer : c) HPV16.

From this list of options HPV16 is the correct answer. Penile cancer is associated with HPV serotypes 16, 18, 31, 33 and 35.
Another risk factor for penile cancers is the uncircumcised male, in which the risk of cancer is much higher than the circumcised male.
A subset of penile cancer is linked to HPV 16 especially basaloid and warty variants. Majority of usual type and verrucous types are not HPV associated. EBV has been reported in some cases but its role in carcinogenesis in unclear.

23) Correct answer : a) Verrucous.

The verruciform penile carcinomas (papillary, warty or verrucous) have a better prognosis. Basaloid, acantholytic and sarcomatoid subtypes have worse prognoses.

24) Correct answer : b) Undifferentiated penile intraepithelial neoplasia.

Penile intraepithelial neoplasia (PeIN) is a premalignant squamous lesion of the penis, with full-thickness dysplasia equivalent to carcinoma in-situ of the epidermis elsewhere. PeIN encompasses and replaces the older terminology of Bowen's disease and erythroplasia of Queyrat. There are two types: differentiated PeIN which shows papillomatous acanthotic epithelium with parakeratosis and is associated with lichen sclerosis, and undifferentiated PeIN which is composed of uniform small basaloid cells and is associated with high-risk HPV infection. Both types of PeIN predisposes to squamous cell carcinoma.

25) Correct answer : c) Zoon's balanitis.

This is a characteristic histological picture for Zoon's balanitis. It is an inflammatory condition which usually occurs in uncircumcised males. The aetiology is unknown.
Balanitis xerotica obliterans (also known as lichen sclerosus et atrophicus) is a chronic and atrophic condition which may be a precursor for cancer. It shows a band-like sclerosis beneath the basement membrane with lymphocytic inflammation beneath sclerosis. Basal cell vacuolation is also a feature.
Peyronie disease manifests as dense fibrosis of dermis and Buck's fascia of penis and causes abnormal curvature.
Psoriasis manifests as scaly plaques and the penis is not a common site.

26) Correct answer : c) Spermatocytic granuloma.

It is also known as vasitis nodosa. This is a result of inflammation and trauma to the epithelium of the vas deferens with leakage of spermatozoa into the interstitium leading to non–caseating granuloma formation.
Mumps is secondary to viral parotitis in adolescent males.
Testicular torsion is usually secondary to trauma and manifests as severe pain and a swollen, haemorrhagic testis.

27) Correct answer : b) Post-operative spindle cell nodule.

This can develop between a few weeks and months after TURP. Grossly, these are red and friable nodules. They are cellular and contain fascicles of minimally pleomorphic spindle shaped cells with extravasated red blood cells. Many mitoses can be seen but they are not abnormal. They can mimic Kaposi's sarcoma. The cells in post-operative spindle cell nodule are positive for keratins, sometimes for actin and are negative for EMA and endothelial markers.

28) Correct answer : b) Scrotal calcinosis.

This is a classical description of scrotal calcinosis. It is thought that this lesion usually develops from ruptured epidermoid cysts in the scrotum.

29) Correct answer : c) Von Hippel-Lindau disease.

Von Hippel-Lindau disease is due to mutation of VHL tumour suppressor gene. It is associated with haemangioblastomas, angiomatosis, renal cell carcinomas, bilateral epididymal cystadenomas, cystadenomas of broad ligament of uterus, pancreatic serous cystadenoma and phaeochromocytoma.

30) Correct answer : d) Angiokeratoma.

Angiokeratoma is a benign vascular lesion typically found in the scrotum, and less commonly in the penis. It is composed of superficial ectatic blood vessels, with overlying acanthotic epidermis and parakeratosis.
There are four type of angiokeratomas:
1. Angiokeratoma of Fordyce, which is often scrotal.
2. Angiokeratoma corporis diffusum, which is associated with Fabry's disease and occur in childhood.
3. Angiokeratoma of Mibelli with multiple lesion in the dorsum of the fingers and toes.
4. Solitary angiokeratoma.

CHAPTER 14

ENDOCRINE PATHOLOGY
Dr Rashpal Flora

1) Which of the following mutations is associated with papillary carcinoma?

   a) BRAF V600E
   b) cKIT D816V
   c) RET M918T
   d) VHL
   e) JAK2

2) A 58-year-old female presents with a unilateral thyroid mass associated with hoarseness. The thyroid is excised and a tumour is identified on the affected side which stains positively for TTF-1, Synaptophysin and Chromogranin. The lesion is negative for Thyroglobulin. The most likely diagnosis is:

   a) Tall cell variant of papillary carcinoma
   b) Anaplastic carcinoma
   c) Medullary carcinoma
   d) Follicular adenoma
   e) Poorly differentiated carcinoma

3) A 60–year-old female presents with virilising symptoms including deepening of the voice and hirsutism. CT scan reveals a 10 cm right sided adrenal mass which is excised. On macroscopic examination the mass shows central necrosis. The tumour is composed of sheets and nests of polygonal cells showing clear and eosinophilic cytoplasm. The tumour stains positively for Calretinin, Melan-A and Synaptophysin. Which would be the important criteria to favour a malignant diagnosis over a benign tumour in this case?

   a) Presence of atypical mitotic figures
   b) Presence of spindle shaped cells
   c) Clear cells comprising >25% of the tumour
   d) Cellular monotony
   e) Large cell nests

4) Which of the following is associated with Multiple Endocrine Neoplasia Type I (Wermer's Syndrome)?

   a) Phaeochromocytoma
   b) Marfanoid habitus
   c) Parathyroid hyperplasia
   d) Mucocutaneous ganglioneuromas
   e) Neurofibromas

Scenario for Questions 5 to 7: The following microscopic image is from a thyroid tumour removed from a 37-year-old female who presented with an asymptomatic, bilateral neck swelling associated with cervical lymphadenopathy.

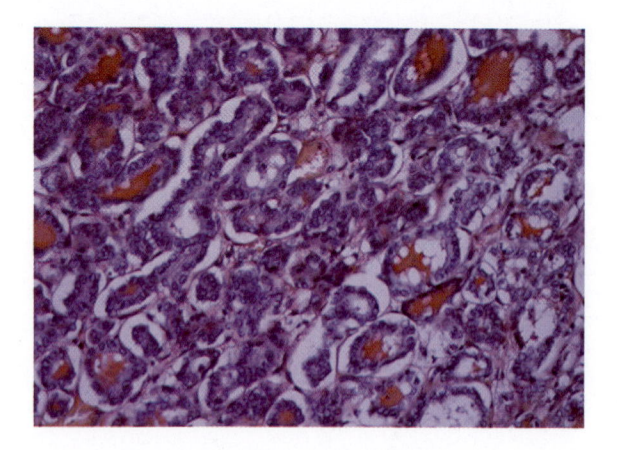

5) Which of the following diagnosis best describes this lesion?

   a) Follicular variant of papillary carcinoma
   b) Medullary carcinoma
   c) Tall cell variant of papillary carcinoma
   d) Diffuse sclerosing variant of papillary carcinoma
   e) Follicular carcinoma

6) Which feature is helpful in differentiating this tumour from a follicular adenoma?

   a) The presence of follicles containing thick colloid
   b) Nuclear atypia
   c) Fibrosis
   d) Presence of cuboidal tumour cells
   e) Presence of hyaline

7) Which criteria are included in the TNM staging of this tumour?

   a) Tumour size
   b) Presence of vascular invasion
   c) Involvement of thyroid capsule
   d) Macroscopic involvement of margins
   e) Presence of capsular invasion

8) A 45-year-old male presents with depression, joint pain, vague abdominal pain and renal calculi. He is diagnosed as having hypercalcaemia and further investigations reveal a single enlarged right inferior parathyroid. He is diagnosed with hyperparathyroidism and is referred for parathyroidectomy. The surgeon reports that he had difficulty in removing the gland which is sent for histological evaluation. Which criteria are helpful in distinguishing a benign from malignant parathyroid tumour?

   a) Invasion of adjacent structures
   b) Loss of staining for parathormone
   c) Clear cells less than 30% of the tumour
   d) Cells showing granular eosinophilic cytoplasm

9-12) Extended matching item questions:

A) Follicular adenoma
B) Follicular carcinoma
C) Hyalinising trabecular adenoma
D) Follicular variant of papillary carcinoma
E) Medullary carcinoma
F) Poorly differentiated (insular) carcinoma
G) Anaplastic carcinoma
H) Atypical follicular adenoma
I) CASTLE
J) Sclerosing mucoepidermoid tumour with eosinophilia

For each of the following microscopic descriptions from surgical specimens of **thyroid tumours**, select the **most appropriate** diagnosis from the list of options above. Each option may be used once, more than once or not at all.

9) An encapsulated follicular patterned lesion with tumour cells showing round/oval nuclei, mild nuclear atypia but neither chromatin clearing nor nuclear pseudoinclusions and no capsular or vascular invasion.

10) A poorly circumscribed tumour composed of nests of polygonal/spindle cells showing granular nuclear chromatin and associated with amyloid deposition.

11) A poorly circumscribed tumour composed of sheets of polygonal and spindle shaped cells showing markedly pleomorphic nuclei, extrathyroidal extension with skeletal muscle invasion and focal positivity for MNF116 and TTF-1.

12) A nested tumour composed of basaloid cells with a high nuclear to cytoplasmic ratio, brisk mitotic activity and foci of necrosis. The tumour stains positively for TTF-1.

13) A 30 year-old lady has a total thyroidectomy following failed medical treatment for hyperthyroidism. The microscopic sections show small follicles with pale scalloped colloid, foci of epithelial hyperplasia and occasional lymphoid aggregates. Increased vascularity is also noted. What is the most likely diagnosis?

   a) Graves' disease
   b) Hashimoto's thyroiditis
   c) De Quervain's thyroiditis
   d) Palpation thyroiditis
   e) Riedel's thyroiditis

14) A 60-year-old lady suffers from tiredness and repeat episodes of fainting. Investigations including CT scans show a 2cm pancreatic head mass. The mass is biopsied during endoscopy. The biopsy shows tumour composed of sheets of polygonal cells with uniform round nuclei and low mitotic index. Tumour cells show granular chromatin. Which of the following immunostains is the lesion most likely to be positive for?

   a) Insulin
   b) Ca19.9
   c) CD20
   d) WT1

15) Which of the following is not a feature of thyroglossal cysts?

a) Cyst lined by squamous epithelium
b) Presence of thyroid follicles
c) Location in the midline anterior to the hyoid bone
d) Cyst lined by respiratory type epithelium
e) Presence of gastric foveolar cells

16) Which of the following diagnoses is associated with a favourable prognosis?

a) Incidental papillary microcarcinoma
b) Tall cell variant of papillary carcinoma
c) Anaplastic carcinoma
d) Poorly differentiated carcinoma
e) Diffuse sclerosing variant of papillary carcinoma

17) A 45-year-old man has a thyroidectomy for bilateral thyroid enlargement and compression symptoms including dysphagia and airway obstruction. The photomicrograph shown below is from a section from the specimen. What is the diagnosis?

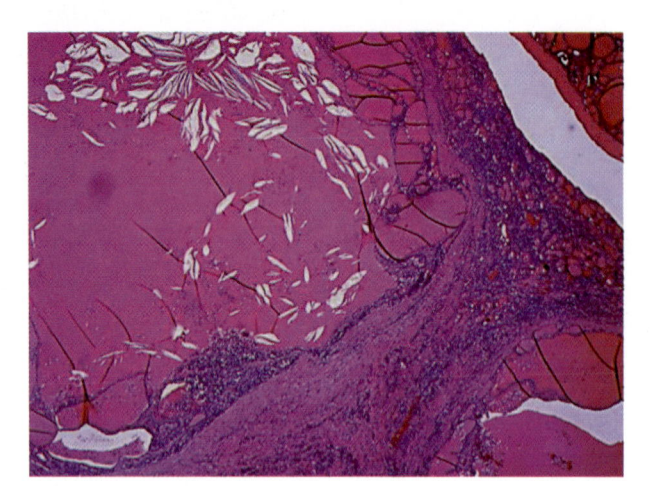

a) Multinodular goitre
b) Follicular adenoma
c) Classic papillary carcinoma
d) C-Cell hyperplasia
e) Anaplastic carcinoma

18-21) Extended matching item questions.

A) Graves' disease
B) Hashimoto's thyroiditis
C) De Quervain's thyroiditis
D) Riedel's thyroiditis
E) Acute thyroiditis
F) Diffuse non-toxic goitre
G) Multinodular goitre
H) C-Cell hyperplasia
I) Subacute lymphocytic thyroiditis
J) Palpation thyroiditis

For each of the following microscopic descriptions from ***thyroidectomy specimens*** please select the **most appropriate** diagnosis. Each option may be used once, more than once or not at all.

18) Diffusely enlarged thyroid showing slight attachment to adjacent structures. Microscopic features which include an inflammatory infiltrate comprising plasma cells, lymphocytes and histiocytes is noted in relation to damaged thyroid follicles. Multinucleate giant cells are seen in relation to extravasated colloid and poorly formed granulomas are observed.

19) An enlarged firm/rubbery thyroid which is adherent to adjacent structures. Microscopic features include atrophy of thyroid follicles with diffuse areas of fibrosis which extend into adjacent structures including skeletal muscle. There is an associated mild to moderate lymphoplasmacytic infiltrate.

20) An enlarged thyroid showing numerous lymphoid follicles including germinal centres associated with follicle atrophy and colloid depletion. Multifocal Hurthle cell change is seen.

21) An enlarged thyroid showing replacement by multiple nodules. The nodules are composed of variably sized follicles, some of which are distended by colloid. Foci of recent and old haemorrhage are seen including cholesterol clefts and aggregates of haemosiderin laden macrophages.

22) A 28 year old male has an adrenalectomy for a unilateral, non-functioning adrenal tumour. The image is a photomicrograph of a section from the tumour. The spindle cells are positive for S100p. NSE is also positive in the lesion. What is the diagnosis?

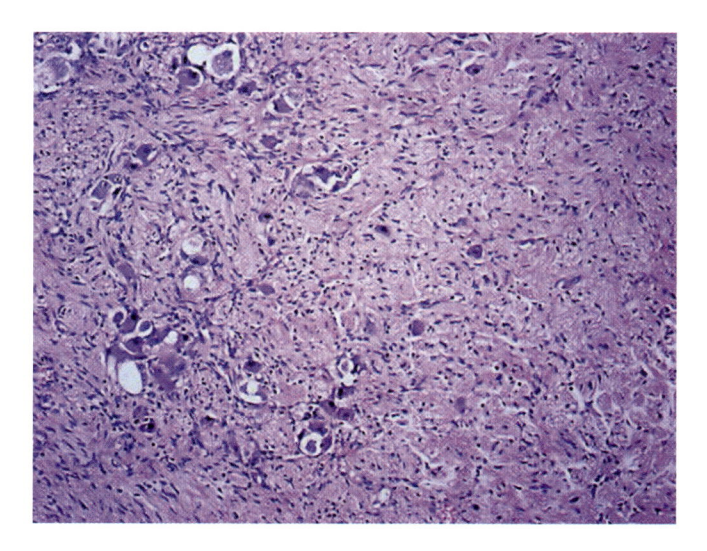

a) Adrenal cortical adenoma
b) Adrenal cortical carcinoma
c) Phaeochromocytoma
d) Neuroblastoma
e) Ganglioneuroma

23) Which of the following are important criteria used to distinguish benign from malignant thyroid follicular tumours (which lack nuclear features of papillary carcinoma)?

a) Mitotic index
b) Haemorrhage

c) Vascular invasion within the capsule
d) Presence of nuclear atypia

24) Which of the following is a characteristic of follicular thyroid lesions of Hurthle cell type?

a) Spontaneous infarction following FNA
b) Inconspicuous nucleoli
c) Widespread nuclear grooves
d) Basophilic cytoplasm

25) Which of the following is a variant of papillary carcinoma of the thyroid?

a) Nested variant
b) Clear cell variant
c) Variant with nodular fasciitis like stroma
d) Pleomorphic variant

26) Which one of the following molecular pathway abnormalities are NOT commonly associated with follicular carcinoma of the thyroid?

a) PAX8/PPARgamma
b) KRAS mutations
c) Activating mutations of NRAS
d) RET mutations

27) A 38 year old woman presents with a unilaterally enlarged thyroid, cervical lymphadenopathy and difficulty swallowing. FNA of the thyroid shows a few thyroid epithelial groups, small amounts of colloid and many inflammatory cells including eosinophils, lymphocytes and histiocytes. Some multinucleated cells are seen. Histiocytic cells show grooved nuclei. Which of the following stains would be most useful in establishing the diagnosis?

a) CD45
b) CD25
c) S100p
d) CD15

28) An adrenal resection specimen reveals a circumscribed tan tumour showing a central fibrotic area. Microscopy reveals nests and sheets of polygonal cells with abundant granular eosinophilic cytoplasm. Which of the following criteria are important in the distinction between a benign and malignant neoplasm?

a) Presence of atypical mitoses
b) Prominent nuclear atypia
c) Diffuse architecture
d) Clear cells <30%of tumour

29) Which of the following is **not** a feature of hyalinising trabecular neoplasm of the thyroid?

a) Presence of nuclear pseudoinclusions
b) Tumour cells aligned perpendicular to trabecula
c) Presence of perinuclear haloes

d) Presence of amyloid

30) Which of the following syndromes is associated with the diffuse sclerosing variant of papillary carcinoma?

a) Familial adenomatous polyposis
b) Cowden's syndrome
c) MEN-1
d) Lynch syndrome
e) Li-Fraumeni syndrome

Correct Answers and Explanations:

1) Correct answer: a) BRAF V600E

BRAF V600E mutation is associated with papillary carcinoma and is also associated with an increased mortality.

2) Correct answer: c) Medullary carcinoma

Medullary carcinoma is a neuroendocrine tumour derived from C-Cells and will therefore be positive for neuroendocrine markers such as Synaptophysin and Chromogranin as well as Calcitonin and TTF-1. The other tumour types are derived from thyroid epithelial cells and will only stain for TTF-1 and Thyroglobulin. The exception is anaplastic carcinoma which may only show focal positivity for epithelial markers, TTF-1 or Thyroglobulin.

3) Correct answer: a) Presence of atypical mitotic figures

The tumour is a cortical neoplasm as ascertained from the description and immunophenotype. The presence of atypical mitotic figures and clear cells comprising <25% of the tumour are components of the Weiss criteria, which favour a malignant cortical tumours over a benign cortical tumour. Other features in the original Weiss criteria include diffuse architecture >one third of the tumour, confluent necrosis, high nuclear grade (Fuhrman grade 3 or4), mitotic activity >5/50 hpf, venous invasion, sinusoidal and capsular invasion.

4) Correct answer: c) Parathyroid hyperplasia

Parathyroid hyperplasia is a component of MEN-1 (Multiple Endocrine Neoplasia Type 1) in addition to pituitary and parathyroid adenomas, pancreatic islet lesions, adrenal cortical hyperplasia and thyroid C-Cell hyperplasia.

5) Correct answer: a) Follicular variant of papillary thyroid carcinoma (FVPTC)

The lesion has a follicular architecture and is composed of cells showing features of papillary carcinoma including nuclear clearing, grooves and crowding. Follicular tumours (adenomas or carcinoma) and medullary carcinomas lack nuclear features of papillary carcinoma. Features of tall cell variant of papillary carcinoma, including columnar cells with height 2-3 times the width of the cell, are not seen. Diffuse sclerosing variant of papillary carcinoma shows diffuse sclerosis with many psammoma bodies.

6) Correct answer: a) The presence of follicles containing thick colloid

Of all the features listed the presence of thick colloid is most helpful in distinguishing the follicular variant of papillary carcinoma from a follicular adenoma. Nuclear atypia, fibrosis and the presence of hyaline material and columnar cells are not specific.

7) Correct answer: a) Tumour size

Tumour size and involvement beyond the thyroid capsule into extra-thyroid structures are features used to stage thyroid tumours in the TNM 7 staging system. The other features mentioned are not in this staging system.

8) Correct answer: a) Invasion of adjacent structures

The presence of thick fibrotic bands in the lesion and invasion of adjacent structures are criteria associated with malignant parathyroid tumours. Other features seen in parathyroid carcinomas include presence of capsular or vascular invasion, high mitotic activity, necrosis, cellular atypia or monotony, and prominent nucleoli.

9) Correct answer: A) Follicular adenoma

The description is of a follicular adenoma. The lesion lacks nuclear features of papillary carcinoma, and the absence of capsular or vascular invasion excludes follicular carcinoma.

10) Correct answer: E) Medullary carcinoma

The question describes medullary carcinoma which can show spindle or polygonal cells with a granular chromatin pattern. Amyloid can also be present (calcitonin deposition).

11) Correct answer: G) Anaplastic carcinoma

Anaplastic carcinoma can show a varied morphology with carcinomatous and sarcomatoid (spindle cells) areas and frequently shows marked nuclear pleomorphism. It is often widely invasive at presentation with extrathyroidal spread.

12) Correct answer: F) Poorly differentiated (insular) carcinoma

Poorly differentiated (insular) carcinoma lacks nuclear features of papillary carcinoma and tumour cells shows a basaloid appearance. Mitotic activity and necrosis are features of insular carcinoma.

13) Correct answer: a) Graves' disease

The microscopic description is typical of Graves' disease including epithelial hyperplasia and pale scalloped colloid. Hashimoto's thyroiditis shows a marked lymphoid infiltrate with follicles showing germinal centre formation. Riedel's thyroiditis shows fibrosis with a lymphoplasmacytic infiltrate and follicle atrophy and features of De Quervain's thyroiditis include granulomatous inflammation. With palpation thyroiditis granulomatous inflammation is seen in relation to ruptured follicles.

14) Correct answer: a) Insulin

The description is of a neuroendocrine tumour likely to be a pancreatic insulinoma and will therefore stain positively for Insulin and neuroendocrine markers (such as chromogranin and synaptophysin). These tumours often have a low mitotic index (grade 1). The fainting episodes described in the clinical history are likely to be secondary to hypoglycaemia.

15) Correct answer: e) Presence of gastric foveolar cells

Gastric foveolar cells are not seen in thyroglossal cysts, whereas squamous cells, respiratory type epithelial cells and thyroid follicles can be seen in these lesions which are located in the midline anterior to the hyoid bone.

16) Correct answer: a) Incidental papillary microcarcinoma

Incidental papillary carcinoma is a common finding in thyroidectomies performed for benign disease. The finding is almost always associated with a very favourable outcome. The other tumour types are associated with a poor prognosis.

17) Correct answer: a) Multinodular goitre

The microscopic shown are those of a multinodular goitre including a nodular architecture with medium-sized and large dilated follicles, features of recent and old haemorrhage (including cholesterol clefts) and fibrosis.

18) Correct answer: C) De Quervain's thyroiditis

The microscopic description is of De Quervain's thyroiditis which is also known as subacute granulomatous thyroiditis.

19) Correct answer: D) Riedel's thyroiditis

The microscopic description is of Riedel's thyroiditis which can show fibrosis and adhesion to adjacent structures. This is a rare form of thyroiditis and results in hypothyroidism. Other symptoms include compression of local structures.

20) Correct answer: B) Hashimoto's thyroiditis

The microscopic description is of Hashimoto's thyroiditis. This is an autoimmune disease which commonly presents in females and can initially present with hyperthyroidism, but eventually results in hypothyroidism.

21) Correct answer: G) Multinodular goitre

The description is typical of a multinodular goitre and not of a tumour or other form of thyroiditis.

22) Correct answer: e) Ganglioneuroma

Features of this adrenal tumour include a mature (schwannian) spindle cell stroma with interspersed ganglion cells seen singly and in nests. Ganglion cells show uniform round nuclei and nucleoli. The tumour presents in older patients, as opposed to ganglioneuroblastoma which tends to be seen in children. Ganglioneuromas can arise from the sympathetic ganglia (mediastinal or retroperitoneal) or the adrenal medulla but can also involve other organs or tissues. They often do not produce any clinical symptoms, but can secrete catecholamines.

23) Correct answer: c) Vascular invasion within the capsule

Capsular and vascular invasion are criteria used to diagnose minimally invasive follicular carcinoma and differentiate it from benign follicular adenoma. The features of haemorrhage, nuclear atypia and mitotic index are non-specific.

24) Correct answer: a) Spontaneous infarction following FNA

Spontaneous infarction following FNA and psammoma body formation are features of Hurthle cell lesions which also have cells with nuclei showing prominent nucleoli and granular eosinophilic (rather than basophilic) cytoplasm. Nuclear grooves are seen in papillary carcinoma.

25) Correct answer: c) Variant with nodular fasciitis like stroma

The macrofollicular variant and variant with nodular fasciitis like stroma are subtypes of papillary carcinoma. The macrofollicular variant is rare and is composed of large follicles which show focal nuclear features of papillary carcinoma. Papillary carcinomas with a nodular fasciitis like stroma are rare and show a spindle cell (myofibroblastic) stroma with intermixed groups of malignant epithelial cells.

26) Correct answer: d) RET mutations

PAX8/PPARgamma, KRAS mutations and activating mutations of NRAS are seen in follicular carcinomas. RAS mutations are the most common. RET mutations are commonly associated with medullary thyroid carcinomas.

27) Correct answer: c) S100p

The description is of Langerhans cell histiocytosis and the neoplastic histiocytic cells express CD68R, CD163 and S100p. This neoplastic disorder of macrophages can be systemic or localised (multifocal or unifocal), often presenting with lymphadenopathy, bone, liver or lung lesions. The thyroid can be affected as well as the thymus.

28) Correct answer: a) Presence of atypical mitoses

The description is of an oncocytic adrenal cortical tumour which is a rare adrenal cortical neoplasm. The criteria used to distinguish benign from malignant oncocytic tumours are different from the Weiss criteria. Major criteria include the presence of atypical mitoses and venous invasion, and minor criteria include large size, confluent necrosis, capsular and sinusoidal invasion.

29) Correct answer: d) Presence of amyloid

Features of hyalinising trabecular neoplasm include tumour cells showing presence of nuclear pseudoinclusions, tumour cells aligned perpendicular to trabeculae and presence of perinuclear haloes.

Hyalinising trabecular neoplasms are a distinct type of thyroid tumour which are mostly benign but can be associated with vascular invasion and metastasis. It is thought that the tumour might be related to papillary carcinoma due to similar morphological features. Amyloid deposition is associated with medullary carcinoma.

30) Correct answer: a) Familial adenomatous polyposis

Diffuse sclerosing variant of papillary carcinoma is associated with familial adenomatous polyposis/Gardner syndrome. This autosomal dominant syndrome is also associated with gastrointestinal polyposis (adenomas) which can progress to colorectal carcinoma as well as fibromatosis (desmoid tumours).

CHAPTER 15

LYMPHORETICULAR PATHOLOGY
Dr Matthew Pugh & Dr Richard Attanoos

1) A 28 year old male presents with a three month history of weight loss, fever and drenching night sweats. A CT scan of the thorax reveals mediastinal lymphadenopathy. A mediastinal lymph node excision was performed and the histology is shown in the following picture.

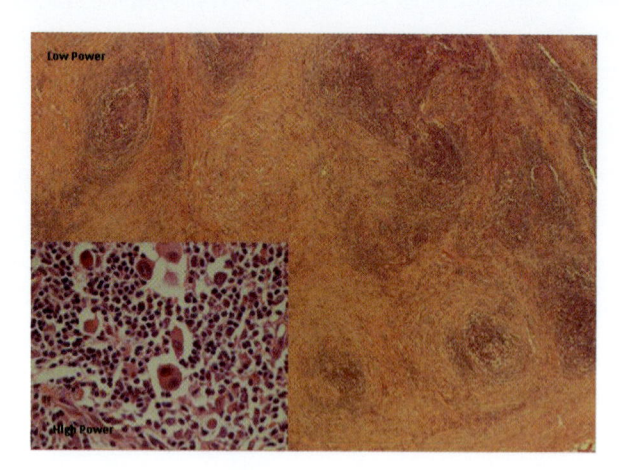

What is the most likely diagnosis?

a) Diffuse large B-cell lymphoma.
b) Anaplastic large cell lymphoma.
c) Burkitt lymphoma.
d) Classical Hodgkin lymphoma, nodular sclerosis subtype.
e) Nodular lymphocyte predominant Hodgkin lymphoma.

2-6) Extended matching item questions:

A) p24 immunohistochemistry
B) Warthin Starry stain
C) Ziehl Neelsen stain
D) CD20 immunohistochemistry
E) Fite-Faraco stain
F) Syphilis serology
G) Giemsa stain
H) Toxoplasma serology
I) Gram stain
J) S100 immunohistochemistry

For each of the following microscopic descriptions of a lymph node, select the most appropriate further investigation from the list above for the **most likely** diagnosis. Each option may be used once, more than once, or not at all.

2) Retained lymph node architecture with florid follicular hyperplasia and macrofollicles. The germinal centres are mitotically active with a starry sky appearance. The mantle zones appear thin and disrupted and occasionally invaginate the germinal centre (follicle lysis).

3) Reactive follicular hyperplasia with monocytoid paracortical expansion. Small clusters of epithelioid histiocytes are seen in the paracortex and encroaching on the germinal centre.

4) Numerous large, pale and foamy histiocytes ("Virchow cells") are seen in the paracortical region and expanding the sinusoids, with no definite granuloma formation or necrosis.

5) Thickened lymph node capsule and florid sinusoidal expansion with plasma cells, lymphocytes and large histiocyte-like cells. The histiocyte-like cells have a large round nucleus and foamy eosinophilic cytoplasm with abundant emperipolesis of plasma cells and lymphocytes

6) Areas of stellate necrosis containing neutrophils with a palisading rim of histiocytes. There is follicular and monocytoid B-cell hyperplasia. No Reed-Sternberg cells are seen.

7-11) Extended matching item questions:

    A) t(11;18)(q32;q21)
    B) t(2;8)(p12 ;q24)
    C) t(2:5)(p23;q35)
    D) Loss of 7q21-32
    E) t(14;18) (q32;q21)
    F) Trisomy 3
    G) t(9;22)(q34;q11)
    H) 13q14.3 deletion
    I) Trisomy 12
    J) t(8;14)(q24;q32)

For each of the following immunoprofiles of a lymphoid neoplasm, decide the most likely diagnosis and select the *most commonly associated* genetic abnormality from the list of options above. Each option may be used once, more than once, or not at all.

7) CD20+, CD3-, CD5+, CD23+, CYCLIN-D1-

8) CD20+, CD3-, CD5+, CYCLIN-D1+

9) CD20+, CD3-, CD5-, Bcl-2+, CD10+, Bcl-6+

10) CD43+, CD30+, CD5+, ALK+

11) CD20+, BCL-2-, CD10+, Ki-67 100%

Scenario for questions 12 -14: A 4 year old boy was admitted to hospital with multiple soft tissue masses on the head and lower limbs. On X-ray, these were associated with multiple destructive punched out lesions of the bone, with a solid sclerotic rim. A biopsy reveals a lesion rich in eosinophils and histiocytoid cells. The histiocytoid cells are oval with grooved, folded and lobulated nuclei. Occasional osteoclast-like giant cells are also seen.

12) Based on the clinical history and biopsy findings, what is the most likely diagnosis.

   a) Lymphoplasmacytic lymphoma.
   b) Rosai-Dorfman disease.
   c) Langerhans cell histiocytosis.
   d) Plasma cell myeloma.

13) What is typical immunoprofile of this lesion?

   a) S100 +, CD1a +, Langerin +.
   b) S100 +, CD1a -, Langerin +.
   c) TdT +, CD138 +, CD56 +.
   d) S100-, CD138+, CD56 +.

14) Which of following is the ultrastructural hallmark of the disease, visible on electron microscopy?

   a) Dutcher body.
   b) Russell body.
   c) Birbeck granule.
   d) Hairy projections.

Scenario for questions 15 -17: A 75 year old male presents with a unilateral testicular mass. A malignant lesion is suspected and an orchidectomy is performed. The histology of the lesion is shown in the following picture. The lesional cells stain positive for CD45 and CD20 on immunohistochemistry.

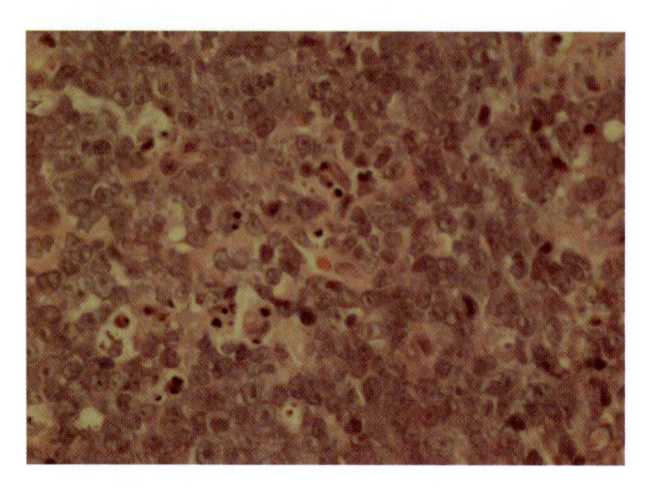

15) What is the most likely diagnosis?

   a) Diffuse large B-cell lymphoma.
   b) Chronic lymphocytic leukaemia/ small lymphocytic leukaemia.
   c) T-cell lymphoblastic lymphoma.
   d) Classical Hodgkin lymphoma.

16) Immunohistochemistry of this lesion yields positive staining for CD10 and BCL-6, and negative staining for MUM-1. Which *one* of the following statements is correct?

   a) This lesion is of germinal centre phenotype.

- b) This lesion is of non-germinal centre phenotype.
- c) This lesion has arisen from blastic transformation from a follicular lymphoma.
- d) This lesion has arisen *de novo* with no preceding low grade lesion.

17) Which of the following features of this malignancy *is not* associated with a poorer prognosis?

- a) *MYC* break.
- b) Germinal centre phenotype.
- c) Age over 60 years at presentation.
- d) EBV infection.
- e) High serum LDH.

Scenario for questions 18 -21: A 72 year old man presents to A&E with a vertebral crush fracture. X-rays show multiple lytic lesions of the spinal column.

A bone marrow trephine is performed, as shown in the picture.

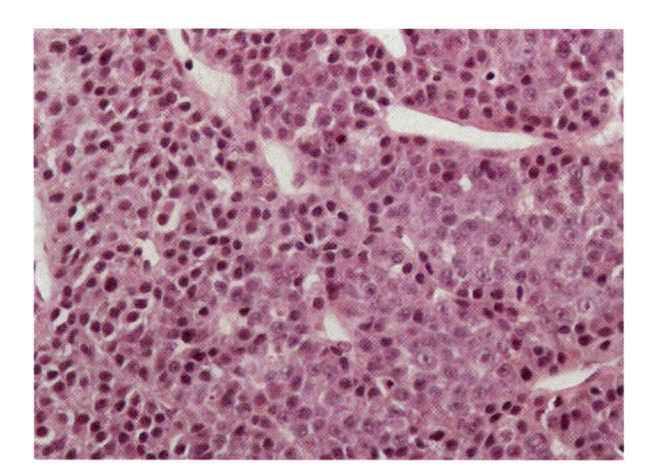

18) What is the likely diagnosis?

- a) Plasma cell myeloma.
- b) Diffuse large B-cell lymphoma.
- c) Classical Hodgkin lymphoma.
- d) Plasmablastic lymphoma.

19) What would be the most likely immunohistochemical profile of this lesion?

- a) CD79a+, CD138+, CD19+, CD56-.
- b) CD79a-, CD138-, CD19+, CD56+.
- c) CD79a+, CD138+, CD19-, CD56+.
- d) CD79+, CD138+, CD19-, CD56-.

20) Which of the following tests *cannot* be used to establish monoclonality in this case?

- a) Kappa/ Lambda light chain immunohistochemistry.
- b) B-cell polymerase chain reaction.
- c) Flow cytometry.
- d) Fluorescent in situ hybridisation for Cyclin-D1.

21) Which of the following *is not* a recognised complication of this disease?

   a) Anaemia.
   b) Renal failure.
   c) Congestive cardiac failure.
   d) Hypocalcaemia.
   e) Recurrent infections.

22) A 56 year old lady is diagnosed with chronic myelogenous leukaemia. Cytogenetic analysis is performed on the lesional cells. What is the most likely cytogenetic abnormality?

   a) t(9;22)(q34;q11.2).
   b) t(1;19)(q23;q13).
   c) t(8;21)(q22;q22).
   d) t(8;14)(q24;q32).

23-27) Extended matching item questions:

   A) Lymphomatoid granulomatosis
   B) Mycosis fungoides
   C) Angioimmunoblastic T-cell lymphoma
   D) Subcutaneous panniculitis-like T-cell lymphoma
   E) Langerhans cell histiocytosis
   F) Diffuse large B-cell lymphoma
   G) Sezary syndrome
   H) Dermatopathic lymphadenitis
   I) Anaplastic large cell lymphoma
   J) Cutaneous follicular lymphoma

For each of the following clinical and microscopic descriptions, select the **most likely** diagnosis from the list of options above. Each option may be used once, more than once, or not at all.

23) A 28 year old female presents with frequent fevers, widespread peripheral lymphadenopathy and multiple ulcerative skin nodules. Biopsy of one of the skin lesions reveals a dermal infiltrate composed of sheets of large, highly pleomorphic atypical cells with abundant cytoplasm and horseshoe nuclei, with an eosinophilic area adjacent to the nucleus. The cells are CD30+ and ALK-1+.

24) A 67 year male presents with multiple skin plaques. A skin biopsy reveals a lichenoid dermal infiltrate of CD4+ lymphoid cells. There is lymphocyte satellitosis, epidermotropism and Pautrier microabscesses. The nuclei of the epidermal infiltrate appear cerebriform.

25) A 45 year old female presents with fever, cough and multiple red nodular skin lesions. A biopsy of one of the skin lesions is performed which reveals a mixed dermal infiltrate primarily composed of small CD3+ lymphoid cells aggregated around blood vessels. Occasional CD20 blastic cells are seen which are also EBV+. There are variable numbers of background plasma cells and histiocytes. There is vasculitis with fibrinoid necrosis present.

26) A 27 year old male with long standing itchy plaques on his torso now presents with inguinal lymphadenopathy. Previous biopsy of the skin lesions showed a non-specific lichenoid dermal

inflammatory infiltrate. A biopsy of the enlarged inguinal lymph nodes is performed which reveals marked paracortical expansion with S100+ and CD1a+ cells, which compress the follicles against the capsule. There is scattered brown pigment present throughout the node. No atypical lymphoid cells are seen.

27) A 72 year old male presents with erythroderma and generalised lymphadenopathy. A skin biopsy reveals a lichenoid dermal infiltrate of monotonous but atypical CD4+ cells with cerebriform nuclei.

28) Which of the following *is NOT* a recognised EBV-induced lymphoma?

   a) NK/T cell lymphoma, nasal type.
   b) Nodular lymphocyte predominant Hodgkin lymphoma.
   c) Classical Hodgkin lymphoma.
   d) Endemic Burkitt lymphoma.
   e) Effusion-associated diffuse large B-cell lymphoma.

Scenario for questions 29 -31: A 68 year man presents with cervical lymphadenopathy. A lymph node excision is performed. On histological examination, there is follicular effacement of the lymph node architecture, largely composed of small lymphoid cells, with aggregates of centroblasts and small lymphoid cells within the follicles. No Reed-Sternberg cells are seen. There is sparse mitotic activity. Initial immunohistochemical tests show the majority of the cells to be CD20+. A diagnosis of follicular lymphoma is suspected.

29) Which of the following *would NOT be* included in the differential diagnosis?

   a) Mantle cell lymphoma.
   b) Marginal cell lymphoma.
   c) Burkitt lymphoma.
   d) Chronic lymphocytic leukaemia/ small lymphocytic lymphoma.

30) Which of the following immunohistochemical markers could be used to highlight the follicular dendritic networks within the follicle?

   a) CD23.
   b) CD15.
   c) CD20.
   d) CD68.

31) Following immunohistochemical testing, the diagnosis of follicular lymphoma is confirmed. Which *one* of the following criteria would indicate WHO grade II disease?

   a) >15 centroblasts per hpf within the germinal centre admixed with centrocytes.
   b) 6-15 centroblasts per hpf within the germinal centre.
   c) A mitotic count of >5 mitoses/ high power field within the germinal centre.
   d) A Ki-67 proliferation index of >50%.
   e) A Ki-67 proliferation index of >90%.
   f) >5 immunoblasts per hpf in the paracortex.

Scenario for questions 32 -33: A 33 year old man with a history of HIV infection presents with generalised lymphadenopathy. A lymph node excision is performed. Light microscopy shows a diffuse plasmacytic infiltrate within the paracortex and occasional plasmablastic cells within the follicle. There is follicular hyperplasia with absent mantle zones. There are prominent hyalinised blood vessels within the follicles and concentric rings of lymphocytes around the germinal centre, giving rise to a 'lollipop' appearance.

32) What is the most likely diagnosis?

   a) Lymphoplasmacytic lymphoma.
   b) Castleman disease.
   c) Plasma cell myeloma.
   d) Kikuchi disease.
   e) Kawasaki disease.

33) Which of the following viruses are associated with this disease?

   a) CMV.
   b) HHV-8.
   c) HSV-1.
   d) HSV-2.
   e) HTLV-1.

34) An 80 year man is incidentally found to have peripheral lymphocytosis. A bone marrow trephine is performed. Which of the following features is *most suggestive* of bone marrow involvement by a lymphoproliferative process?

   a) A left shift in the cellular population.
   b) Paratrabecular aggregates of lymphoid cells.
   c) Scattered interstitial lymphoid follicles.
   d) Trilineage hematopoiesis.

Correct answers and explanations:

1) Correct answer: d) Classical Hodgkin Lymphoma, nodular sclerosis subtype

This picture shows an effaced lymph node with broad fibrotic bands separating nodular aggregates of small lymphoid cells, admixed with Reed-Sternberg (RS) cells. In the context of this characteristic history, the most likely diagnosis is classical Hodgkin lymphoma (cHL), nodular sclerosis subtype. Classical Hodgkin lymphoma accounts for 25% of all lymphomas in western countries, with a bi-modal peak of incidence in young adulthood (15-35 years) and in late life. The lesional cell in Hodgkin lymphoma, the Reed-Sternberg cell, typically accounts for less than 1% of the overall cellular elements. The RS cell is a large cell which is sometimes bi-nucleated, and has a prominent eosinophilic nucleolus. The lacunar variant of RS cell is often seen in the nodular sclerosis subtype of cHL. The typical immunophenotype of the RS cell is CD30+, CD15+ and CD20- . The nodular sclerosis subtype of cHL is the most common subtype.

Diffuse large B-cell lymphoma usually appears as diffuse sheets of mitotically active blast cells. Anaplastic large cell lymphoma (ALCL) is typically composed of large, atypical pleomorphic cells, with variable numbers of 'hallmark cells'. Although a variant of ALCL can mimic classical Hodgkin lymphoma, this is much less common. Burkitt lymphoma classically appears as sheets of monotonous medium sized lymphoid cells with frequent mitoses and a starry sky appearance. In nodular lymphocyte predominant Hodgkin lymphoma (NLPHL), there is nodular effacement of the lymph node with small lymphoid cells admixed with the lesional LP cells. Fibrosis is not typically a feature of this lymphoma, which accounts for only 5% of Hodgkin lymphoma. LP cells are CD20+, CD30- and CD15- on immunostaining.

2) Correct answer: A) p24 immunohistochemistry

Florid follicular hyperplasia with macrofollicles, attenuated mantle zones and follicle lysis is the typical appearance seen in HIV-related benign lymphadenopathy. The high proliferation index in the germinal centre with subsequent increased apoptosis and tingible body macrophage formation, gives rise to the starry sky appearance. Patients with HIV infection may develop lymphadenopathy due to lymphoma, Kaposi sarcoma, multicentric Castleman disease or infection, however, the most common cause for lymphadenopathy in this population is a reactive process related directly to the virus, as described here. HIV infection can be demonstrated with immunostaining for the p24-gag viral capsid protein, which is usually localised to the follicular dendritic cells.

3) Correct answer: H) Toxoplasma serology

The classic triad of histological features seen in the lymph node in Toxoplasmosis are:
- Reactive follicular hyperplasia
- Small clusters of epithelioid macrophages which encroach onto the germinal centre
- Reactive monocytoid B-cell hyperplasia

Toxoplasmosis is caused by the protozoal parasite *Toxoplasma gondii*. Most infected individuals are asymptomatic, however, a subset of individuals will present with an acute febrile illness. Lymphadenopathy is a feature of the later stages of the acute infection, most commonly affecting the posterior cervical region. Immunohistochemical or PCR studies do not have a high sensitivity for the detection of the organism, and *Toxoplasma* serology is best used for confirmation of the diagnosis.

4) Correct answer: E) Fite-Faraco stain

Paracortical and sinusoidal accumulation of large, foamy macrophages, or Virchow cells, are typically seen in lepromatous leprosy. Granulomas and necrosis are not typical features. Leprosy is a slowly progressive infection caused by *Mycobacterium leprae* that primarily affects the skin and peripheral nerves, but can involve the lymph nodes. The presence of the acid-fast organisms packed into macrophages can be demonstrated with Fite-Faraco or Wade-Fite special stain.

5) Correct answer: J) S100 immunohistochemistry

Florid sinusoidal expansion with large foamy histiocyte-like cells with a round nucleus and moderately prominent nucleoli, is characteristic of Rosai-Dorfman disease. The histiocyte-like cells phagocytose plasma cells and lymphocytes, which are present within the cytoplasm of the cell. This process is known as emperipolesis. A thickened lymph node capsule, a non-specific feature of chronic lymph node disease, is sometimes present. Rosai-Dorfman disease is thought to represent a reactive proliferation of histiocyte-like cells, which stain positive for S100 and negative for CD1a. It is most common in children but can affect all age groups. Around 90% of patients present with multiple, massive, painless cervical lymph nodes, but other lymph nodes can be affected including axillary, inguinal, para-aortic and mediastinal lymph nodes.

6) Correct answer: B) Warthin-Starry stain

Suppurative necrotising granulomatous inflammation with follicular and monocytoid B-cell hyperplasia is seen in Cat Scratch lymphadenitis, which is the most common specific cause of suppurative granuloma in western populations. The causative agent is the bacterium *Bartonella henselae*, which is usually introduced following a cat bite. Patients usually present several weeks later with axillary, cervical or submandibular lymphadenopathy. Lymphogranuloma Venereum caused by *Chlamydia trachomatis* can also give a similar histological appearance, however, this is less common. Both *Bartonella henselae* and *Chlamydia trachomatis* are highlighted on Warthin-Starry special staining.

7) Correct answer: H) 13q14.3 deletion

Chronic lymphocytic leukaemia/ small lymphocytic lymphoma (CLL/SLL) is a low grade B-cell neoplasm (CD20+, CD3-) which aberrantly expresses CD5, which is normally a T-cell marker. Cyclin-D1 is negative in CLL/SLL, distinguishing it immunohistochemically from mantle cell lymphoma. CD23 and CD43 can also be positive in CLL/SLL. About 80% of CLL/SLL cases show cytogenetic abnormalities on FISH, the most common of which is the 13q14.3 deletion, which is present in 50% of cases.

8) Correct answer: A) t(11;18)(q32;q21)

Mantle cell lymphoma is a low grade B-cell neoplasm (CD20+, CD3-) which aberrantly expresses the T-cell marker CD5, in addition to Cyclin-D1 and CD43. Cyclin-D1 belongs to the G1 cyclins and plays a key role in cell cycle regulation during the G1/S transition by cooperating with cyclin-dependent kinases (CDKs). This protein is overexpressed in mantle cell lymphoma as a result of the t(11;18)(q32;q21) translocation, which is present in nearly all cases of mantle cell lymphoma.

9) Correct answer: E) t(14;18)(q32;q21)

Follicular lymphoma is a low grade B-cell neoplasm (C20+, CD3-) which co-expresses BCL-2 and the germinal centre antigens, BCL-6 and CD10. Disrupted and expanded follicular dendritic networks are highlighted on CD21 and CD23 immunostaining. The t(14;18)(q32;q21) translocation is present in 90% of grade 1-2 follicular lymphomas. The translocation involves the *BCL-2* (B cell leukaemia/lymphoma 2) and the *IgH* (immunoglobulin heavy chain) genes, resulting in BCL-2 overexpression and subsequent suppression of apoptosis.

10) Correct answer: C) t(2:5)(p23;q35)

Anaplastic large cell lymphoma (ALCL) is a highly aggressive T-cell lymphoma. Immunohistochemically, tumour cells typically show membrane and Golgi CD30 positivity, ALK-1 positivity and usually CD43 positivity. There is frequently loss of pan T-cell markers, such as CD3. CD2, CD5 and CD4 are positive in a significant number of cases. The most frequent translocation in ALCL is t(2:5)(p23;q35), between the *ALK* gene on chromosome 2 and the *nucleophosmin (NPM)* gene on chromosome 5. The *NPM-ALK* chimeric gene encodes a constitutively activated tyrosine kinase that has been shown to be a potent oncogene.

11) Correct answer: J) t(8:14)(q24;q32)

Burkitt lymphoma is a B-cell lymphoma (CD20+, CD3-), with a very short doubling time. Nearly all of the cells are proliferating, therefore, Ki-67 is approximately 100%. BCL-6 and CD10 are usually expressed, whilst BCL-2 is usually negative. The most common cytogenetic abnormality in Burkitt lymphoma is the *MYC* translocation, t(8;14)(q24;q32). The *MYC* gene produces an oncogenic transcription factor that affects diverse cellular processes involved in cell growth, cell proliferation, apoptosis and cellular metabolism.

12) Correct answer: c) Langerhans cell histiocytosis

The key histological feature of Langerhans cell histiocytosis (LCH) is the Langerhans cell, an oval cell with grooved, folded, indented and lobulated nuclei, fine chromatin, inconspicuous nucleoli and thin nuclear membranes. Other cells usually present include variable numbers of eosinophils, histiocytes including osteoclastic multinucleated giant cells, neutrophils and small lymphocytes. Plasma cells are less common. LCH usually affects children, and can present as single site disease, multiple site disease within a single system, or as multisystem disease. Bone and adjacent soft tissue, in particular, skull, vertebrae, pelvic bones, ribs and femur, are preferentially affected in the solitary form of the disease. Multifocal disease is largely confined to the bone and soft tissue, whilst multisystem disease typically affects the skin, bone, liver, spleen and bone marrow. Whilst plasma cell myeloma can present with multiple lytic bony lesions, myeloma is a plasma cell neoplasm which usually affects older patients.

13) Correct answer: a) S100 +, CD1a +, Langerin +

LCH cells are classically S100, CD1a and Langerin positive, in addition to vimentin and CD68. CD45, B-cell, and T-cell markers (except CD4), are usually positive. In lymph node biopsies, LCH cells can be distinguished from the histiocytoid cells of Rosai-Dorfman on immunohistochemistry, which are S100 positive, but CD1a and Langerin negative. TdT is a marker of lymphoblastic neoplasms, CD138 is a plasma cell marker, and CD56 is marker of neoplastic plasma cells, NK cells, and some T-cell lymphomas.

14) Correct answer: c) Birbeck granule

The ultrastructural hallmark of LCH is the Birbeck granule, a tennis racquet shaped organelle, with as zipper like appearance, 200-400nm long, and 33nm wide. Langerin positivity on immunohistochemical staining confirms the presence of the Birbeck granule, obviating the need for electron microscopy. A Dutcher body is a PAS positive intranuclear pseudoinclusion seen on light microscopy and are usually found in plasma cell myelomas and lymphoplasmablastic lymphoma. Russell bodies are eosinophilic, large, homogenous immunoglobulin-containing inclusions usually found in a plasma cell undergoing excessive synthesis of immunoglobulin. Hairy projections are a feature of hairy cell leukaemia.

15) Correct answer: a) Diffuse large B-cell lymphoma

The picture shows replacement of the normal testicular tissue with diffuse sheets of large blastic lymphoid cells. The blastic cells are primarily immunoblastic with a prominent central nucleolus, admixed with occasional centroblastic cells, with vesicular chromatin and peripherally arranged nucleoli. Immunostaining for CD20 was positive in this case, indicating a B-cell blastic lymphoma. The most common B-cell blastic neoplasm is diffuse large B-cell lymphoma, which is the most common malignant testicular tumour in men over 50 years old.

T-lymphoblastic lymphoma is composed of sheets of blastic cells, however, this neoplasm is of T-cell lineage, and CD20 would therefore be negative. Chronic lymphocytic leukaemia/ small lymphocytic leukaemia is of B-cell lineage, but is typically composed of small non-blastic lymphoid cells forming pseudofollicles. The lesional cells of Hodgkin lymphoma, the Reed-Sternberg cells, are large cells with a prominent large nucleolus, however, these cells are larger and typically account for less than 1% of the cellular population. Although, B-cells may be present alongside the Reed-Sternberg cells, the Reed-Sternberg cells themselves are negative for B-cell markers and positive for CD30 and CD15. T-lymphoblastic lymphoma, chronic lymphocytic leukaemia/ small lymphocytic leukaemia and classical Hodgkin lymphoma rarely present in the testis.

16) Correct answer: a) This lesion is of germinal centre phenotype

Diffuse large B-cell lymphomas can be immunophenotypically divided into 'germinal centre phenotype' and 'non-germinal centre phenotype' using BCL-6, CD10 and MUM-1 immunostaining as follows:

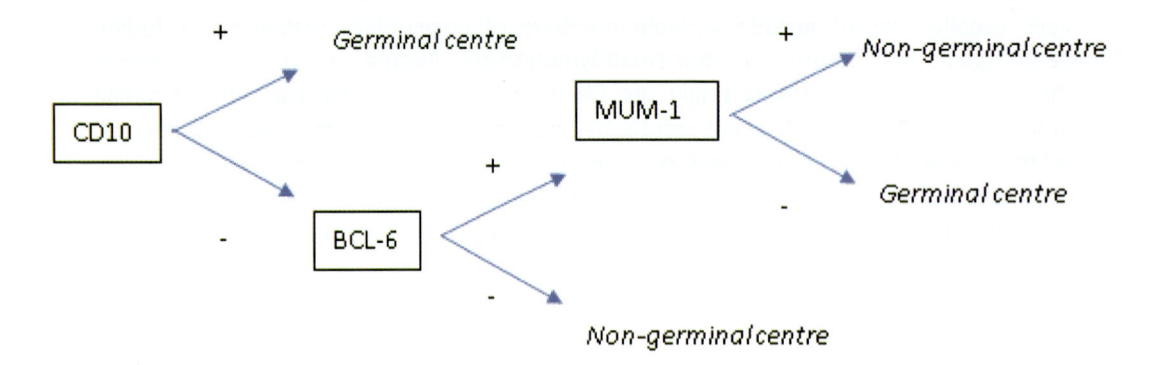

This lesion is therefore of germinal centre phenotype. Diffuse large B-cell lymphoma can arise *de novo*, or from a pre-existing low grade B-cell lymphoma. This distinction is usually based on the clinical history, and cannot be determined by the immunophenotype.

17) Correct answer: b) Germinal centre phenotype

The expression of BCL-6 and CD10 have been linked to a more favourable outcome, therefore germinal centre phenotype is not associated with a poorer prognosis. A worse clinical outcome has been associated with EBV positive diffuse large B-cell lymphoma compared to EBV negative diffuse large B-cell lymphoma. The presence of a MYC gene break is associated with a markedly poorer outcome. The following clinical and biochemical parameters are also associated with a worse clinical outcome:
- Age>60
- Poor performance status (ECOG≥2)
- Advanced Ann Arbor stage (III-IV)
- Extra-nodal involvement (≥2 sites)
- High serum LDH (>normal)

18) Correct answer: a) Plasma cell myeloma

The picture shows aggregates of atypical plasmacytoid cells. Typically, plasma cells only account for 1% of the bone marrow nucleated cell population, and do not typically form aggregates or clusters. This would indicate an abnormal plasmacytoid cell proliferation. The differential diagnosis for neoplastic plasmacytoid population includes:
- Plasma cell myeloma
- Plasmablastic lymphoma
- Lymphoplasmacytic lymphoma
- Plasmacytoid marginal zone lymphoma

Plasmablastic cells indicative of plasmablastic lymphoma are not prominent in this picture. There are no diffuse sheets of large blastic cells present to suggest diffuse large B-cell lymphoma. No Reed-Sternberg cells are present to suggest classical Hodgkin lymphoma, and bone marrow involvement is uncommon (~5%). The most appropriate diagnosis is therefore plasma cell myeloma.

Plasma cell myeloma is a multifocal plasma cell neoplasm primarily involving the bone marrow. Myeloma is associated with serum or urine M protein and organ related impairment (CRAB: hypercalcaemia, renal insufficiency, anaemia, lytic bone lesions).

19) Correct answer: c) CD79a+, CD138+, CD19-, CD56+

CD79a and CD138 are positive in both normal plasma cells and neoplastic plasma cells. CD19 is not expressed in myeloma plasma cells, in contrast to normal plasma cells which show CD19 positivity. CD56, normally a natural killer cell marker, is aberrantly expressed in myeloma cells in 67-79% of cases. Furthermore, myeloma cells may also aberrantly express CD117, CD20, CD52 and CD10. Some cases also express Cyclin-D1 which is associated with a lymphoplasmacytic appearance morphologically.

20) Correct answer: d) Fluorescent in situ hybridisation for Cyclin-D1

Plasma cell myeloma is characterised by a clonal proliferation of plasma cells. Determination of clonality in mature B-cells is possible as during development, B-cells undergo immunoglobulin VDJ gene rearrangement and subsequent somatic hypermutation. This leaves normal B-cells with a unique genetic fingerprint, which is inherited during clonal proliferation in neoplastic processes. A number of molecular techniques take advantage of this in order to determine clonality in B-cells. In a normal or reactive plasma cell population, there should be variable expression of both kappa and lambda light chains. Expression of primarily kappa or lambda on immunohistochemistry suggests a clonal proliferation. PCR is a molecular technique used to amplify single or multiple genes, and is used to look for immunoglobulin heavy chain gene rearrangements in order to determine clonality. Flow cytometry is a laser based technology with can detect specific antigens on the surface of cells. This can be used to identify kappa and lambda light chain antigens in order to determine clonality. Fluorescent *in situ* hybridisation is a cytogenetic technique which can be used to detect translocations involving the CCND1 gene, but this does not indicate clonality.

21) Correct answer: d) Hypocalcaemia

Hypercalcaemia, not hypocalcaemia, is a feature of plasma cell myeloma secondary to bony destruction and release of calcium into the blood. Anaemia is a recognised complication of myeloma and results from destruction of haematopoietic tissue within the bone marrow. Myeloma cells produce and secrete immunoglobulin free light chains which are toxic to the renal tubules. Furthermore, hypercalcaemia, volume depletion, hyperuricemia and hyperviscosity can contribute towards acute kidney injury. The resulting renal failure can have a further deleterious effect on red cell production, as erythropoietin production is diminished. Proteinaemia can also result in amyloid deposition, which may further affect the kidney. When amyloid is deposited in the heart, congestive cardiac failure and sudden arrhythmic death can ensue.

22) Correct answer: a) t(9;22)(q34;q11.2)

The t(9;22)(q34;q11.2) translocation present in 90-95% of chronic myelogenous leukaemia (CML) cases is a reciprocal translocation that results in the Philadelphia (Ph) chromosome. The Ph chromosome fuses sequences on the *BCR* gene on chromosome 22 with regions of the *ABL1* gene on chromosome 9. The resultant oncogene produces a constitutively active tyrosine kinase which affects several signal transduction pathways, resulting in the leukaemic phenotype of CML.

23) Correct answer: I) Anaplastic large cell lymphoma

Anaplastic large cell lymphoma is a T-cell lymphoma which is characteristically positive for CD30 and ALK-1, although ALK-1 negative variants exist. The characteristic lesional cells are large and pleomorphic, with horseshoe or kidney shaped nuclei with an eosinophilic region adjacent to the nucleus. These cells are referred to as 'hallmark' cells. Most patients present with stage III-IV disease with peripheral and/or abdominal lymphadenopathy and extranodal infiltrates. Skin, bone, soft tissues, lung and liver are common extranodal sites of involvement. Primary cutaneous anaplastic large cell lymphoma is a separate disease entity with a similar morphology, which is usually CD30 positive and ALK-1 negative.

24) Correct answer: B) Mycosis fungoides

Mycosis fungoides (MF) is an epidermotropic, primary cutaneous T-cell lymphoma characterised by infiltrates of small to medium sized lymphoid cells with cerebriform nuclei. The typical phenotype is CD2+, CD3+, TCRβ+, CD5+, CD4+ and CD8-. Clinically, MF lesions evolve through three stages referred to as patches, plaques and tumours. Lesions at various stages can co-exist.

The histological appearance varies with the stage of the lesion:
· **Patch lesions** show superficial band-like or lichenoid infiltrates, mainly consisting of lymphocytes and histiocytes. Few atypical cells with cerebriform nuclei are seen and are usually confined to the epidermis where they colonise the basal layer as haloed cells, either singly or in a continuous linear distribution. When the lymphoid cells are present in a continuous linear distribution, this is known as lymphocyte satellitosis.
· **Plaques** are characterised by more pronounced epidermotropism and intraepidermal collections of atypical lymphocytes may be present, known as Pautrier microabscesses.
· **Tumour stage lesions** show a more diffuse dermal infiltrate. The tumour cells increase in number and size, with variable numbers of small, medium and large cerebriform cells.
Histological transformation is defined by the presence of >25% large lymphoid cells in the dermal infiltrate, usually in the tumour stage. These large cells may be CD30+.

25) Correct answer: A) Lymphomatoid granulomatosis

Lymphomatoid granulomatosis is an angiocentric, angio-destructive lymphoproliferative disease involving extranodal sites. The cellular infiltrate is composed of relatively few lesional EBV positive blastic B-cells, admixed with large numbers of reactive T-cells and variable numbers of plasma cells and histiocytes. Vascular changes are usually prominent with vascular wall infiltration and necrosis. Pulmonary involvement is present in 90% of patients, producing chest symptoms such as cough, dyspnoea and chest pain. Other sites commonly involved include brain, kidney, liver and skin.

26) Correct answer: H) Dermatopathic lymphadenitis

Dermatopathic lymphadenitis is a reactive condition which occurs in lymph nodes draining areas in which there has been skin irritation or disruption. This may be due to a benign or neoplastic skin process. Lymph nodes show massive paracortical expansion with Langerhans cells (CD1a+ and S100+) and interdigitating cells (CD1a- and S100+). This contrasts with Langerhans cell histiocytosis, in which the Langerhans cells are present in the sinuses. Melanin pigment is often seen. Patients with neoplastic skin conditions such as mycosis fungoides can develop dermatopathic lymphadenitis and lymph nodes may show lymphomatous involvement in addition to reactive changes.

27) Correct answer: G) Sezary syndrome

Sezary syndrome is a rare cutaneous T-cell lymphoma defined by the triad of:
- Erythroderma
- Generalised lymphadenopathy
- Clonal neoplastic T-cells with cerebriform nuclei in the skin

Furthermore, one or more of the following criteria are required for the diagnosis:
- Absolute Sezary cell count >1000 cells/mm^3
- An expanded CD4+ population, with a CD4/CD8 ratio >10
- Loss of one or more T-cell markers.

Overall, the histological appearances of Sezary syndrome are similar to mycosis fungoides, however, the cellular infiltrate is usually more monotonous and epidermotropism can be absent. Sezary cells are usually cutaneous lymphocyte antigen (CLA)+, CCR4+, CD2+, CD3+, TCRβ+, CD5+, CD4+ and CD8-.

28) Correct answer: b) Nodular lymphocyte predominant Hodgkin Lymphoma.

The Epstein-Barr virus (EBV) is a gamma herpes virus that establishes a mostly harmless latent infection in B cells in over 95% of the human population, but has also been linked to a number of lymphoid and non-lymphoid neoplasms. Lymphoid lesions commonly associated with EBV include:
- Infectious mononucleosis
- Post-transplant lymphoproliferative disease (PTLD)
- Classical Hodgkin lymphoma (particularly lymphocyte deplete classical Hodgkin lymphoma and mixed cellularity classical Hodgkin lymphoma)
- Non-Hodgkin lymphoma:
- Burkitt lymphoma (endemic 100%; sporadic 20%)
- Nasal-type NK/T cell lymphoma (>95%)
- Angioimmunoblastic T cell lymphoma
- Lymphomatoid granulomatosis (>95%)
- CNS lymphoma in AIDS (>95%)
- Primary effusion lymphoma
- Plasmablastic lymphoma

The LP cells of nodular lymphocyte predominant Hodgkin lymphoma are consistently negative for EBV.
The EBV encoded early RNA (EBER) antigen is the most sensitive marker for the detection of EBV in cells using *in situ* hybridisation.

29) Correct answer: c) Burkitt lymphoma

The description is that of a low grade non-Hodgkin B-cell lymphoma with a nodular or follicular architecture. The main differential includes:
- Follicular lymphoma
- Mantle cell lymphoma
- Marginal cell lymphoma
- Chronic lymphocytic leukaemia/small lymphocytic lymphoma

Although Burkitt lymphoma can infiltrate germinal centres resulting in a follicular appearance, the typical appearance is that of sheets of medium sized cells. The cells are nearly all mitotically active and there is a corresponding high rate of apoptosis, giving rise to a 'starry sky appearance'.

30) Correct answer: a) CD23

Follicular dendritic cells form a network of antigen presenting cells that surround the lymphoid cells of the follicles. CD23 and CD21 can be used to highlight follicular dendritic cell networks. In follicular lymphoma, as neoplastic cells infiltrate the follicles, the networks become expanded and disrupted. CD15 is a marker for myeloid cells and Hodgkin cells. CD20 is a B-cell marker. CD68 is a histiocytic marker.

31) Correct answer: b) 6-15 centroblasts per hpf within the germinal centre

Follicular lymphoma is graded according to the number of centroblasts within the follicle as follows:

WHO follicular lymphoma grading system

Grade 1-2 (low grade)	**0-15 centroblasts per hpf**
Grade 1	*0-5 centroblasts per hpf*
Grade 2	*6-15 centroblasts per hpf*
Grade 3 (high grade)	**>15 centroblasts per hpf**
Grade 3A	*Centrocytes present*
Grade 3B	*Solid sheets of centroblasts*

32) Correct answer: b) Castleman disease

There are two distinct morphological forms of Castleman disease, vascular hyaline type and plasma cell type. Vascular hyaline type is characterised by large abnormal follicles with marked vascular proliferation and hyalinisation. There is concentric layering of the lymphocytes at the periphery of the follicle giving rise to an onion skin appearance. Vessels are arranged perpendicular to the concentric rings of lymphocytes giving a 'lollipop' appearance to the follicle. The interfollicular zone shows prominent hyperplastic vessels and an admixture of plasma cells, immunoblasts and eosinophils. The main features of the plasma cell variant described in this case of Castleman disease include a diffuse plasma cell proliferation in the interfollicular zone with Russell bodies, deposition of acidophilic amorphous material with fibrin in the follicles, and variable degrees of vascular changes as seen in the hyaline vascular type. The vascular hyaline type is typically unicentric, usually involving the mediastinal lymph nodes. The plasma cell variant can be either unicentric or multicentric. The multicentric form is strongly associated with HIV infection, and as in this case, can show diminished or absent mantle zones. Patients often present with generalised lymphadenopathy in cases of multicentric Castleman disease.

Kikuchi disease is a necrotising lymphadenitis characterised by apoptotic debris without neutrophils in areas of necrosis and typically presents with cervical lymphadenopathy. Kawasaki disease (mucocutaneous lymph node syndrome) is characterised by geographical areas of necrosis with karyorrhectic debris with fibrin and neutrophils, with fibrin thrombi adjacent to areas of necrosis. Plasma cell myeloma usually manifests in the bone marrow and is associated with increasing age. Lymphoplasmacytic lymphoma is a low grade B-cell lymphoma which can present with the symptoms of Waldenstrom macroglobulinaemia. There is paracortical involvement, or architectural effacement, with a mixture of small lymphoid cells, mature plasma cells and cells with intermediate features. Patients are typically older and present with generalised lymphadenopathy.

33) Correct answer: b) HHV-8

HHV-8, as well as EBV can be found in multicentric Castleman disease, particularly those with HIV infection. Nearly all cases in those with HIV infection are found to have co-infection with HHV-8 in lesional tissue, whilst around 40% of cases are infected with HHV-8 in non-HIV patients.

34) Correct answer: b) Paratrabecular aggregates of lymphoid cells

Paratrabecular aggregates of lymphoid cells are almost always a pathological feature, suggestive of a lymphoproliferative process involving the bone marrow. Occasional interstitial lymphoid follicles may reflect a neoplastic process, but are also seen in reactive processes. Trilineage haematopoesis refers to the normal production of the lymphoid, myeloid and erythroid lineage cells seen in haematopoetic tissue. The term "left shift" means that a particular population of cells is "shifted" towards more immature precursors, and usually refers to the neutrophil series. A left shift in the neutrophil series alone is usually indicative of infection, inflammation or necrosis. If there is a left shift in the red cell and neutrophil series, this is termed a leukoerythroblastic reaction, and may reflect a more sinister problem.

CHAPTER 16

NEUROPATHOLOGY
Dr Leslie R Bridges

1) Which of the following are types of prion disease (more than one answer applies)?

   a) Kuru
   b) Fatal familial insomnia
   c) Guamanian ALS-parkinsonism-dementia syndrome
   d) Neurolathyrism

2) A 60-year-old man presented with dementia and ataxia and died of his disease within six months of onset. What is the likely diagnosis?

   a) Alzheimer's disease
   b) Parkinson's disease
   c) Creutzfeldt-Jakob disease
   d) Mercury poisoning

3-5) Extended matching item questions: For each of the following descriptions of *neurodegenerative disorders*, select the **most appropriate** diagnosis from the list of options above. Each option may be used once, more than once, or not at all.

   A) Alzheimer's disease
   B) Parkinson's disease
   C) Creutzfeldt-Jakob disease
   D) Huntington's disease
   E) Friedreich's ataxia

3) The patient had tremor, slow movement and rigidity. The autopsy brain showed depigmentation of the substantia nigra and locus coeruleus. Microscopy showed Lewy bodies.

4) The patient was demented. The autopsy brains showed atrophy. Microscopically there were senile plaques and neurofibrillary tangles.

5) The patient showed dementia and ataxia. Microscopically there were proteinaceous deposits in the brain (immunopositive for prion protein, PrP).

Scenario for questions 6 and 7: The patient presented in middle age with involuntary movements (chorea) and dementia. The patient's father and sister were similarly affected. The autopsy brains showed atrophy of the caudate nucleus and putamen. Histologically there were intranuclear intraneuronal inclusions.

6) What is the diagnosis?

   a) Wilson's disease
   b) Huntington's disease
   c) Progressive supranuclear palsy

d) Multiple systems atrophy

7) The inclusions were immunopositive with antibodies against which of the following proteins?

a) Ubiquitin and huntingtin
b) Ubiquitin and tau
c) Ubiquitin and alpha-synuclein
d) Ubiquitin and ceruloplasmin

8) A four-year-old boy presented with headache and ataxia. Magnetic resonance imaging (MRI) showed a cystic tumour in the cerebellum. Histologically the tumour comprised elongated fibrillated cells with Rosenthal fibres and a low mitotic rate. Molecular studies showed a BRAF fusion. What is the likely diagnosis?

a) Schwannoma
b) Medulloblastoma
c) Pilocytic astrocytoma
d) Ependymoma

9-16) Extended matching item questions:

A) Diffuse astrocytoma
B) Oligodendroglioma
C) Ependymoma
D) Anaplastic astrocytoma
E) Anaplastic oligodendroglioma
F) Anaplastic ependymoma
G) Subependymoma
H) Glioblastoma
I) Pleomorphic xanthoastrocytoma

For each of the following descriptions of a ***primary brain tumour***, select the **most appropriate** diagnosis from the list of options above. Each option may be used once, more than once, or not at all.

9) A primary brain tumour characterised by pleomorphic cells, pseudopalisading necrosis, capillary proliferation and frequent mitoses. Typically lacks isocitrate dehydrogenase (IDH) mutation.

10) A primary brain tumour characterised by pleomorphic cells, pericellular reticulin and a low mitotic rate. Typically lacks IDH mutation but often shows BRAF mutation.

11) A primary brain tumour with fairly uniform clear cells and low mitotic rate. Typically has both IDH mutation and 1p19q co-deletion.

12) A primary brain tumour with fairly uniform fibrillated cells and low mitotic rate. Typically has IDH mutation but not BRAF fusion or 1p19q codeletion.

13) A primary brain tumour with clear cells, mini-gemistocytes and high mitotic rate. Sometimes shows necrosis and capillary proliferation. Typically has IDH mutation and 1p19q codeletion.

14) A tumour which grows from the ventricles and has fibrillated cells arranged in perivascular pseudo-rosettes and a low mitotic rate. Immunostaining for EMA shows dot-like positivity. Typically lacks IDH mutation and 1p19q codeletion.

15) A very low-grade tumour arising from the ventricles comprising fibrillated cells with extensive anuclear zones. Can be an incidental finding on brain scan or at post-mortem.

16) A primary brain tumour comprising pleomorphic fibrillary cells and a high mitotic rate lacking necrosis and capillary proliferation. Typically has IDH mutation but not 1p19q codeletion.

17) The following are WHO grade I tumours (more than one answer applies):

   a) Subependymoma
   b) Pilocytic astrocytoma
   c) Ganglioglioma
   d) Ependymoma

18) The following are WHO grade II tumours (more than one answer applies):

   a) Diffuse astrocytoma
   b) Oligodendroglioma
   c) Ependymoma
   d) Pilocytic astrocytoma

19) The following are WHO grade III tumours (more than one answer applies):

   a) Pleomorphic xanthoastrocytoma
   b) Anaplastic astrocytoma
   c) Anaplastic oligodendroglioma
   d) Anaplastic oligoastrocytoma

20) The following are WHO grade IV tumours (more than one answer applies):

   a) Pleomorphic xanthoastrocytoma
   b) Glioblastoma
   c) Supratentorial primitive neuroectodermal tumour (sPNET)
   d) Dysembryoplastic neuroepithelial tumour (DNET)

21) Small round blue cell tumours of the CNS include which of the following (more than one answer applies)?

   a) Medulloblastoma
   b) Supratentorial primitive neuroectodermal tumour (sPNET)
   c) Pineoblastoma
   d) Subependymoma

22) Isocitrate dehydrogenase (IDH) mutation is typically NOT found in which of the following?

   a) Diffuse astrocytoma
   b) Oligodendroglioma
   c) Anaplastic astrocytoma

d) Glioblastoma

23) Chromosome 1p19q codeletion occurs commonly in which of the following tumours (more than one answer applies)?

   a) Oligodendroglioma
   b) Anaplastic oligodendroglioma
   c) Astrocytoma
   d) Anaplastic astrocytoma

24) Signs of raised intracranial pressure include which of the following (more than one answer applies)?

   a) Diplopia due to VIth cranial nerve palsy
   b) Uncal herniation with posterior cerebral artery territory infarction
   c) Cerebellar tonsillar herniation with medullary compression
   d) Discrete patches of demyelination throughout the white matter

25) Alzheimer's disease is characterised by which of the following features (more than one answer applies)?

   a) Extracellular deposits of beta-amyloid
   b) Neurofibrillary tangles containing hyperphosphorylated tau
   c) Kuru plaques
   d) Lewy bodies

26) Which of the following descriptions is the best fit for the figure below?

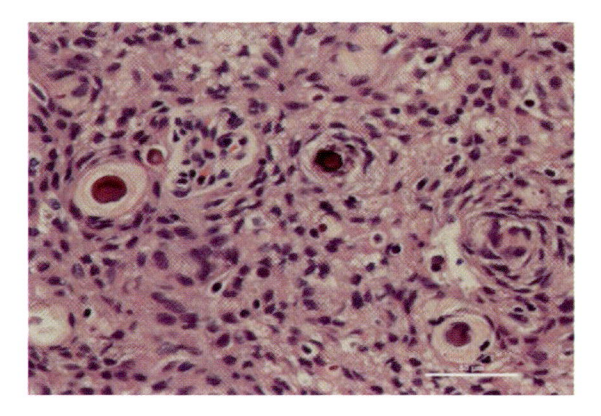

   a) Spindle-shaped cells arranged in whorls typical of a meningioma
   b) Spindle-shaped cells arranged in interlacing fascicles typical of a schwannoma
   c) Spindle-shaped cells arranged in perivascular patterns typical of a haemangiopericytoma
   d) Pleomorphic spindle-shaped cells typical of a gliosarcoma

27) Which of the following descriptions is the best fit for the figure below?

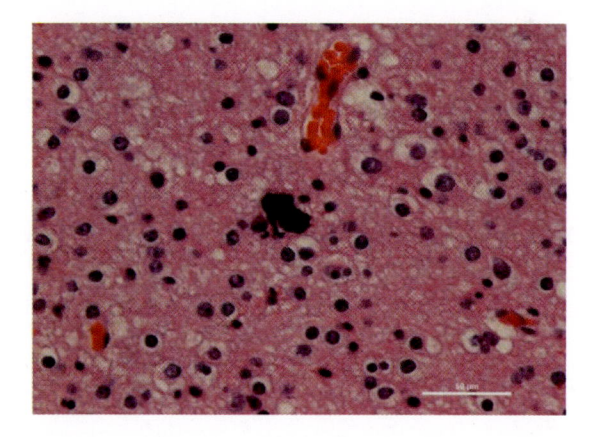

a) Cells with clear cytoplasm, distinct cytoplasmic borders and rounded nuclei typical of an oligodendroglioma
b) Spindle-shaped cells and microcalcification typical of a psammomatous meningioma
c) Pleomorphic clear cells typical of metastatic renal carcinoma
d) Physalliferous cells typical of a chordoma

28) Which of the following is a demyelinating disease due to proven direct viral infection?

a) Multiple sclerosis
b) Progressive multifocal leukoencephalopathy
c) Cerebral Whipple's disease
d) Rabies

29) In which of the following diseases is spread within and/or between nerve cells thought to play a role (more than one answer applies)?

a) Rabies
b) Herpes zoster infection
c) Creutzfeldt-Jakob disease
d) Alzheimer's disease

30) Which of the following are recognised common causes of subarachnoid haemorrhage (more than one answer applies)?

a) Cerebral trauma
b) Rupture of a berry aneurysm
c) Rupture of a middle meningeal artery
d) Rupture of a cerebral micro-aneurysm

31) Which of the following features indicate an atypical meningioma (WHO grade II)?

a) Meningioma with 4 mitoses per 10 high-power fields
b) Meningioma with 20 mitoses per 10 high-power fields
c) Meningioma with necrosis but no other worrying features
d) Meningioma with CEA-immunopositive, PAS-positive secretory globules

32-35) Extended matching item questions: For each of the following descriptions, select the **most appropriate** diagnosis from the list of options above. Each option may be used once, more than once, or not at all.

- A) Craniopharyngioma, adamantinomatous type
- B) Craniopharyngioma, papillary type
- C) Granular cell tumour of the neurohypophysis
- D) Pituitary carcinoma
- E) Germinoma
- F) Silent corticotroph adenoma
- G) Null-cell pituitary adenoma
- H) Prolactinoma
- I) Sparsely-granulated growth hormone cell adenoma

32) A suprasellar epithelial tumour with stellate reticulum, "wet" keratin and calcification seen mainly in young people.

33) A suprasellar malignant tumour with large CD117-immunopositive cells and reactive lymphocytes.

34) A sellar tumour, typically large and locally aggressive which contains but does not secrete ACTH.

35) A chromophobe pituitary adenoma characterised by large round cytoplasmic inclusions (fibrous bodies).

36-39) Extended matching item questions:

- A) Epidural (extradural) haematoma
- B) Subdural haematoma
- C) Subarachnoid haemorrhage
- D) Intracerebral haemorrhage
- E) Intraventricular haemorrhage

For each of the following descriptions of an ***intracranial haemorrhage***, select the **most appropriate** diagnosis from the list of options above. Each option may be used once, more than once, or not at all.

36) Caused by rupture of bridging veins between arachnoid mater and dura mater

37) Caused by tearing of the middle meningeal artery

38) Haemorrhage into the parenchyma of the brain

39) Haemorrhage into the space between arachnoid mater and pia mater

40) Which of the following types of multiple sclerosis (MS) are correctly linked to their descriptions (more than one answer applies)?

- a) Charcot-type – classical MS
- b) Devic's disease – neuromyelitis optica
- c) Marburg type – concentric sclerosis
- d) Balo's disease – acute MS

41) Haemorrhagic infarcts are typically caused by which of the following conditions (more than one answer applies)?

    a) Embolism
    b) Venous infarction
    c) Thrombosis and occlusion of an end-artery
    d) Rupture of a cerebral microaneurysm

42) Which of the following is the best description of the structure of an established cerebral abscess (from the centre outwards)?

    a) Necrosis, granulation tissue, collagenous capsule and reactive brain tissue
    b) Necrosis, reactive brain tissue, granulation tissue and collagenous capsule
    c) Necrosis, granulation tissue, reactive brain tissue and collagenous capsule
    d) Granulation tissue, necrosis, collagenous capsule and reactive brain tissue

43) Which of the following statements are true about the Arnold-Chiari malformation (more than one answer applies)?

    a) Also known as the Chiari type I malformation
    b) Also known as the Chiari type II malformation
    c) Features include lumbosacral myelomeningocoele, shallow posterior fossa and hydrocephalus
    d) Features include syringomyelia, skeletal abnormalities and hydrocephalus

Correct Answers and Explanations:

1) Correct answers: a) Kuru and b) Fatal familial insomnia

Kuru is a form of prion disease due to ritual cannibalism in the Fore tribe of New Guinea.

Fatal familial insomnia is an autosomal dominant form of prion disease characterised by thalamic atrophy and sleep disturbance due to mutation in the prion (PRNP) gene on chromosome 20.

Guamanian amyotrophic lateral sclerosis-parkinsonism-dementia syndrome in the Chamorro people is suspected of being due to neurotoxicity caused by the cycad seed.

Neurolathyrism is a form of spastic paraplegia prevalent in some areas of India and Africa due to neurotoxicity caused by ingestion of the Lathyrus chickpea.

2) Correct answer: c) Creutzfeldt-Jakob disease

Creutzfeldt-Jakob disease (CJD) occurs at an incidence of about one case per million population per year. Although rare, it should always be considered in rapidly progressive dementias. CJD is usually sporadic but is sometimes genetic (due to mutation in the PRNP gene) or iatrogenic (e.g. spread via dura mater grafts from other patients with prion disease).

3) Correct answer: B) Parkinson's disease

The patient has the classical clinical triad of Parkinson's disease (tremor, bradykinesia and rigidity). Histologically there are rounded eosinophilic intraneuronal inclusions (immunopositive for alpha-synuclein) known as Lewy bodies. There are some rare genetic causes of Parkinson's disease but it is usually sporadic.

4) Correct answer: A) Alzheimer's disease

The histological hallmarks of Alzheimer's disease are senile plaques (extracellular deposits of beta-amyloid) and neurofibrillary tangles (intracellular accumulations of hyperphosphorylated tau).

5) Correct answer: C) Creutzfeldt-Jakob disease

Patients with CJD show diffuse or synaptic patterns of deposition of prion protein (PrP). Typically, there is spongiform degeneration of grey matter and sometimes small (kuru type) plaques.

6) Correct answer: b) Huntington's disease

Huntington's disease is an autosomal dominant disease of high penetrance due to mutation of the huntingtin gene (coding for huntingtin protein) on chromosome 4. The mutation is an expanded CAG repeat sequence (usually there are 9 to 37 CAG repeats; in disease there are 37 to 100 or more).
Wilson's disease is an autosomal recessive disease caused by mutation in the ATPase gene (ATP7B) on chromosome 13. The ATPase enzyme normally transports copper into bile and incorporates it into ceruloplasmin. In Wilson's disease copper accumulates abnormally in liver and brain.
Progressive supranuclear palsy causes parkinsonism and paralysis of upward gaze and is associated with the H1 haplotype of the tau gene. Neurons and glia in the basal ganglia, brainstem and cerebral cortex accumulate hyperphosphorylated tau protein.
Multiple systems atrophy is a sporadic disease causing olivopontocerebellar atrophy (OPCA), Shy-Drager syndrome or striatonigral degeneration. Histologically there are glial cytoplasmic inclusions (GCIs) containing alpha-synuclein.

7) Correct answer: a) Ubiquitin and huntingtin

The expanded CAG repeat in the huntingtin gene is translated into an expanded polyglutamine (polyQ) tract in the huntingtin protein. This accumulates in neuronal nuclei. Ubiquitin is part of the ubiquitin proteosome system (UPS) and is commonly tagged onto proteins that accumulate in neurodegenerative diseases, labelling them for disposal.

8) Correct answer: c) Pilocytic astrocytoma

Pilocytic astrocytoma (WHO grade I) occurs usually in children in the posterior fossa or optic chiasma, sometimes in association with neurofibromatosis type 1 (NF1). Genomic duplication involving BRAF and the KIAA1549 genes creates a fusion protein with B-Raf kinase activity and is a useful biomarker for the tumour. Pilocytic astrocytoma lacks mutation of the isocitrate dehydrogenase (IDH) genes and is molecularly distinct from diffuse astrocytoma.

9) Correct answer: H) Glioblastoma

Glioblastoma (WHO grade IV) is the commonest and one of the most malignant forms of primary brain tumour. Treatment with radiotherapy and chemotherapy is palliative only; patients often survive for only a few months after diagnosis.

10) Correct answer: I) Pleomorphic xanthoastrocytoma

Pleomorphic xanthoastrocytoma (PXA, WHO grade II) typically occurs in a superficial location in the brain. The cells contain lipid and are surrounded by reticulin (collagen IV). Despite cellular pleomorphism, the tumours are low-grade. About 60% have BRAF V600E mutations.

11) Correct answer: B) Oligodendroglioma

Oligodendroglioma (WHO grade II) is characterised by clear cells (giving rise to a "fried egg" or "frog-spawn" appearance), fine branching "chicken-wire" capillaries and perineuronal satellitosis.

12) Correct answer : A) Diffuse astrocytoma

Diffuse astrocytoma (WHO grade II) comprises fibrillated cells arranged in sheets and microcysts. Isocitrate dehydrogenase (IDH) mutation is common in both diffuse astrocytoma and oligodendroglioma. Codeletion 1p19q is seen commonly in oligodendroglioma but not astrocytoma.

13) Correct answer: E) Anaplastic oligodendroglioma

Anaplastic oligodendroglioma (WHO grade III) shows increased cellularity, pleomorphism and mitotic rate compared to oligodendroglioma (WHO grade II). The presence of IDH mutation and 1p19q codeletion in the majority of anaplastic oligodendroglioma helps distinguish it from glioblastoma.
Codeletion of chromosome arms 1p and 19q is found in high frequency in oligodendroglioma (WHO grade II) and anaplastic oligodendroglioma (WHO grade III) and is associated with better prognosis and response to chemotherapy. It can be detected by fluorescent in-situ hybridisation (FISH) on paraffin sections.
Somatic mutation of IDH1 is seen in the majority of diffuse astrocytomas and oligodendrogliomas. The common IDH1 mutation can be detected in paraffin sections using a antibody specific for the mutant protein. IDH2 mutations are much less common but have a similar effect to IDH1 mutations. Possession of an IDH mutation is prognostically favourable.

14) Correct answer: C) Ependymoma

Ependymoma (WHO grade II) is a relatively common CNS tumour in childhood and occurs in the posterior fossa and spinal-cord as well as supratentorially. Although low-grade it can disseminate throughout the craniospinal axis.

15) Correct answer: G) Subependymoma

Subependymoma (WHO grade I) may be seen as a firm, well-demarcated 1 to 2 cm diameter nodule bulging into the floor of the fourth ventricle or lateral ventricle. Subependymoma is often asymptomatic but can cause brainstem compression.

16) Correct answer: D) Anaplastic astrocytoma

The distinction between anaplastic astrocytoma (WHO grade III) and glioblastoma (WHO grade IV) can be difficult in a small biopsy. Although capillary proliferation and pseudopalisading necrosis are features of glioblastoma and not anaplastic astrocytoma, these features may be absent in a small biopsy. In such a situation immunostaining for the common IDH1 mutation may be helpful (positive staining makes anaplastic astrocytoma more likely than glioblastoma).

17) Correct answers: a) Subependymoma, b) Pilocytic astrocytoma and c) Ganglioglioma

Other WHO grade I tumours in the WHO classification of tumours of the CNS include myxopapillary ependymoma, dysembryoplastic neuroepithelial tumour (DNET), paraganglioma of the spinal-cord, pineocytoma, schwannoma, neurofibroma, meningioma, haemangioblastoma and craniopharyngioma.

18) Correct answers: a) Diffuse astrocytoma, b) Oligodendroglioma and c) Ependymoma

Other WHO grade II CNS tumours include pleomorphic xanthoastrocytoma, mixed glioma (oligoastrocytoma), central neurocytoma, atypical meningioma and haemangiopericytoma. WHO grade I and WHO grade II tumours are considered "low-grade".

19) Correct answers: b) Anaplastic astrocytoma, c) Anaplastic oligodendroglioma and d) Anaplastic oligoastrocytoma.

Other WHO grade III CNS tumours include anaplastic ependymoma, anaplastic ganglioglioma, anaplastic meningioma and anaplastic haemangiopericytoma.

20) Correct answers: b) Glioblastoma and c) Supratentorial primitive neuroectodermal tumour (sPNET)

Other WHO grade IV tumours include gliosarcoma (a variant of glioblastoma), pineoblastoma, ependymoblastoma and atypical teratoid/rhabdoid tumour (AT/RT). WHO grade III and WHO grade IV tumours are considered "high-grade".

21) Correct answers: a) Medulloblastoma, b) Supratentorial primitive neuroectodermal tumour (sPNET) and c) Pineoblastoma

These are all primitive neuroectodermal tumours (PNETs) comprising malignant neural-type cells (synaptophysin-immunopositive) with scanty cytoplasm and hyperchromatic pleomorphic nuclei. They are named according to their location in the CNS (i.e. medulloblastoma in the cerebellum, supratentorial PNET in the cerebrum and pineoblastoma in the pineal). Although of similar morphology, there are molecular differences between these entities as well as differences in behaviour and response to treatment.

22) Correct answer: d) Glioblastoma

IDH mutation is a molecular signature of diffuse astrocytoma, oligodendroglioma, anaplastic astrocytoma, anaplastic oligodendroglioma, oligoastrocytoma and anaplastic oligoastrocytoma. IDH mutation is unusual in glioblastomas. When it does occur in glioblastoma it suggests that the glioblastoma may have arisen from a previous low-grade glioma. Such glioblastomas are called "secondary" glioblastomas. Glioblastomas arising *de novo* are called "primary" glioblastomas.

23) Correct answers: a) Oligodendroglioma and b) Anaplastic oligodendroglioma

Chromosome 1p19q codeletion is the molecular signature of oligodendroglial tumours. It is seen in the majority of oligodendrogliomas and anaplastic oligodendrogliomas and in fewer but still significant numbers of oligoastrocytomas and anaplastic oligoastrocytomas. It is not generally seen in astrocytomas and anaplastic astrocytomas. Chromosome 1p19q codeletion generally denotes favourable prognosis and response to treatment.

24) Correct answers: a) Diplopia due to VIth cranial nerve palsy, b) Uncal herniation with posterior cerebral artery territory infarction and c) Cerebellar tonsillar herniation with medullary compression

Other effects of raised intracranial pressure include IIIrd nerve palsies due to uncal (transtentorial) herniation, cingulate (subfalcine) herniation and brainstem midline (Duret) haemorrhage.

25) Correct answers: a) Extracellular deposits of beta-amyloid and b) Neurofibrillary tangles containing hyperphosphorylated tau

Beta-amyloid results from deranged metabolism of the beta-amyloid precursor protein (APP). The APP gene is on chromosome 21. Mutation in the gene or increased dose of the gene (as occurs in trisomy 21, Down's syndrome) are recognised causes of Alzheimer's disease, although the disease is usually sporadic. According to the amyloid cascade hypothesis of Alzheimer's disease, amyloid deposition is primary and the other features (such as intracellular accumulation of hyperphosphorylated tau) are secondary effects.

26) Correct answer: a) Spindle-shaped cells arranged in whorls typical of a meningioma

The figure shows cells arranged in whorls. Other recognised features of meningiomas include intranuclear inclusions and calcified psammoma bodies. Meningiomas usually show diffuse positivity for EMA. Schwannomas are usually are immunopositive for S100 and haemangiopericytomas for CD34. Gliosarcoma (WHO grade IV) is a variant of glioblastoma in which the usual reactive vascular proliferation progresses to frank neoplasia. Gliosarcoma shows a biphasic pattern of malignant gliomatous and sarcomatous elements.

27) Correct answer: a) Cells with clear cytoplasm, distinct cytoplasmic borders and rounded nuclei typical of an oligodendroglioma

The figure shows microcalcification typical of oligodendroglioma. Microcalcification can also be seen in other low-grade gliomas, subependymoma, craniopharyngioma, pineocytoma and meningioma.

28) Correct answer:b) Progressive multifocal leukoencephalopathy

Progressive multifocal leukoencephalopathy (PML) is due to infection by the SV40 virus. This polyoma virus usually lies dormant in the body but can be reactivated, particularly in immunosuppressed patients, such as those with AIDS.
Multiple sclerosis is a demyelinating disease usually attributed to autoimmunity and multiple genetic factors. Although viral triggers have been imputed, there is no proven association with any virus.
The gut infection Whipple's disease (due to the Gram-positive bacillus Tropheryma whipplei) sometimes affects the brain. Histologically there are clusters of macrophages containing PAS-positive diastase-resistant material. CNS Whipple's disease is an important diagnosis to make as it can be treated with antibiotics. Confirmation is available by EM (which shows bacteria in macrophages) and specific PCR.
Rabies is a direct viral disease but it affects grey matter rather than white matter (i.e. it is not a demyelinating disease).

29) Correct answers : a) Rabies, b) Herpes zoster infection, c) Creutzfeldt-Jakob disease and d) Alzheimer's disease

In rabies the initial infection by the lyssavirus in skin due to an animal bite travels to the brain by the peripheral nerves.
In herpes zoster infection, the varicella zoster virus (VZV) is dormant in dorsal root ganglia. On activation it can spread out into the skin by peripheral nerves (or rarely into the spinal-cord and brain).
A nerve to nerve spread of prions has been proven in experimental animals and is thought to occur also in human prion disease. Prion disease arises from misfolding of the cellular form of the PrP protein. In recent years, misfolded proteins in other neurodegenerative diseases (including beta-amyloid, tau and alpha-synuclein) have been imputed to spread from nerve to nerve in a "prion-like" fashion. Although spread within the brain may well be a common mechanism for these diseases, as yet only PrP disease is proven to be transmissible between humans.
The Braak staging of Alzheimer's disease shows a topographical pattern suggestive of neuroanatomical spread of disease from medial temporal lobe (stages I and II) to limbic structures (stages III and IV) to cerebral neocortex (stages V and VI).

30) Correct answers: a) Cerebral trauma and b) Rupture of a berry aneurysm

Subarachnoid haemorrhage is a common component of intracranial haemorrhage due to head trauma in road traffic accidents and assaults.
Berry aneurysms commonly arise from the anterior communicating, middle cerebral and posterior communicating arteries of the circle of Willis. They are due to congenital weakening of the arterial wall, causing progressive dilatation with time, particularly in patients with hypertension. They are the commonest cause of spontaneous subarachnoid haemorrhage.
Rupture of the middle meningeal artery due to trauma is a cause of extradural haematoma.
Rupture of cerebral micro-aneurysms is a cause of spontaneous intracerebral haemorrhage.

31) Correct answer:a) Meningioma with 4 mitoses per 10 high-power fields

Atypical meningioma (WHO grade II) is defined as having 4 or more mitoses per 10 high-power fields.
Anaplastic meningioma (WHO grade III) is defined as having 20 or more mitoses per 10 high-power fields.
An alternative definition of atypical meningioma (WHO grade II) require three or more of the following: increased cellularity, small cells with high nucleocytoplasmic ratio, prominent nucleoli, sheet like growth and necrosis. Necrosis alone is insufficient for the diagnosis.
Meningioma with CEA-immunopositive material indicates the so-called "secretory" subtype of meningioma (WHO grade I). Such tumours may produce measurable levels of carcinoembryonic antigen in the serum.

32) Correct answer: A) Craniopharyngioma, adamantinomatous type

The adamantinomatous type of craniopharyngioma (WHO grade I) has a bimodal age distribution with peaks in children aged 5 to 15 years in adults aged 45 to 60 years.
Papillary craniopharyngioma (WHO grade I) comprises well differentiated squamous epithelium lacking surface maturation. It occurs mainly in adults aged 40 to 55 years.

33) Correct answer: E) Germinoma

Germinoma of the CNS is a similar to seminoma of the testis and dysgerminoma the ovary. Germinoma and other germ-cell tumours of the CNS occur in the midline in the suprasellar and pineal regions.

34) Correct answer: F) Silent corticotroph adenoma

Corticotroph pituitary adenomas sometimes retain but do not secrete ACTH and are therefore endocrinological "silent". They tend to be more locally aggressive and more likely to recur than corticotroph adenomas with endocrine effects.

35) Correct answer: I) Sparsely-granulated growth hormone cell adenoma

Growth hormone cell adenomas can be densely granulated (eosinophil) or sparsely granulated (chromophobe). Fibrous bodies (pale spherical cytoplasmic inclusions containing intermediate filaments immunopositive for low molecular weight cytokeratin) are diagnostic of the sparsely granulated type.

36) Correct answer: B) Subdural haematoma

Virtually always due to trauma. Typically occurs in assaults (including non-accidental injury in children) and road traffic accidents.

37) Correct answer: A) Epidural (extradural) haematoma

The middle meningeal artery passes beneath the pterion, the thinnest part of the skull, just behind the temple. Trauma here (e.g. due to being struck on the side of the head by a cricket ball) can cause rupture of the artery with consequent rapid accumulation of haematoma. This is a neurosurgical emergency requiring removal of the haematoma and ligation of the artery.

38) Correct answer: D) Intracerebral haemorrhage

Haemorrhage into the substance of the brain can be primary (e.g. due to spontaneous rupture of a cerebral microaneurysm) or secondary (e.g. due to trauma from an assault or road traffic accident).

39) Correct answer: C) Subarachnoid haemorrhage

Subarachnoid haemorrhage can be either primary (typically due to spontaneous rupture of a berry aneurysm on an artery at the base of the brain) or secondary (e.g. due to trauma from an assault or road traffic accident).
Intraventricular haemorrhage can be primary (e.g. in premature infants, due to weakness of capillary walls in the subventricular zone) or secondary (typically caused by extension from an intracerebral haemorrhage).

40) Correct answers : a) Charcot-type – classical MS and b) Devic's disease – neuromyelitis optica

The cause of classical MS (Charcot-type) is still unknown, although genetic, autoimmune and viral factors have been implicated.
Neuromyelitis optica (Devic's disease) is a rare form of MS affecting the optic nerves and spinal cord which is caused in the majority of cases by an autoimmune attack on aquaporin 4 (an astrocyte protein).
Balo's disease (concentric sclerosis) is a rare form of MS characterised by alternating rings of demyelination and myelination in the white matter.
Acute MS (Marburg-type) is usually fatal within a few months. The associated oedema is pronounced and sometimes simulates a neoplasm.

41) Correct answers: a) Embolism and b) Venous infarction

In embolism to a cerebral artery (e.g. from a heart valve vegetation) the cerebral tissue in the territory of the occluded vessel dies. If the embolus then breaks up and moves on the restored blood flow into the dead tissue resulting in a haemorrhagic infarct as blood seeps into the dead tissue.

Venous infarction causes death of brain tissue due to obstruction of venous outflow but the arterial input is still patent. Blood can seep into the dead tissue, causing a haemorrhagic infarct.

Thrombosis and occlusion of an end-artery in the brain usually causes pale infarction since there is no route for blood to seep into the dead tissue.

Rupture of a cerebral artery microaneurysm is a cause of cerebral haemorrhage as opposed to haemorrhagic infarction. Haemorrhage is where blood under pressure ploughs into and displaces previously healthy brain parenchyma whereas haemorrhagic infarction is where blood seeps into already-dead brain tissue.

42) Correct answer : a) Necrosis, granulation tissue, collagenous capsule and reactive brain tissue

A brain abscess commences as a focus of suppurative cerebritis (days 1 - 2).
Then confluent central necrosis occurs (days 3 - 4).
This is followed by a granulation tissue response around the necrosis (days 5 – 7).
Eventually, concentric layers of collagenous tissue then gliotic reactive brain tissue form (by 14 days).

43) Correct answers: b) Also known as the Chiari type II malformation and c) Features include lumbosacral myelomeningocoele, shallow posterior fossa and hydrocephalus

The Chiari type II (also known as Arnold Chiari) malformation includes features of lumbosacral myelomeningocoele, shallow posterior fossa and hydrocephalus.
The Chiari type I malformation includes features of syringomyelia, skeletal abnormalities and (sometimes) hydrocephalus.
The Chiari type III malformation is similar to the Chiari type II malformation but also includes cerebello-encephalocoele.

CHAPTER 17

PAEDIATRIC AND PLACENTAL PATHOLOGY
Dr Iona Jeffrey & Dr Ruth Nash

1) A four year old presents with an upper abdominal mass. Biopsy reveals the following picture. What is the most likely diagnosis?

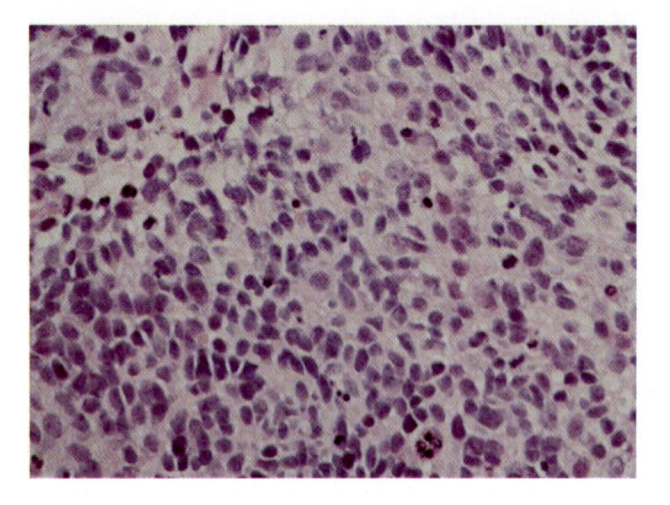

a) Wilm's tumour (nephroblastoma)
b) Desmoplastic small round cell tumour
c) Embryonal rhabdomyosarcoma
d) Primitive neuroectodermal tumour
e) Poorly differentiated neuroblastoma

2 – 6) Extended matching item questions:

A) Wilm's tumour
B) Precursor B acute lymphoblastic leukaemia/lymphoma
C) Precursor T acute lymphoblastic leukaemia/lymphoma
D) Desmoplastic small round cell tumour
E) Embryonal rhabdomyosarcoma
F) Primitive neuroectodermal tumour (PNET)
G) Poorly differentiated neuroblastoma
H) Well differentiated neuroblastoma
I) Hepatoblastoma

For each of the following descriptions select the most appropriate diagnosis from the list of options above. Each option may be used once, more than once or not at all.

2) Fourteen year old boy with a mass in the thoracic wall composed of groups of undifferentiated small round blue cells with clumped chromatin and virtually no cytoplasm, divided by thin fibrous septa and showing occasional rosette like structures; crisp membranous positivity on CD99 immunostaining.

3) Three month old infant with an upper abdominal mass composed of trabeculae of medium sized cells with round, relatively uniform nuclei, eosinophilic cytoplasm and distinct cell membranes that are α fetoprotein positive on immunostaining; between the trabeculae there are sinusoids lined by CD34 positive endothelium; scattered extramedullary haemopoietic cells are present but virtually no stroma.

4) A four year old girl with a mass growing below the bladder urothelium composed of small round and oval cells with hyperchromatic nuclei and very little cytoplasm among which are occasional larger cells with elongated tails of eosinophilic cytoplasm. The tumour shows variable cellularity, accentuated under the mucosal surface, and patchy myxoid change in the stroma.

5) A fourteen year old girl with a thoracic mass composed of sheets of small blue cells with nuclei containing multiple small basophilic nucleoli and a thin crescent of basophilic cytoplasm; nuclear positivity for TdT and cytoplasmic positivity for CD3 and CD5 on immunostaining.

6) A sixteen year old boy with an abdominal mass composed of thick collagenous septa between which are nests of hyperchromatic ovoid small cells with very little cytoplasm; immunostaining reveals tumour cells are cytokeratin, desmin and NSE positive.

7 – 10) Extended matching item questions:

- A) Mesenchymal hamartoma
- B) Mesoblastic nephroma
- C) Clear cell sarcoma
- D) Ganglioneuroma
- E) Yolk sac tumour
- F) Phaeochromocytoma
- G) Ganglioneuroma
- H) Rhabdoid tumour

For each of the following descriptions of an abdominal mass in a child, select the most appropriate diagnosis from the list of options above. Each option may be used once, more than once or not at all.

7) Three year old child with cellular renal tumour composed of a diffuse growth of small cells with stippled chromatin and small amounts of clear cytoplasm arranged in nests and cords. There is a prominent branching vascular network. The tumour is vimentin and BCL2 positive, and WT1, cytokeratin and CD99 negative.

8) Nine month old infant with large renal tumour composed of cells with vesicular nuclei, prominent nucleoli and scattered eosinophilic cytoplasmic inclusions. There is extensive haemorrhage and necrosis. Cells show dot like perinuclear cytoplasmic staining with vimentin with absence of INI1 staining in nuclei.

9) Four month old infant with upper right abdominal mass composed of loose, myxoid, spindle cell stroma with prominent cystic spaces and intermixed biliary ductular epithelial elements and foci of extramedullary haemopoiesis.

10) Seven year old boy underwent excision post chemotherapy of a paravertebral thoracic mass, which showed stroma composed of wavy spindle cells arranged in bundles among which are large cells with plentiful eosinophilic cytoplasm and a round nucleus with a single prominent nucleolus. There is no small blue cell component.

11) A 10 year old girl presents with abdominal pain and rectal bleeding. Laparotomy reveals an ileocolic intussusception that cannot be reduced. The lead point of the intussusception is a mass composed of sheets of lymphoid cells with round nuclei and clumped chromatin and small basophilic nucleoli with minimal cytoplasm. There are interspersed prominent tingible body macrophages. The lymphoid cells are CD20, CD10, CD79A and BCL6 positive on immunostaining; TdT stain is negative. KI-67 stains almost 100% of the cells. Which of the following is the most likely diagnosis?

    a) Ileal follicular lymphoid hyperplasia
    b) B cell lymphoblastic lymphoma
    c) Diffuse large B cell lymphoma
    d) Follicular lymphoma
    e) Burkitt lymphoma
    f) Classical Hodgkin's lymphoma

12) For which one of these commonly biopsied tumours of childhood is the availability of fresh tissue currently essential for determining treatment and prognosis?

    a) Wilms' tumour
    b) Rhabdomyosarcoma
    c) Primitive neuroectodermal tumour (PNET)
    d) Neuroblastoma
    e) Anaplastic large cell lymphoma

13) A 3 year old child is found to have a renal mass. Biopsy reveals the following picture. What is the likely diagnosis?

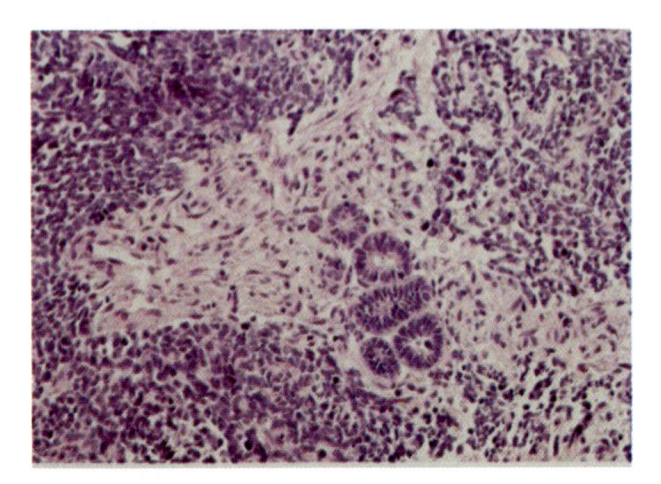

    a) Teratoma
    b) Clear cell sarcoma of the kidney
    c) Adenofibroma
    d) Mesoblastic nephroma
    e) Wilms' tumour

14) A 2 year old child is found to have a renal mass. Biopsy reveals the following picture. What is the most likely diagnosis?

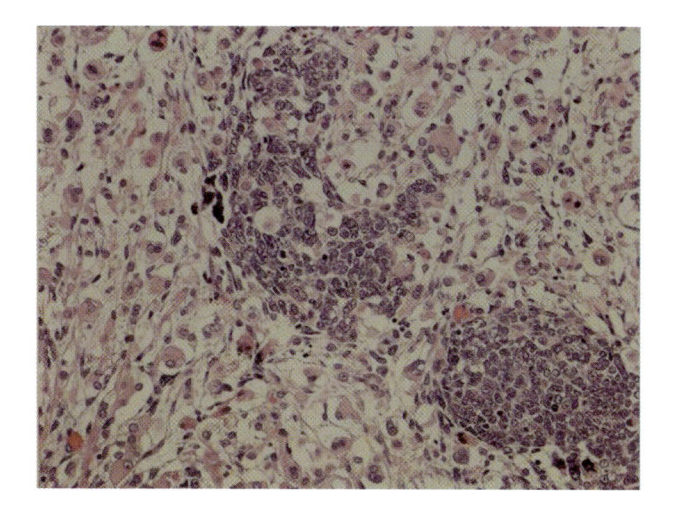

a) Teratoma
b) Clear cell sarcoma of the kidney
c) Adenofibroma
d) Mesoblastic nephroma
e) Wilms' tumour
f) Anaplastic Wilms' tumour

15) From the list below, which is NOT essential to determine the local staging of a resected post chemotherapy Wilms' tumour?

a) Appearance of renal capsule or fibrous pseudocapsule
b) Intrarenal vascular invasion
c) Viable tumour in the renal sinus including vessels and ureteric wall
d) Chemotherapy effect in lymph nodes
e) Completeness of excision of viable tumour
f) Completeness of excision of necrotic tumour

16 – 19 Extended matching item questions:

A) Congenital infantile fibrosarcoma
B) Alveolar rhabdomyosarcoma
C) Neuroblastoma
D) Ewing's sarcoma
E) Desmoid fibromatosis
F) Primitive neuroectodermal tumour (PNET)

For each of the following genetic changes, select the most likely tumour diagnosis from the list of options above. Each option may be used once, more than once or not at all.

16) PAX3 or 7-FKHR fusion transcript

17) EWS-FLI1 fusion transcript

18) 1p deletion; 17p gain.

19) ETV6-NTRK3 fusion transcript

20) A newborn infant has not passed meconium for 48 hours. Imaging reveals a large sigmoid colon, contracted rectum and biopsy of the rectal mucosa and submucosa (rectal suction biopsy) reveals large nerves in the submucosa, no ganglion cells and marked acute inflammation throughout the biopsy. Which of the following is the most appropriate diagnosis?

    a) Necrotising enterocolitis
    b) Hirschsprung's enterocolitis
    c) Infective colitis
    d) Hirschsprung's disease
    e) Colonic atresia
    f) Congenital anorectal malformation

21) Which of the following statements about Hirschsprung's disease is NOT correct?

    a) Often associated with RET mutations
    b) Higher incidence in babies with trisomy 21 than chromosomally normal infants
    c) Only involves the rectum and colon
    d) Calretinin immunohistochemistry may be useful in diagnosis
    e) May be hereditary

22) A 3 month old child is undergoing a Duhamel procedure for operative treatment of Hirschsprung's disease. At operation, seromuscular biopsies are taken from the outside of the descending colon in an area that is believed to be proximal to the aganglionic segment. Frozen section reveals ganglion cells in the intermyenteric plexus but also occasional nerves. What should you tell the surgeon?

    a) The biopsy shows Hirschsprung's disease and anastomosis needs to be more proximal to the site from which the biopsies were taken.
    b) Ganglion cells present so an anastomosis at this site should function satisfactorily.
    c) Ganglion cells present but the definitive anastomosis requires to be more proximal.

23) A non-functioning kidney is resected from a 1 year old child. Macroscopically the resected kidney is large with multiple thin walled subcapsular cysts. On cross section, the centre of the kidney shows firm pale fibrous tissue with no definite pelvis. Microscopically the cysts are lined by flattened epithelium. The rest of the kidney is fibrotic with only occasional tubules and glomeruli. Occasional foci of cartilage and groups of haemopoietic cells are present. Towards the centre of the kidneys there are scattered primitive collecting ducts surrounded by concentric immature mesenchyme. What is the most likely diagnosis?

    a) Cystic nephroma
    b) Multicystic dysplasia
    c) Cystically partially differentiated nephroma
    d) Reflux nephropathy
    e) Polycystic kidney disease

24) A lobe of lung is resected from a 1 month old male infant following antenatal diagnosis of cystic lung lesion on ultrasound scan. The lobe shows multiple thin walled cysts ranging from 5 to 42mm in maximum dimension with no solid masses. Histologically the cysts are lined with columnar epithelium, some of which is ciliated, with occasional groups of mucous secreting cells. The cyst walls are composed of fibrous tissue, and focally, small islands of cartilage. What is the likely diagnosis:

a) Congenital pulmonary airway malformation (CPAM)
b) Pleuroblastoma
c) Congenital emphysema
d) Cystic bronchopulmonary dysplasia
e) Extralobar sequestration

Scenario for questions 25 & 26: A 4 year old child presents with a single raised pink lesion on the scalp.

25) The biopsy from the scalp lesion shows a diffuse dermal infiltrate of large cells with plentiful eosinophilic cytoplasm and lobulated nuclei, some of which contain grooves, scattered granulomas and a mixed inflammatory infiltrate with prominent numbers of eosinophils. What is the most likely diagnosis?

a) Langerhans cell histiocytosis
b) Juvenile xanthogranuloma
c) Epithelioid sarcoma
d) Deep granuloma annulare
e) Spitz naevus

26) Which one of the following immunostains is most likely to help make a definitive diagnosis?

a) CD68
b) CD45
c) CK8/18
d) CD1a
e) S100

27) A 1 year old child has a 2cm soft lump removed from the upper arm. Microscopy shows that it is composed of interlacing connective tissue septa, mature adipose tissue and areas of loose immature mesenchymal tissue. What is the diagnosis?

a) Inclusion body fibromatosis
b) Fibrous hamartoma of infancy
c) Desmoid type fibromatosis
d) Nodular fasciitis
e) Myofibromatosis
f) Inflammatory myofibroblastic tumour
g) Congenital/ infantile fibrosarcoma

28) A mass is resected from the lateral aspect of the neck of an eleven year old child. Microscopy reveals a tract extending through the dermis and subcutaneous tissue lined partly by granulation tissue and partly by squamous or ciliated columnar epithelium. There are large subepithelial lymphoid aggregates in the wall. What is the most likely diagnosis?

a) Branchial cleft sinus
b) Reactive lymph node
c) Thyroglossal duct sinus
d) Discharging epidermal cyst

PLACENTA QUESTIONS

29) A 28 year old mother delivers a markedly growth restricted baby at 34 week gestation. Placental histology reveals the following picture. What is the main lesion shown in this photomicrograph?

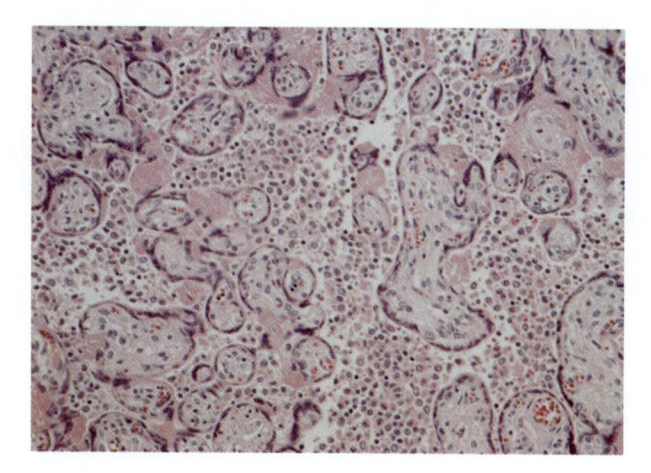

a) Infarction
b) Perivillous fibrin deposition
c) Chronic histiocytic intervillositis
d) Intervillous thrombosis
e) Chronic villitis
f) Fetal thrombotic vasculopathy

30 – 33) Extended matching item questions:

A) Chorangioma
B) Chorangiomatosis
C) Old placental infarct
D) Recent intervillous thrombus
E) Excavating placental haematoma
F) Subchorionic thrombus

For each of the following descriptions of a *placental mass*, select the **most appropriate** diagnosis from the list of options above. Each option may be used once, more than once, or not at all.

30) Brown mass with a laminated appearance located just under the fetal surface.

31) Dark red mass with a glassy surface in the middle part of the placenta

32) A large expansile nodular lesion at one edge of the placenta resulting in elevation of the fetal surface with a firm homogenous dark red cut surface.

33) A pale firm lesion demarcated from the adjacent villous tissue and extending from the maternal surface through half the thickness of the placental disc.

34) A baby with body weight on the 3rd centile is delivered at 36 weeks gestation. Which of the following placental findings would NOT explain the growth restriction?

a) Massive perivillous fibrin deposition
b) Chronic villitis
c) Necrotising funisitis
d) Fresh marginal haemorrhage

35) With which placental lesion can this decidual vascular abnormality be associated?

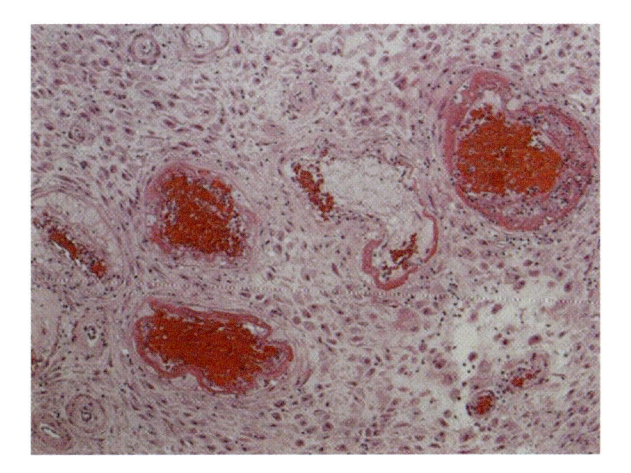

a) Infarction
b) Marginal haemorrhage
c) Intervillous thrombi
d) Fetal thrombotic vasculopathy
e) Subchorionic thrombus

36) A placenta from a baby delivered by emergency caesarean for abnormal CTG with fetal bradycardia in labour shows lymphocytes and plasma cells in many of the villi with sclerosis of the stroma of affected areas. Of the following, which is the most likely aetiology?

a) Herpes simplex infection
b) Group B Streptococcal infection
c) Rubella infection
d) Cytomegalovirus (CMV) infection
e) Enteroviral infection

37) A non-oedematous baby weighing 2046g is stillborn at 38 weeks gestation. The placenta is large, weighing 900g, and friable, with very large, congested and focally thrombosed chorionic vessels on the fetal surface and small cysts on the cut surface of the villous tissue. Which of the following lesions would explain the macroscopical findings in the placenta and the death of the baby?

a) Fetal thrombotic vasculopathy
b) Molar change in the placenta
c) Placental oedema
d) Chorangiomatosis
e) Placental mesenchymal dysplasia

38) In a hysterectomy specimen for massive postpartum haemorrhage two days after delivery, which of the following pathological findings would NOT explain the cause of the bleeding?

a) Necrotic tissue and blood clot lining the uterine cavity
b) Placental tissue in the myometrium
c) Partly thrombosed vessels in the myometrium
d) Acute inflammation of the myometrium

39 – 42) Extended matching item questions:

A) Acute chorioamnionitis
B) Subacute chorioamnionitis
C) Fetal thrombotic vasculopathy
D) Placental infarction
E) Acute villitis
F) Placental oedema
G) Massive perivillous fibrin deposition

For each of the following clinicopathological scenarios, select the most likely placental abnormality from the list above. Each option may be used once, more than once or not at all

39) 34 weeks gestation. Mother with hypertension and proteinuria and a growth restricted baby

40) 37 weeks gestation. Mother with diabetes and a placenta showing groups of completely sclerotic villi alternating with normally vascularised villi.

41) 16 weeks gestation. Mother with baby who has congenital cardiac abnormalities and history of recent amniocentesis.

42) 25 weeks gestation. Mother with history of prolonged rupture of membranes.

Correct answers and explanations:

1) Correct answer: e) Poorly differentiated neuroblastoma

Many cancers in children are morphologically different from typical carcinomas and sarcomas of adults. Like the example illustrated, they are composed of cells that recapitulate embryonic tissues, with darkly staining nuclei and very little cytoplasm. These are collectively referred to as the 'small round blue cell tumours of childhood'. They include tumours mimicking embryonic neural tissue (neuroblastoma) and skeletal muscle (rhabdomyosarcoma) as well as lymphomas and leukaemias and tumours of specific organs that have a small round blue cell component such as nephroblastoma (synonym Wilms' tumour), arising in the kidney and hepatoblastoma arising in the liver. Diagnostic morphological findings may not be present on light microscopy in small biopsies of these tumours and often require immunostaining and genetic testing in order to arrive at a precise conclusion.

In this case, the presence of pink fibrillary stroma (neuropil) allows diagnosis of a neuroblastoma. A few cells have vesicular nuclei with a single nucleolus indicating early differentiation towards ganglion cells. This change is present in less than 5% of the cells, so this is a *poorly differentiated* neuroblastoma. In *differentiating neuroblastomas* more than 5% of the tumour cells show ganglionic nuclear changes. *Undifferentiated neuroblastomas* have no obvious neuropil or ganglionic differentiation.

2) Correct answer: F) Primitive neuroectodermal tumour (PNET)

PNETs derive from neural crest cells and most occur in children, adolescents and young adults. Peripheral PNETs i.e. those occurring outside the CNS are now thought to be the same as Ewing's sarcoma because they have the same phenotype and chromosome translocation. PNETs show more neural differentiation the Ewing's sarcoma which is undifferentiated. Crisp membranous staining with CD99 is a typical finding in this family of tumours.

3) Correct answer: I) Hepatoblastoma

The description is typical of an epithelial type hepatoblastoma. This is an uncommon malignant tumour of very young children. The histological appearance of hepatoblastomas is highly variable dependent upon the constituent cell types. These include epithelial cells resembling foetal liver, small darkly staining embryonic cells and mesenchymal elements including spindle cells and osteoid. Extramedullary haemopoiesis is common and small amounts of bile pigment may be present in the tumour. One cell type may predominate or the tumours may be of mixed type. Hepatoblastomas are more common in infants and young children with familial adenomatous polyposis.

4) Correct answer: E) Embryonal rhabdomyosarcoma

These malignant tumours arise from primitive mesenchymal cells and most patients are less than 10 years old. Most common sites are the skeletal muscle of the head and neck and the genitourinary tract. Histological appearances vary from an almost undifferentiated small blue cell tumour to those containing considerable numbers of cells that resemble fetal skeletal muscle. Embryonal rhabdomyosarcomas are desmin positive and show *nuclear* positivity for Myo-D1 and myogenin.

5) Correct answer: C) Precursor T-acute lymphoblastic leukaemia / lymphoma

The cellular morphology is typical of lymphoblasts and CD3 and CD5 positivity is indicative of T cell origin. Nuclear positivity for TdT is indicative of an immature phenotype. The disease may present with a mass, typically in the thorax, or with leukaemia, the difference depending upon the degree of bone marrow involvement. Although currently precursor T-acute lymphoblastic leukaemia is considered to be the same as precursor T-acute lymphoblastic lymphoma, some recent studies have shown variation in degree of expression of some genes, dependent upon whether presentation is with a tissue mass or leukaemia.

6) Correct answer: D) Desmoplastic small round cell tumour

The description of dense fibrous stroma surrounding clusters of undifferentiated blue cells that show trilinear expression of epithelial, muscle and neural markers is characteristic of these highly malignant tumours. Although originally described in the abdomen, they are now known to occur at many sites throughout the body. Diagnosis is confirmed by demonstration of a unique chromosomal translocation that results in an EWS-WT1 fusion transcript.

7) Correct answer: C) Clear cell sarcoma (of the kidney)

The description is of a classic form of clear cell sarcoma of the kidney (CCSK). It is important to note that this is a **renal** tumour and must not be confused with clear cell tumour of soft tissue that is a completely different entity, mainly occurring in adults. CCSK needs to be distinguished from a stromal Wilms' tumour because of its resistance to conventional Wilms' tumour therapy. It is more common in boys and has a propensity to metastasise to bone (indeed it was originally described as 'bone-metastasizing renal tumor of childhood'). Variants include epithelioid, spindle cell, cystic and myxoid patterns. The prominent branching vascular pattern and lack of WT1 and cytokeratin staining are helpful distinguishing features. Unlike mesoblastic nephroma, a spindle cell renal tumour of infancy with a low propensity to metastasise, CCSK is rare in the first year of life.

8) Correct answer: H) Rhabdoid tumour

Rhabdoid tumours are highly malignant and may be difficult to diagnose especially in small biopsies especially in examples showing extensive necrosis and lacking cytoplasmic inclusions. These tumours are characterised by abnormalities of the INI1 gene on chromosome 22. As a result, unlike other childhood renal tumours, INI1 nuclear staining is negative. Rhabdoid tumours may occur at other sites including the CNS, liver and soft tissue. Most patients are less than 2 years old.

9) Correct answer: A) Mesenchymal hamartoma

This is a cystic lesion of the liver that usually presents as an abdominal mass before 2 years of age although examples in older children and even in adults have been reported. It is benign but often very large. A history of a rapidly growing mass is not uncommon, due to sudden accumulation of fluid within cystic spaces. The microscopical appearance of myxoid and fibrous tissue with bile ducts, small vessels and cysts are lined by epithelium or vascular endothelium is typical.

10) Correct answer: D) Ganglioneuroma

Ganglioneuromas are benign neural tumours that may be primary or due to secondary maturation of a neuroblastic malignancy, the latter often after chemotherapy. The stroma is composed of S100 positive Schwann cells that need to be distinguished from fibrillary neuropil seen in neural tumours that retain an immature, neuroblastic component. Tumours showing Schwannian stroma and ganglion cells, but also groups of neuroblasts, are referred to as ganglioneuroblastomas and are **not** benign. A firm diagnosis of ganglioneuroma should not be made on a needle biopsy. A resected specimen requires wide sampling to exclude the presence of neuroblastic elements before it is concluded that the tumour is a benign ganglioneuroma.

11) Correct answer: e) Burkitt lymphoma

The description of diffuse sheets of lymphoid cells in not compatible with either follicular hyperplasia or follicular lymphoma (the latter is extremely rare in children).

The cellular morphology is typical of lymphoid blasts but the absence of nuclear TdT is indicative of a more mature phenotype than precursor lymphoblastic lymphoma. CD20 and CD79a positivity are indicative of a B cell neoplasm. The cellular morphology, immunostaining pattern and high mitotic rate with numerous tingible body macrophages are typical of Burkitt lymphoma. The cell size and description would be against diffuse large B cell lymphoma or classical Hodgkin's lymphoma.

12) Correct answer: d) Neuroblastoma

Genetic studies are currently essential to decide optimum treatment for a neuroblastoma patient and give information about prognosis and are best done on fresh frozen tissue. If fresh tissue is not available it may be necessary to re-biopsy a patient to obtain material for genetic studies, even if the histological diagnosis is known. The potential translocations associated with alveolar rhabdomyosarcoma are also helpful for correct treatment stratification but generally can be done on paraffin-processed material. PNETs and anaplastic large cell lymphomas can show chromosomal rearrangements and these genetic changes may be useful for diagnosis, but at the present time are not essential for determining treatment or prognosis.

13) Correct answer: e) Wilms' tumour (nephroblastoma)

The photomicrograph shows the typical triphasic appearance of classical Wilms' tumour, the three elements being: i) a small undifferentiated blue cell component (known as blastema) ii) epithelial tubules many of which appear embryonic and iii) stroma. The proportion of each component in an individual tumour is very variable. The epithelial elements may show squamous or mucinous differentiation and skeletal muscle, and occasionally adipose tissue cartilage and/or bone, may be present in the stromal component. Cystic change is not uncommon. Hence the gross and histological appearances of Wilms' tumours are very variable.

14) Correct answer: f) Anaplastic Wilms' tumour (nephroblastoma)

Although this looks completely different from the example in Question 13, it is also a Wilms' tumour, an example of this neoplasm showing skeletal muscle differentiation in the stromal component. There is also marked enlargement and hyperchromasia of some of the cells and abnormal mitotic figures, features that are indicative of anaplasia. If this change is present diffusely through the tumour, which can only be assessed in resected neoplasms not biopsies, it signifies a worse prognosis.

The variety of metaplastic changes in the epithelium and stroma in Wilms' tumour means it is referred to as 'the great mimicker'. In young children, primary renal teratomas are virtually unknown, most apparent examples being due to misdiagnosis of Wilms' tumour. Of the other renal tumours listed, mesoblastic nephroma is an infiltrating spindle cell tumour of infants that occurs in classical and cellular forms and adenofibroma is a very rare benign renal tumour, definitive diagnosis of which should not be made without the opinion of a specialist children's renal tumour pathologist.

15) Correct answer: b) Intrarenal vascular invasion (is NOT essential)

Currently in the UK and Europe, local staging of Wilms' and other childhood renal tumours is determined on resected kidneys, usually after chemotherapy, according to the criteria of SIOP (Societe Internationale D'oncologie Pediatrique) as endorsed by the Royal College of Pathologists. Based on histopathological findings, tumours are divided into stages I to IV but are also histologically stratified with regards to risk (low, intermediate or high). Thus the morphological characteristics must be recorded as well as all the variables listed and most are important in risk stratification, but only a, c, d, e and f determine the tumour ***stage***.

Stage I tumours are surrounded by the renal capsule or an intact pseudocapsule. Viable tumour may infiltrate but does not extend through the capsule or pseudocapsule into the perirenal fat (a). Necrotic tumour outside the capsule or in the renal sinus does not upstage the tumour *provided* it does not reach the resection margins (f).

Stage II : Viable tumour is present in the perirenal fat or in the renal sinus soft tissue, including in vessels and pelvis, or adjacent organs or vena cava, but this is completely excised (c, e). Necrotic tumour must also be fully excised (f)

Stage III: Tumour, viable or non-viable, is present at a resection margin (e,f). Tumour rupture before or during an operation irrespective of other criteria upstages the tumour to a III. Tumour in a lymph node even if showing necrosis and chemotherapy effect also upstages the lesion to a III (d).

Stage IV tumours have haematogenous metastasis or positive lymph nodes out with the abdomino-pelvic region.

Morphology of childhood renal tumours and extent of necrosis are important for assigning risk. Low risk tumours are cystic partly differentiated nephroblastoma, completely necrotic tumours and mesoblastic nephroma. Intermediate risk tumours are epithelial, stromal or mixed Wilms' tumours including those showing focal anaplasia and/or regression (65-99% necrosis). High risk tumours are rhabdoid tumour and clear cell sarcoma of kidney, Wilms' tumours with diffuse anaplasia and blastemal type Wilms' tumours after chemotherapy.

16) Correct answer: B) Alveolar rhabdomyosarcoma

PAX-3-FKHR (also known as PAX-3-FOX01) results from translocation and fusion of the PAX 3 gene on chromosome 2 and a forkhead gene on chromosome 13. PAX-7-FKHR (also known as PAX-7-FOX01) involves the PAX-7 gene on chromosome 1 and the same forkhead gene on chromosome 13. This genetic mutation is diagnostically useful in some cases of rhabdomyosarcoma, especially in young children with a solid form of embryonal rhabdomyosarcoma. Failure to identify the translocation favours embryonal rather than alveolar rhabdomyosarcoma although 20% of alveolar rhabdomyosarcomas lack either of these forkhead gene translocations.

17) Correct answer: F) Primitive neuroectodermal tumour (PNET)

The EWS-FLI1 fusion transcript results from rearrangement of genetic material between the EWSR1 gene on chromosomes 22 and the *FLI1* gene on chromosome 11. The protein produced from the EWSR1-FLI1 fusion gene is called EWS/FLI, and functions as an aberrant transcription factor. It is likely to be of aetiological importance in causation of this malignancy by disrupting normal cellular growth controls. It is the most common mutation in both PNETs and Ewing's sarcoma.

18) Correct answer: C) Neuroblastoma

MYCN amplification, 1p deletion, 17p gain, although not thought to *cause* neuroblastoma, are associated with a poorer prognosis and neuroblastoma patients whose tumours are MYCN amplified are given more intensive chemotherapy.

19) Correct answer: A) Congenital/infantile fibrosarcoma

ETV6-NTRK3 gene fusion results from the translocation of genetic material between chromosomes 12 and 15 and its presence is diagnostically useful in congenital /infantile fibrosarcoma. This is a rare fibrous tissue neoplasm of early childhood and, in spite of its name, is not always recognised at birth. It exhibits rapid growth and extensive local invasion but unlike fibrosarcoma in adults, metastases are infrequent. It is rarely fatal although is relatively insensitive to chemotherapy requiring extensive surgery up to and including amputation. This gene fusion is also found in cellular mesoblastic nephroma, a spindle cell tumour of the infantile kidney that may also be congenital and is rarely diagnosed after the first year of life. Secretory breast cancer shares the same genetic marker.

20) Correct answer: b) Hirschsprung's enterocolitis

Hirschsprung's disease is congenital absence of ganglion cells in the rectum and varying lengths of bowel proximal to the rectum. In short segment disease no ganglion cells are present in the submucosal and intermyenteric nerve plexuses in the rectum and varying lengths of the sigmoid colon. In long segment disease the aganglionic segment can extend to the splenic or hepatic flexure. In a small number of cases there is total aganglionosis of the colon. Very rarely part or all of the small intestine is involved. The latter is lethal in the absence of an intestinal transplant.

The lack of ganglion cells in the biopsy described is typical of Hirschsprung's disease but there is also acute inflammation. This justifies a diagnosis of Hirschsprung's enterocolitis. None of the other options would be expected to be associated with a lack of ganglion cells. Hirschsprung's associated enterocolitis usually presents in the newborn period with failure to pass meconium, abdominal distension and foul smelling stool. It is a major cause of morbidity and can also occur following operative treatment of Hirschsprung's disease if the bowel re-obstructs.

21) Correct answer: c) Only involves the rectum and colon (is INCORRECT)

Hirschsprung's disease is genetically heterogeneous with several genes playing a role in its pathogenesis. However the main susceptibility is RET gene mutation(s) on chromosome 10 (a). An affected child may inherit a causative gene (e). Some congenital syndromes confer an increased risk of Hirschsprung's disease, including trisomy 21 (b). The length of the affected segment is very variable and rarely the whole intestine may be aganglionic, so (c) is incorrect.

Acetylcholinesterase staining in frozen sections is useful for diagnosis, showing an increased thickness and number of nerve fibrils in the lamina propria, sometimes with abnormal nerve fibrils in the lamina propria. However this is not essential for diagnosis. The essential diagnostic criterion is a demonstration of absence of ganglion cells after study of adequate material, but the presence of abnormally large submucosal nerves is a useful pointer. In biopsies of rectum that do not include the muscularis propria, this usually means at least 50 serial sections of a biopsy with plentiful submucosa have to be examined. Recently calretinin immunohistochemistry in routine paraffin sections has been suggested as a satisfactory replacement for acetylcholinesterase enzyme histochemistry, with staining being absent in areas of aganglionosis (d). However experience with calretinin staining is still relatively limited.

22) Correct answer: c) Anastomosis needs to be more proximal

The presence of a large nerve in the myenteric plexus indicates that the site from which the biopsy has been taken is in so called 'transitional zone'. This is a variable length of bowel between that which is totally aganglionic and that which is morphologically normal. Transitional zone is characterised, as described here, by the presence of ganglion cells but also large hypertrophied nerves of the type that are typically seen in biopsies from aganglionic bowel. Transitional zone bowel may or may not function normally so a definitive anastomosis should be made with bowel that appears fully ganglionic.

23) Correct answer: b) Multicystic dysplasia

When unilateral, this is generally a sporadic developmental abnormality resulting from obstruction of the urinary collecting system during development. Bilateral multicystic dysplasia is part of several congenital syndromes but in these cases, the large cysts tend to be more centrally located. The presence of foci of cartilage, although not always seen, is a useful diagnostic pointer to renal dysplasia as are the presence of abortive collecting ducts surrounded by immature mesenchyme.

Reflux nephropathy is due to incompetence of the vesicoureteric junction and if this is present *in utero* prior to 36 weeks gestation, the kidney may show features of renal dysplasia. However cysts are not a prominent feature and the pelvicalyceal system is present and dilated with segmental scarring of renal tissue over dilated calyces.

Cystic partially differentiated nephroblastoma is a lesion in the Wilms' tumour spectrum with variable sized cysts replacing the kidney and showing embryonic elements in the septa between cysts that do not form solid nodules. Cystic nephroma is a benign tumour which, although macroscopically similar to cystic partially differentiated nephroblastoma, shows only mature elements in the septa. There is dispute as to whether this is at the benign end of the Wilms' spectrum or should be regarded as a separate lesion.

24) Correct answer: a) Congenital pulmonary airway malformation

Congenital (cystic) pulmonary airway malformations, previously called congenital cystic adenomatoid malformations, are of varying types. The description here is of a type I lesion which is the most common and usually confined to one lobe. It is generally an isolated finding with no other congenital malformations. Type II has smaller cysts, up to 2cm in diameter with no mucous secreting cells or mural cartilage, although bands of striated muscle may be present below the epithelium. Type II is commonly found in babies with other congenital malformations. Type III is very much less common and may occupy one or more lobes and is solid and rubbery with small cysts reminiscent of immature lung at the canalicular stage of development with thick fibrous septa between cysts. Congenital cystic pulmonary malformations have to be distinguished from childhood forms of pulmonary blastoma (pleuropulmonary blastoma) which, at the benign end of the spectrum, frequently has large cysts and may be similar to a congenital cystic pulmonary airway malformation Type 1. The difference is the presence of collections of primitive cells below the epithelium in the pulmonary blastoma.

Pulmonary sequestrations are composed of lung tissue lacking a connection to the bronchial tree and usually having an anomalous systemic arterial supply. They may be within the lung (intralobar) or outside the lung (extralobar). They may show features indistinguishable from congenital pulmonary airway malformation in which case only the systemic arterial supply allows the diagnosis of sequestration.

25) Correct answer: a) Langerhans cell histiocytosis

The large cells with plentiful eosinophilic cytoplasm and indented nuclei are the characteristic of Langerhans cell histiocytosis (LCH). An infiltrate of eosinophils is a typical finding and granulomas composed of epithelioid macrophages may be prominent in such lesions.

Cutaneous juvenile xanthogranuloma differs in that the lesional cells are more typical macrophages with foamy cytoplasm and these lesions contain scattered multinucleate giant cells of Touton type, the latter characterised by a ring of nuclei around an island of central eosinophilic cytoplasm and surrounded by a rim of peripheral lipid laden foamy cytoplasm.

Another histiocytic skin lesion that needs to be distinguished from LCH is superficial granuloma annulare that shows a central area of degenerate collagen in the dermis surrounded by palisaded histiocytes and a lymphocytic infiltrate that is largely perivascular. Epithelioid sarcoma tends to arise in older adolescents and young adults and consists of largely subcutaneous nodules of epithelioid type cells around areas of central necrosis. Spitz nevi arise in the junctional zone of the epidermis and the cells are arranged in large nests of epithelial cells with vesicular round or oval nuclei, frequently with giant cells in superficial parts of the lesion.

26) Correct answer: d) CD1a

Unlike ordinary histiocytes, Langerhans cells show positive immunostaining with CD1A. The lesional cells will also stain with S100 but this is not as specific as CD1A. Many reactive cells in the lesion will be CD45 or CD68 positive.

27) Correct answer: b) Fibrous hamartoma of infancy

This is a spot diagnosis. Although the mesenchyme is immature and the edge of these lesions is infiltrative, as its name suggests, the lesion is benign. Even if not fully excised it rarely recurs. Diagnostic difficulty can come with older lesions in which the mesenchymal component is less marked. Inclusion body fibromatosis is confined to the digits and desmoid type fibromatosis, although occurring in children, only has one tissue element as does the rare congenital fibrosarcoma. As its name suggests, inflammatory myofibroblastic tumour has a prominent chronic inflammatory cell infiltrate. Infantile myofibromatosis has a zonal pattern with a peripheral area of plump eosinophilic spindle cells and a central component of thin walled branching vessels (so called haemangiopericytomatous pattern). None of the lesions listed apart from fibrous hamartoma of infancy contain adipose tissue.

28) Correct answer: a) Branchial cleft sinus

This is a developmental abnormality with varying appearances. Branchial cleft sinuses can show a heavy lymphoid infiltrate including formation of follicles with germinal centres but the presence of an epithelial component and lack of a defined capsule or subcapsular sinus differentiates them from lymph nodes.

Thymic cysts are also found in the lateral aspect of the neck but require demonstration of thymic tissue in their walls for diagnosis. Thyroglossal duct remnants present in the midline and thyroid acini may be present in the wall. In the absence of thyroid tissue thyroglossal duct lesions may resemble branchial sinuses, being distinguished by the site in the neck from which the lesion has been excised.

29) Correct answer: c) Chronic histiocytic intervillositis

The photomicrograph shows clusters of monomorphic mononuclear cells within the intervillous space with a small amount of perivillous fibrin. These cells are CD68 positive monocytes/macrophages and this is the typical appearance of massive chronic histiocytic intervillositis. This lesion can be seen in placental malaria but in such cases malarial pigment is seen within the macrophages. In the absence of malaria this is an idiopathic condition that is well recognised to be associated with recurrent miscarriage, growth restriction and stillbirth. It is not associated with infection but may have an underlying immunological pathogenesis and can recur in future pregnancies.

30) Correct answer: F) Subchorionic thrombus

The appearance and colour of mass lesions in the placenta depends on their cause but also their duration. Different types of lesions typically occur in different parts of the placenta. The description in this question is typical of a subchorionic thrombus. As suggested by its name, subchorionic thrombi are located below the chorion on the fetal surface and, depending on size, may extend partly or completely through the thickness of the placental disc. The lesions are typically laminated and, histologically, no villi are included in the thrombus. Small subchorionic thrombi are not uncommon but massive thrombi that extend through the placental disc to the maternal aspect separate the chorionic plate from the villous tissue over a large area and can result in intrauterine death.

31) Correct answer: D) Recent intervillous thrombus

These lesions are located in the centre of the villous tissue and as they age may show lamination. Lesions that are older still appear as a laminated white plaque in the villous tissue. As with subchorionic thrombi, histology reveals no villi in the centre of the thrombus. They are composed of a mixture of maternal and foetal blood and would appear to develop from localised rupture of villous syncytiotrophoblast over a dilated foetal villous capillary. When multiple they may result in significant foetomaternal haemorrhage but small isolated lesions are not of pathological significance.

32) Correct answer: A) Chorangioma

Chorangiomas unlike most other placental masses, are typically well circumscribed spherical lesions usually protruding from the foetal surface of the placenta and commonly located at its edge. They are composed of capillary and larger vascular channels and intervening stroma, surrounded by trophoblast. The cut surface is very variable dependent on the degree of degenerative changes, thrombosis and infarction occurring within the lesion. Although regarded by some as benign tumours, they are probably best regarded as hamartomas arising as a result of malformation of the primitive angioblastic tissue of the placenta. Most chorangiomas are small but large lesions more than 4cms in diameter may result in foetal hydrops, polyhydramnios and foetal anaemia.

33) Correct answer: C) Old placental infarct

Placental infarcts are localised areas of ischemic necrosis. They are typically triangular with the base on the maternal surface. Their appearance varies with their duration; fresh placental infarcts are dark red but can be difficult to distinguish before the placenta is fixed, although the infarcted tissue is firmer than the surrounding viable villi. With time, the colour changes from dark red to dull red to brown and finally, after at least two weeks, to white. Most infarcts are due to necrosis and/or thrombosis of a maternal spiral artery in the basal decidua that supplies the affected area. Some cases result from rupture of spiral arteries and retroplacental haemorrhage, the latter stripping the placenta away from the decidua thus depriving the villous tissue of a maternal blood supply.

34) Correct answer: d) Fresh marginal haemorrhage (is INCORRECT)

a) Perivillous fibrin deposition in the placenta is very common and it is only significant if it is of a massive degree, present in a large percentage of several sections. The percentage of the placenta that has to be involved for diagnosis varies with different authors but is of the order of 50%. It is characterised by amorphous eosinophilic material surrounding and separating villi and completely obliterating the intervillous space in affected areas. The entrapped villi in established lesions lose their covering syncytiotrophoblast and there is progressive stromal fibrosis. Cytotrophoblast cells persist and are scattered in the fibrin. The cause of massive perivillous fibrin deposition is often unknown and there is dispute as to whether loss of syncytiotrophoblast is the primary problem or if this is secondary to fibrin deposition. Some mothers have thrombophilias and occasional cases have followed maternal Coxsackie viral infection. Massive perivillous fibrin deposition results in a significant reduction in functioning villous tissue hence fetal growth restriction or even IUD is a frequent consequence. The lesion may recur in future pregnancies.

b) Chronic villitis is characterised by an infiltrate of chronic inflammatory cells within groups of villi throughout the placenta, accompanied by varying degrees of stromal fibrosis. As a result, inflamed villi are rendered non functional and if a sufficient proportion of the placenta is involved, intrauterine growth restriction may develop. Chronic villitis may be caused by a wide variety of pathogens that reach the placenta via the maternal bloodstream or result from spread from pelvic infections. However in many cases of chronic villitis no specific infection is identified and these cases are referred to a chronic non-specific villitis (synonym villitis of unknown aetiology). Some of these cases may be due to a subclinical undiagnosed intrauterine infection but it is suggested that a proportion of cases have an underlying immunological pathogenesis due to a maternal immune reaction against trophoblast antigens. In cases that are not due to infection, chronic non-specific villitis may recur in future pregnancies.

c) Necrotising funisitis is a form of chronic inflammation of the umbilical cord in which concentric bands of necrotic, often calcified, inflammatory debris surround the umbilical vessels. Secondary occlusion of umbilical vessels can lead to intrauterine growth restriction. Necrotising funisitis can be seen in association with a wide variety of infections although in most case by the time the placenta is delivered, the infecting pathogen can no longer be identified. Necrotising funisitis used to be considered to be pathognomonic of syphilis but this is no longer the case.

The other condition in the list (d) does not cause growth restriction. Marginal haematomas referred to as subchorionic haematomas in the obstetric literature appear as a crescentic mass of blood clot at the placental edge. They may extend onto the maternal surface of the placenta but, unlike retroplacental haemorrhages, do not indent the surface of the placental disc. Marginal haemorrhages may cause antepartum vaginal bleeding and, occasionally, early labour but do not affect placental function or result in intrauterine growth restriction. The cause of marginal haemorrhage is often unknown although they may be associated with implantation of part of the placenta into the lower uterine segment.

35) Correct answer: a) Infarction

The photomicrograph shows the classical picture of a maternal spiral artery vasculopathy with fibrinoid necrosis of the walls of these vessels and presence of foamy macrophages. This lesion is characteristic of maternal preeclampsia and occlusion of necrotic vessels results in placental infarction. These necrotic vessels may also rupture resulting in retroplacental haemorrhage.

In the early part of normal pregnancies, maternal spiral arteries in the decidua and myometrium below the placental disc are invaded by trophoblast cells. As a result, the muscle of their walls degenerates and is replaced by fibrinoid material, a process known as 'normal physiological conversion'. As a result, the lumen of converted spiral arteries dilates and allows a vast increase in maternal blood flow to the intervillous space in the placenta. Furthermore, loss of muscle means that these vessels do not narrow in response to factors that cause maternal peripheral vasoconstriction. In some mothers, particularly in first pregnancies, this change in the spiral arteries does not occur in some placental bed spiral arteries or is incomplete. Vessels that retain a muscular wall limit maternal blood flow to the placenta and are at risk of the development of a vasculopathy, of the type shown. This vasculopathy does not occur in spiral arteries that have undergone normal physiological conversion.

Marginal haemorrhage is thought to be venous in origin, intervillous thrombi result from breaks in the villous syncytiotrophoblast and the cause of subchorionic thrombus is often unknown but is not the result of pathological changes in maternal spiral arteries.

36) Correct answer: d) Cytomegalovirus infection

Herpes Simplex, Rubella and Enterovirus may cause chronic villitis but infection with these pathogens is not usually associated with the presence of plasma cells. In contrast, plasma cells in the inflammatory infiltrate in chronic villitis are commonly seen in association with Cytomegalovirus infection. The latter typically also shows well formed or degenerate viral inclusions. Group B streptococcal infection more commonly causes acute chorioamnionitis than villitis. In cases where the infection has spread to the placenta via the maternal blood stream it may result in villitis, but the inflammation is acute not chronic.

37) Correct answer: e) Placental mesenchymal dysplasia

The description given is typical of this unusual placenta abnormality. Histologically, the characteristic finding is markedly enlarged stem villi that show overgrowth of fibroblastic and mesenchymal stromal tissue, oedema and cistern formation, the latter giving the appearance of cysts on naked eye examination. Large stem villi may also show proliferation of dilated, abnormally branching, sinusoidal blood vessels especially at their edges. The stem villi are covered with trophoblast showing no evidence of hyperplasia. In contrast to stem villi, terminal villi are generally normal although can show some immaturity or ischaemic changes. The lesion may be associated with Beckwith-Wiedemann syndrome, a congenital overgrowth syndrome due to a genetic abnormality affecting the short arm of chromosome 11. Even in cases not associated with Beckwith-Wiedemann syndrome and chromosome 11 abnormalities, the lesion is likely to have a genetic aetiology.

Fetal thrombotic vasculopathy may be associated with thrombosis of large chorionic vessels but does not show the other features mentioned.

Molar change in the placenta is generally diagnosed earlier, in a pregnancy without a formed fetus, and is not associated with large chorionic vessels.

Non-specific placental oedema is usually seen in situations when the baby is oedematous and involves all villi and lacks the structural changes of placental mesenchymal dysplasia.

Chorangiomatosis is diagnosed by the presence of multiple nodules of chorangioma-like tissue but does not cause the changes described in the stem villi.

38) Correct answer: a) Necrotic tissue and clot lining the uterine cavity (is INCORRECT)

Placental tissue in the myometrium, without intervening decidua, is a typical finding in placenta accreta, increta and percreta, collectively known as morbidly adherent placenta. The three categories refer to the depth of invasion of the myometrium of villous tissue. In placenta accreta villi are present in the superficial myometrium, in placenta increta villi extend deeply into the myometrium and in placenta percreta villi extend through into the uterine serosa. These lesions can cause antepartum vaginal spotting but also massive postpartum haemorrhage.

Partly thrombosed vessels in the myometrium with parts of the myometrial spiral arteries still containing blood is suggestive of subinvolution of the placental bed, a condition that is well known to be associated with massive postpartum haemorrhage.

Acute inflammation in the superficial layers adjacent to the uterine cavity would be expected following delivery but acute inflammation extending well into the myometrium is suggestive of a superadded infection.

Necrotic tissue lining the uterine cavity and old blood clot in the uterine lumen would be expected after any delivery and *per se* would not be satisfactory explanation for massive postpartum haemorrhage.

39) Correct answer: D) Placental infarction

The clinical description is typical of preeclampsia and this is characteristically associated with placental infarction.

40) Correct answer: C) Foetal thrombotic vasculopathy

Groups of sclerotic villi in an otherwise well vascularised placenta, in the absence of chronic inflammation, are indicative of thrombosis of a foetal stem vessel or chorionic vessel supplying the affected part of the placenta. In some cases, the occluding vessel cannot be demonstrated. The cause of thrombosis of foetal vessels in the placenta may not be identified but is associated with familial thrombophilias and maternal diabetes.

41) Correct answer: A) Acute chorioamnionitis

Any invasive procedure, such as amniocentesis, carries the risk of introduction of pathogens into the gestational sac that can result in infection and acute inflammation of the amnion and chorion.

Subacute chorioamnionitis would defer to a more longstanding inflammatory process, often with pathogens of low virulence.

Acute villitis is more likely to result from haematogenous spread of infection to the placenta than from a complication of amniocentesis.

42) The most likely correct answer is: B) Subacute chorioamnionitis

Prolonged rupture of membranes predisposes to ascending infection. Infection with virulent pathogens such as E.coli or group B streptococci usually results in early onset of labour and rapid delivery. However prolonged rupture of membranes may predispose to recurrent bouts of infection with low grade pathogens resulting in chronic inflammation of the membranes characterised by the presence of mononuclear inflammatory cells and nuclear debris as well as neutrophil polymorphs.

CHAPTER 18

GENERAL PATHOLOGY
Dr. Charanjit Kaur

1) Hypertrophy is an increase in the size of cells and thus the organ. It can be physiological or pathological and is the result of:

   a) Increased production of cellular proteins
   b) Increased water content in the cell
   c) Inadequate nutrition Influx of calcium

2) Atrophy is reduction in cell size of an organ or tissue resulting from a decrease in the cell size and number. Atrophy can also be physiologic and pathologic. Possible causes of atrophy include :

   a) Decreased workload or disuse
   b) Loss of innervation
   c) Diminished blood supply
   d) Inadequate nutrition
   e) Loss of endocrine stimulation
   f) Pressure
   g) All of the above

3) Metaplasia is a reversible change in which one differentiated cell type is replaced by another cell type. Metaplasia can involve:

   a) Epithelial cells
   b) Mesenchymal cells
   c) Red blood cells
   d) White blood cells
   e) Epithelial and mesenchymal cells

   4-10) Extended matching item questions:

   A) Coagulation necrosis
   B) Liquefactive necrosis
   C) Gangrenous necrosis
   D) Caseous necrosis
   E) Fat necrosis
   F) Fibrinoid necrosis
   G) Avascular necrosis

The morphological appearance of necrosis is the result of denaturation of intracellular proteins and enzymatic digestion of lethally injured cells. For each of the following patterns of necrosis, select the **most appropriate** aetiopathogenesis.

4) Areas of fat destruction due to trauma or action of lipases particularly pancreatic lipases in acute pancreatitis.

5) Form of necrosis seen in immune reactions involving blood vessels when antigen-antibody complexes are deposited in the wall of the arteries.

6) Friable white necrosis which comprises amorphous granular debris of lysed cells with granulomatous inflammation.

7) Ischaemic necrosis of tissues with superadded bacterial infection, resulting in abscess formation..

8) Ischaemic cell death, with de-naturing of cellular proteins, but preservation of cellular structures.

9) Cell death due to localized infections, by bacteria or fungi, resulting in a homogenous viscous mass.

10) Necrosis of bone due to ischaemia, with dead bony trabeculae containing empty lacunae.

11) Tightly regulated processes maintain blood in a fluid state in normal vessels and also permit formation of a clot at the site of vascular injury. Select the three components that regulate both haemostasis and thrombosis:

    a) Oxygen, water and blood
    b) Endothelium, platelets and coagulation cascade
    c) Fibrinogen, red blood cells and endothelium
    d) Factor V, fibrin and platelets
    e) Thrombin, prostaglandins and neutrophils

12) Which of the following statements is true?

    a) Thrombophilia is due to deficiency of factor VIII.
    b) Red infarcts occur with end arterial occlusions in solid organs.
    c) Antiphospholipid antibodies induce a hypercoagulable state in vivo but in vitro they inhibit coagulation.
    d) Leiden mutation is a genetic defect of the factor VI gene, resulting in a hypercoagulable state.

13-16) Extended matching item questions:

    A) Immediate (type I) hypersensitivity
    B) Antibody mediated (type II) hypersensitivity
    C) Immune complex mediated (type III) hypersensitivity
    D) Cell mediated (type IV) hypersensitivity

Match the prototypic disorders to the **associated** hypersensitivity immune reaction.

13) Contact dermatitis, multiple sclerosis, type I DM, rheumatoid arthritis, inflammatory bowel disease, tuberculosis

14) Anaphylaxis, bronchial asthma, allergies

15) Autoimmune hemolytic anaemia, Goodpasture's syndrome

16) Systemic lupus erythematosus, serum sickness, Arthus reaction

17-23) Extended matching item questions:

A) AL (amyloid light chains)
B) AA (serum amyloid associated protein)
C) A beta2 microglobulin (beta2 microglobulin)
D) ATTR (transthyretin)
E) A Cal (calcitonin)
F) AIAPP (islet amyloid peptide)
G) AANF (atrial natriuretic factor)

Amyloidosis is a systemic disease related to abnormal protein folding and possibly abnormalities in the immune system. Amyloid is not a chemically distinct entity. Several biochemical forms exist. For each of the following underlying disorders, select the **most appropriate** associated biochemical subtype of amyloid.

17) Chronic inflammatory conditions

18) Medullary carcinoma of thyroid

19) Systemic senile amyloidosis

20) Chronic renal failure

21) Isolated atrial amyloidosis

22) Type 2 diabetes

23) Multiple myeloma

24) Which one of the following statements is incorrect:

   a) MSH2, MLH1 and MSH6 are tumor suppressor genes.
   b) Hereditary non polyposis colon cancer is an autosomal dominant condition caused by inactivation of DNA mismatch repair genes.
   c) Familial adenomatous polyposis is an autosomal dominant disorder caused by mutation of the adenomatous polyposis coli (APC) tumour suppressor gene.
   d) Mutations in p16 and INK4A are associated with a familial predisposition to melanomas.
   e) MEN-1 and MEN-2 are associated with mutations in BRCA1 and BRCA2 genes.

25) Which one of the following statements about p53 are false?

   a) It is a tumour suppressor gene.
   b) It is commonly known as the "guardian of the genome".
   c) Inherited mutations of the gene are associated with Li-Fraumeni syndrome.
   d) p63 and p73 are collaborators of p53.
   e) p53 is independent of MDM2 gene.

26) Specific chromosomal abnormalities have been identified in most leukaemias and lymphomas. In Burkitt's lymphoma which of the following translocations is /are present?

a) 8; 14
   b) 14; 18
   c) 11; 22
   d) 9; 22
   e) 8; 21

27) The paraneoplastic syndrome of inappropriate antidiuretic hormone secretion is most commonly associated with which one of the following underlying cancers?

   a) Breast cancer
   b) Renal carcinoma
   c) Gastric carcinoma
   d) Small cell carcinoma of lung
   e) Pancreatic carcinoma

28) Cancer cachexia is due to which one of the following:

   a) Reduced basal metabolic rate
   b) Increased basal metabolic rate
   c) Starvation
   d) Increased exertion
   e) Increased breathlessness
   f) Nutritional demands of the tumour(s)

29) Metastasis from the following malignant tumours are well recognised, except for:

   a) Renal cell carcinoma
   b) Bronchogenic carcinoma
   c) Colorectal carcinoma
   d) Melanoma
   e) Basal cell carcinoma

30) In septic shock the tissue hypoperfusion is due to:

   a) Fluid loss
   b) Peripheral vasodilatation due to endothelial injury
   c) Failure of myocardial pump
   d) Anaphylaxis

31) The apple green birefringence seen with polarized light and Congo red stain is due to:

   a) Beta -pleated structure of amyloid
   b) Light chains in amyloid
   c) Transthyretin in amyloid
   d) Beta - macroglobulin in amyloid
   e) Serum amyloid associated protein

32) Sago spleen is the morphological appearance of spleen involved with:

   a) Follicular non-Hodgkin's lymphoma

b) Infective endocarditis
c) Tuberculosis
d) Amyloidosis
e) Malaria

33) In cardiac amyloidosis, amyloid deposition may occur in:

a) Sub-endocardium
b) Between myocardial fibers
c) Conduction system
d) Coronary arteries
e) All of the above

34) T cell immunodeficiency in HIV infection is due to:

a) Infection of B cells with HIV virus and direct cytopathic effect.
b) Loss of immature precursors of CD8+ cells due to direct infection of thymic progenitor cells.
c) Fusion of infected and uninfected T cells which leads to ballooning and cell death.
d) All of the above.

35) What type of microscopic necrosis would you expect to see in a cerebral infarction?

a) Liquefactive necrosis
b) Caseation necrosis
c) Coagulative necrosis
d) Fibrinoid necrosis
e) Avascular necrosis

36) Giant cell arteritis is vasculitis involving all of the following, except:

a) Lymphatics
b) Small caliber blood vessels
c) Medium sized blood vessels
d) Large blood vessels

37) C reactive protein, fibrinogen and serum amyloid A (SAA) protein are:

a) Cytokines
b) Coagulation factors
c) Carcinogens
d) Acute phase reactants
e) Immune complexes

38) Prostatic gland hyperplasia is an example of:

a) Compensatory hyperplasia
b) Physiological hyperplasia
c) Pathological hyperplasia
d) None of the above

39) Left ventricular hypertrophy is an example of:

   a) Compensatory hypertrophy
   b) Physiologic hypertrophy
   c) Pathological hypertrophy
   d) None of the above

40) Which of the following statements on dysplasia are correct?

   a) Only used to describe change in epithelial cells
   b) Identified via changes in architectural atypia only
   c) Can be a premalignant lesion
   d) All of the above

41) Which of the following statements about Marfan syndrome are correct?

   a) It is an autosomal recessive disorder
   b) It results from an inherited defect in extracellular glycoprotein fibrillin
   c) It results from an inherited defect in synthesis or structure of fibrillar collagen
   d) It is manifested by changes in the skeleton only

42) The type II major histocompatibility complex is present on which of the following cells:

   a) Red blood cells
   b) Neutrophils
   c) B lymphocytes
   d) T lymphocytes
   e) Eosinophils

43) Which of the following statements are **incorrect**?

   a) Burton's agammaglobulinaemia is an X-linked disorder.
   b) X-linked agammaglobulinaemia is an acquired immunodeficiency syndrome.
   c) X-linked agammaglobulinaemia is characterized by failure of B cell precursors to develop into mature B cells.
   d) B cell are absent or markedly reduced in the circulation in X linked agammaglobulinaemia.

44) Which of the following statements about primary immunodeficiency disorders are **incorrect**?

   a) Common variable immunodeficiency disorders are related to antibody deficiency.
   b) B cell numbers are normal or near normal in common variable immunodeficiency disorders.
   c) Severe combined immunodeficiency group of disorders have defects in both humoral and cellular immunity.
   d) Severe combined immunodeficiency group of disorders have defects in cellular immunity.
   e) DiGeorge syndrome is related to a defect in cellular Immunity.

45) Acquired immunodeficiency syndrome (AIDS) is caused by human immunodeficiency virus (HIV). The following facts about HIV are true:

   a) HIV is a retrovirus which targets immune system only.

b) HIV has selective tropism for CD4+ T cells.
c) HIV has selective tropism for CD8+ T cells.
d) Loss of CD8+ T cells is because of infection of the cells and the direct cytopathic effects of the replicating virus.

46) Which of the following statements regarding genetic predisposition to cancer are correct?

a) Autosomal dominant inherited cancer syndromes are those in which the inherited mutation is usually a point mutation occurring in a single allele of a tumour suppressor gene.
b) Hereditary nonpolyposis colon cancer is an autosomal dominant condition caused by inactivation of DNA mismatch repair gene.
c) Carriers of mutant retinoblastoma tumor suppressor gene have increased susceptibility of developing osteosarcoma.
d) Carriers of mutant retinoblastoma tumor suppressor gene have increased susceptibility of developing retinoblastoma.
e) Familial adenomatous polyposis is an autosomal disorder caused by mutation of the adenomatous polyposis coli (APC) tumour suppressor gene.
f) All of the above.

47) Which of the following viruses are **not known** to be oncogenic?

a) HTLV-1
b) Varicella zoster
c) HPV
d) EBV
e) HBV
f) HCV

48) Which of the following may be useful in typing of cancers?

a) Tumour antigens
b) Enzymes
c) Hormones
d) Mucins
e) All of the above

49) Which of the following substances can embolise and lead to infarction following occlusion of the vessel?

a) Thrombus
b) Tumour
c) Air/gas
d) Fat
e) Amniotic fluid
f) Cholesterol
g) All of the above

50) What are the possible etiologies of the following pathology:

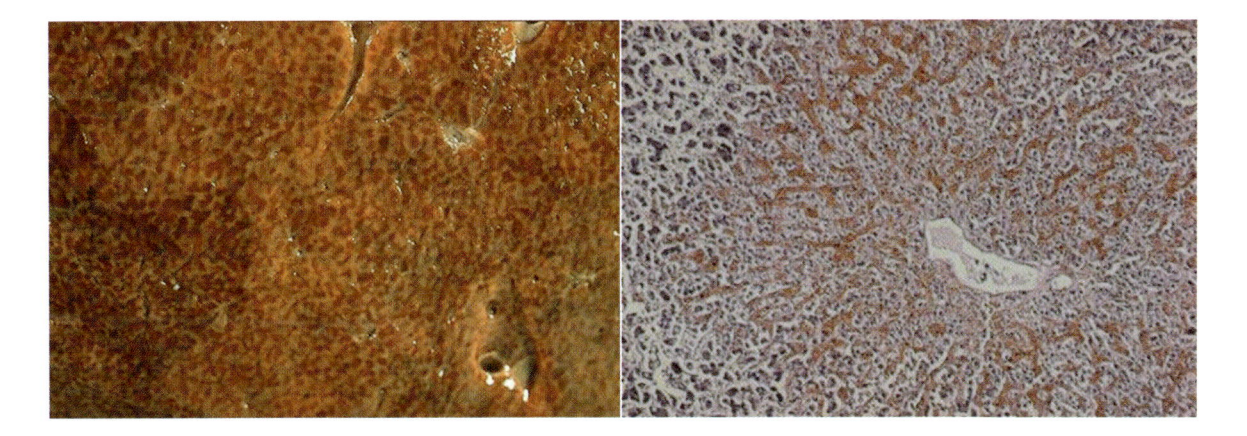

a) Cancer
b) Congestive cardiac failure
c) Paracetamol overdose
d) Carbon monoxide poisoning

Correct answers and explanations:

1) Correct answer: a) Increased production of cellular proteins

In the myocardium, hypertrophy is induced by mechanical sensors triggered by increased workload leading to production of growth factors which increase the synthesis of muscle proteins responsible for the hypertrophy.

Eventually, the hypertrophy reaches a limit beyond which enlargement of muscle mass is no longer able to compensate for the increased burden and regressive changes set in, leading to lysis of myofibrillar contractile proteins. This illustrates how an adaptation to stress can progress to cell injury if stress is not relieved.

2) Correct answer: g) All of the above

Atrophy results from decreased protein synthesis and increased protein degradation in cells.

Reduced metabolic activity of the cell will reduce protein synthesis.

Degradation of cellular proteins occurs mainly by the ubiquitin-proteasome pathway. In catabolic conditions like cancer, nutritional deficiency and disuse, the ubiquitin ligases get activated. Consequently, ubiquitin binds to cellular proteins which then get degraded in proteasomes.

Autophagy may also lead to atrophy in which the starved cell 'eats' its own components to survive.

3) Correct answer: e) Epithelial and mesenchymal cells

Metaplasia results from reprogramming of stem cells in normal tissues or undifferentiated mesenchymal cells in connective tissue. It is not due to change in the phenotype of differentiated cells. External stimuli like cytokines, growth factors and extracellular matrix promote expression of genes that drive cells towards a specific differentiation pathway. Examples of metaplasia include:

- Columnar respiratory epithelium to squamous epithelium in smokers
- In Barrett's oesophagus from squamous to columnar type epithelium
- Connective tissue metaplasia (e.g. formation of cartilage, bone or adipose tissue) in tissues that normally do not contain these elements.

4) Correct answer: E) Fat necrosis

5) Correct answer: F) Fibrinoid necrosis

6) Correct answer: D) Caseous necrosis

7) Correct answer: C) Gangrenous necrosis

8) Correct answer: A) Coagulation necrosis

9) Correct answer: B) Liquefactive necrosis

10) Correct answer: G) Avascular necrosis

11) Correct answer: b) Endothelium, platelets and coagulation cascade

The interaction of endothelium, platelets and coagulation factors regulates haemostasis and thrombosis. Normally endothelial cells exhibit antiplatelet, anticoagulant and fibrinolytic properties. Endothelium can be activated by trauma, infectious agents, haemodynamic forces, plasma mediators and cytokines to acquire procoagulant properties and stimulate thrombosis.

12) Correct answer : c) Antiphospholipid antibodies induce a hypercoagulable state *in vivo* but *in vitro* they inhibit coagulation.

Thrombophilia is a hypercoagulable state and one of its causes is increased levels of factor VIII. Deficiency of Factor VIII is haemophilia i.e. increased bleeding tendencies.

Red infarcts are related to venous obstruction in loose tissues with dual circulations while white infarcts occur in solid organs with occlusion of the end arterial circulation.

In vivo, the autoantibodies induce a hypercoagulable state due to activation of platelets, complement, coagulation factors and endothelial injury through interaction with certain catalytic domains. However, *in vitro*, in absence of platelets and endothelial cells the auto-antibodies interfere with phospholipids and thus inhibit coagulation.

Leiden mutation involves the Factor V gene, and is the commonest inherited cause of hypercoagulopathy. It results in protein C resistance to inactivation of the Factor Va protein.

13) Correct answer: d) Cell mediated (type IV) hypersensitivity

14) **Correct answer: a) Immediate (type I) hypersensitivity**

15) **Correct answer: b) Antibody mediated (type II) hypersensitivity**

16) **Correct answer: c) Immune complex mediated (type III) hypersensitivity**

17) **Correct answer: b) AA (serum amyloid associated protein)**

18) **Correct answer: e) A Cal (calcitonin)**

19) **Correct answer: d) ATTR (transthyretin)**

20) **Correct answer: c) A beta2 microglobulin (beta2 microglobulin)**

21) **Correct answer: g) AANF (atrial natriuretic factor)**

22) **Correct answer: f) AIAPP (islet amyloid peptide)**

23) **Correct answer: a) AL (amyloid light chains)**

24) **Correct answer: e) MEN-1 and MEN-2 are associated with mutations in BRCA1 and BRCA2 genes.**

MSH2, MLH1 and MSH6 are tumour suppressor genes (these are DNA mismatch repair genes associated with HNPCC).
Hereditary non - polyposis colon cancer is an autosomal dominant condition caused by inactivation of DNA mismatch repair genes.
Familial adenomatous polyposis is an autosomal dominant disorder caused by mutation of the adenomatous polyposis coli (APC) tumour suppressor gene.
Mutations in p16 and INK4A are associated with familial predisposition to melanomas.
MEN 1 and 2 are caused by mutations in genes that encode menin transcription factor and RET \tyrosine kinase respectively.

25) **Correct answer: e) p53 is independent of MDM2 gene**

The statement is false as MDM2 and MDMX genes regulate the degradation of p53. Mutations in MDM2 thereby cause functional of loss of p53 in tumours without p53 mutations i.e. sarcomas. All other statements about p53 are true.

26) **Correct answer: a) 8;14 involves c-myc gene in Burkitt's lymphoma**

14 ;18 involves bcl2 gene in follicular lymphoma
11; 22 involves FLI1 gene in Ewing sarcoma
9; 22 involves ABL gene in chronic myeloid leukaemia.
8; 21 involves AML gene in acute leukaemias (AML and ALL)

27) **Correct answer: d) Small cell carcinoma of lung**

Endocrinopathies which are paraneoplastic syndrome are due to ectopic hormone production by the cancer cells. Syndrome of inappropriate antidiuretic hormone is due to ectopic production of antidiuretic hormone by small cell carcinoma of lung. This induces hyponatraemia due to inappropriate ADH secretion.

Other paraneoplastic endocrinopathies associated with lung cancer include:

- Cushing syndrome due to ACTH production.
- Hypercalcaemia due to parathormone and parathyroid hormone related peptide.
- Hypocalcaemia due to calcitonin
- Gynaecomastia due to gonadotropins
- Carcinoid syndrome due to serotonin and bradykinin.

28) Correct answer: b) Increased basal metabolic rate

Increased basal metabolic rate in cancer patients is due to production of TNF, tumour necrosis factor by the macrophages and tumour cells which mobilise fat and suppress the appetite. Other cytokines like IL-1 (interleukin -1), interferon gamma and leukaemia inhibitory factor also act in similar ways to TNF. Additionally, other factors like proteolysis - inducing factor and lipid - mobilizing factor, produced by tumours, increase the catabolism of muscle and fat.

29) Correct answer: e) Basal cell carcinoma

Basal cell carcinomas are locally destructive but rarely metastasize. Dissemination of cancer may occur by direct seeding in body cavities, lymphatic and haematogenous spread.

30) Correct answer: b) Peripheral vasodilatation due to endothelial injury

Septic shock is associated with severe haemodynamic and haemostatic derangements due to overwhelming microbial infections. Endotoxins and other microbial products lead to complement activation, endothelial activation and cytokine cascade. Complement activation results in microvascular thrombosis i.e. disseminated intravascular coagulation with consequent tissue ischaemia. With endothelial activation there is vasodilatation, increased permeability and decreased perfusion. Systemic effects of cytokines lead to diminished myocardial contraction and metabolic abnormalities. The net outcome is shock leading to multi-organ failure.

31) Correct answer: a) Beta - pleated structure of amyloid

Whatever the mechanism and biochemical composition of the amyloid, there is abnormal folding of the protein which is deposited as fibrils in extracellular tissue. These fibrils have a beta - pleated structure (composed of 4-6 fibrils wound around one another) which is common to all the biochemical subtypes of amyloid. Molecules of Congo red align at regular intervals along the fibrils which gives the birefringence with polarized light.

32) Correct answer: d) Amyloidosis

Sago spleen is the fine granular appearance due to deposition of amyloid limited to the splenic follicles in the white pulp. When large confluent deposits of amyloid develop in the sinuses and connective tissue framework of the red pulp the gross appearance is called lardaceous spleen.

33) Correct answer: e) All of the above

Amyloid deposition may begin focally in the sub-endocardium and intra-myocardium. This may result in cardiomyopathy and involvement of the conduction system may lead to electrocardiographic abnormalities.

34) Correct answer: c) Fusion of infected and uninfected T cells which leads to ballooning and cell death.

HIV virus infects T cells and causes direct cytopathic effect. This leads to loss of immature precursors of CD4+ cells due to direct infection of thymic progenitor cells. Fusion of infected and uninfected T cells leads to ballooning and cell death.

35) Correct answer: a) Liquefactive necrosis

Under normal circumstances liquefactive necrosis is seen in situations where there is ischaemic necrosis i.e. infarction with superadded bacterial infection. However, cerebral parenchyma is unique in that despite the absence of superadded infective organisms, liquefactive necrosis is seen with ischaemic infarction rather than coagulative necrosis which would be expected in such a scenario.

36) Correct answer: a) Lymphatics

Giant cell arteritis is a chronic, typically granulomatous inflammation of large to medium to small arteries. It affects mainly the arteries in the head especially the temporal arteries but may also affect the vertebral and ophthalmic arteries. Lesions may also arise in the aorta. Involvement of the ophthalmic artery can lead to blindness. The characteristic granulomatous reaction seen in giant cell arteritis supports the theory of immune aetiology and thus the basis of therapeutic response to steroids.

37) Correct answer: d) Acute phase reactants

Acute phase reactants / proteins are mostly synthesized in the liver. The plasma concentration of these proteins increases many-fold in response to inflammatory stimuli. Synthesis of C-reactive protein, fibrinogen and serum amyloid A protein in the hepatocytes is regulated by cytokines IL-6 (for CRP and fibrinogen) and IL-1 or TNF (for SAA).

38) Correct answer: c) Pathological hyperplasia

Hyperplasia is an increase in organ/tissue function via an increase in cell **number** and may be seen in any tissue capable of cell division. It also increases organ/tissue size. It is mediated predominantly by growth factors and cytokines and often increases the risk of neoplasia.

Hyperplasia may be:

- Physiological hyperplasia (due to hormonal stimulation, e.g. female breast at puberty and cyclical endometrial changes).
- Compensatory hyperplasia, e.g. liver regeneration following partial hepatectomy.
- Pathological Hyperplasia due to excessive hormonal stimulation, e.g. prostatic enlargement leading to prostatism and multinodular colloid goitre leading to hyperthyroidism.

39) Correct answer: c) Pathologic hypertrophy

Hypertrophy is an increase in tissue/organ function via an increase in cell **size.** It occurs on its own in tissues incapable of cell division i.e. muscle and neurons but is also often seen together with hyperplasia in cells that can undergo cell division. Also, it often results in an increase in the size of tissue/organ. It is mediated by growth factors and cytokines.

Hypertrophy may be:
- Physiological hypertrophy (hormonal hypertrophy, e.g. uterine myometrium during pregnancy)
- Compensatory hypertrophy, e.g. skeletal muscle in athletes
- Pathological hypertrophy, e.g. Left ventricular hypertrophy due to increased workload on the heart via hypertension or aortic stenosis. It is an initial compensation mechanism which is effective but eventually becomes insufficient and results in cardiac failure.

40) Correct answer: c) Can be a premalignant lesion

Dysplasia is commonly used to describe change in epithelial cells, but can also be used in haematopoietic cells (eg. myelodysplasia). It is identified via changes in both architecture and cytological atypia. It can also be a precursor to frank neoplasia (premalignant lesion).

41) Correct answer: b) It results from an inherited defect in extracellular glycoprotein fibrillin

Mendelian disorders result from defects involving single genes leading to the formation of an abnormal protein or a reduction in the output of the gene product. Several types of proteins may be affected in these disorders, e.g. enzymes, enzyme inhibitors, extracellular proteins such as collagen, fibrillin and cell membrane proteins such as dystrophin. Marfan's syndrome is an autosomal dominant disorder which results from an inherited defect in fibrillin, a major component of microfibrils found in extracellular matrix. Many of the clinical manifestation of Marfan syndrome can be explained by changes in the mechanical properties of extracellular matrix resulting from abnormalities of fibrillin.
Ehlers-Danlos syndrome is another Mendelian disorder of connective tissues resulting from a defect in the conversion of procollagen to collagen (fibrillar collagen).

42) Correct answer: c) B lymphocytes

Type I MHC complex is present in all nucleated cells, while type II MHC complex (also known as human leukocyte antigen complex) is present on B cells (and other antigen presenting cells) and is fundamental to recognition of antigens by the T cells. The MHC/HLA is linked to many autoimmune diseases including transplant rejection. The genes encoding MHC molecules are clustered on chromosome 6. The HLA system is highly polymorphic i.e. there are many alleles of each MHC gene and each individual inherits one set of these alleles that is different from the alleles in most other individuals which form the basis of organ transplant rejection.

43) Correct answer: b) X-linked agammaglobulinaemia is an acquired immunodeficiency syndrome

X-linked agammaglobulinaemia is not an acquired immunodeficiency syndrome, but is a primary immunodeficiency syndrome affecting the humoral arm of adaptive immunity. It is an X-linked disorder almost always seen in males but sporadic cases have been described in females. Due to the failure of B cell precursors development into mature B cells, B cells are absent or markedly decreased in circulation and the serum levels of all classes of immunoglobulins are depressed. This leads to recurrent bacterial infections of the respiratory tract after six months of age when the circulating maternal immunoglobulins are depleted. Almost always, the causative organisms are Haemophilus influenzae, Streptococcus pneumoniae or Staphylococcus aureus.

44) Correct answer: d) Severe combined immunodeficiency group of disorders have defects in cellular immunity (note this is an INCORRECT statement)

Answer d) is an incorrect statement, as severe combined immunodeficiency disorders (SCID) is a group of disorders related to defects in both humoral and cell mediated immunity thus the affected persons are susceptible to a range of pathogens including Candida, P.jiroveci, Pseudomonas, CMV, Varicella and a host of bacteria.

Common variable immunodeficiency (CVI) is a heterogeneous group of disorders (both sporadic and inherited forms occur). The B cell number in these disorders is normal or near normal however, the B cells are not able to differentiate into plasma cells. Thus the clinical manifestations of CVI are caused by antibody deficiency and hence resemble X- linked agammaglobulinaemia. Both intrinsic B cell defects and abnormalities in T helper cell mediated activation of B cells may account for the antibody deficiency in this disease.

DiGeorge Syndrome (thymic hypoplasia) is a T cell deficiency that results from failure of development of the third and fourth pharyngeal pouches which give rise to the thymus gland. Due to the absence of cell mediated immunity caused by low T cell numbers, there is a predisposition to certain viral and fungal infections.

45) Correct answer: b) HIV has selective tropism for CD4+ T cells

HIV retrovirus infects cells of the immune system (CD4+ T cells, macrophages, dendritic cells) and of the central nervous system. The B cells also are activated, presumably via different pathways. CD4+ T cell reduction occurs due to direct infection and cytopathic effects of the replicating virus.

46) Correct answer: f) All of the above

All the given statements are correct. Hereditary predisposition exists for a number of cancers in addition to environmental influences. In the autosomal dominant inherited cancer syndromes the inherited mutation is usually a point mutation in a tumour suppressor gene which greatly increases the risk of developing a tumour. Carriers of the RB tumour suppressor gene have a 10,000-fold increased risk of developing retinoblastoma and a second cancer, particularly osteosarcoma. FAP is linked to mutation in APC tumour suppressor gene. HNPCC is linked to DNA instability due to inactivation of DNA mismatch repair gene.

47) Correct answer: b) Varicella zoster

Varicella (VZ), influenza and HSV are not oncogenic viruses.

Many DNA and RNA viruses have proven to be oncogenic:
- HTLV-1 (human T - cell leukaemia virus type 1) has tropism for CD4+ T cells which is the target for neoplastic transformation and hence leukaemogenesis.
- HPV- high risk subtypes 16 and 8 cause squamous cell carcinomas of oropharyngeal and anogenital sites including cervix.
- EBV- Burkitt lymphoma.
- HBV and HCV- hepatocellular carcinoma.

48) Correct answer: e) All of the above

A number of tumour markers have been described which can contribute towards detection and typing of cancer.

Clinically useful tumour markers include:

- **Hormones :**

HCG in trophoblastic tumours and non-seminomatous testicular tumours
Calcitonin in medullary carcinoma
Catecholamines in phaeochromocytoma
Ectopic hormones in paraneoplastic syndromes.

- **Onco-fetal antigens (tumour antigens):**
    α-fetoprotein in hepatocellular cancer, non-seminomatous testicular tumours
    CEA (carcinoembryonic antigen) in carcinomas of colon, pancreas, lung and stomach

- **Isoenzymes:**
    Prostatic acid phosphatase in prostate cancer
    Neuron specific enolase in small cell cancer of lung and neuroblastoma

- **Mucins and glycoproteins:**
    CA-125 in ovarian cancer
    CA19-9 in colon and pancreatic cancer
    CA 15-3 in breast cancer.

49) Correct answer: g) All of the above

All of the elements have the propensity to embolise into the systemic circulation. An embolus is defined as a detached circulating intravascular solid, liquid or gaseous mass, and is carried until it blocks the vessel lumen to cause infarction

50) Correct answer: b) Congestive cardiac failure

This is a gross and a microscopic picture of nutmeg liver. Nutmeg liver is seen in the irreversible stage of end stage cardiac failure. The paler areas seen macroscopically correspond to the residual viable areas and the darker areas correspond to the markedly congested sinusoids in the vicinity of necrotic hepatocytes.

CHAPTER 19

AUTOPSY AND FORENSIC PATHOLOGY
Dr Fiona Scott and Dr Charanjit Kaur

1) An 82 year old man is found dead at his nursing home by his carer. On examination the body shows features of rigor mortis. Which of the following statements is true regarding rigor mortis?

   a) The onset is accelerated in cold temperatures
   b) It is due to build-up of ATP (adenosine triphosphate) in the muscles
   c) The hands and feet are involved first, followed by the limbs and face
   d) Violent activity before death will speed up the onset of rigor mortis

2) Under the Human Tissue Act (2004), which of the following is recognised as 'relevant material'?

   a) Urine
   b) Sweat
   c) Serum
   d) Sperm cells

3) Biochemical examination of vitreous humour is preferentially used for assessment of:

   a) Insulin in possible overdose
   b) Glucose in diabetic ketoacidosis
   c) Potassium levels to estimate interval between death and post mortem
   d) All of the above

4) Post mortem hypostasis is also called 'Livor mortis'. Which of the following statements are true of livor mortis?

   a) A blue/purple discolouration of the skin occurring where blood has settled after death
   b) The onset of livor mortis is approximately 12 hours after death
   c) In the early stages of livor mortis the discoloured skin will not blanch when pressed
   d) Bilateral conjunctival haemorrhages are typically seen

5) Which of the following options is true regarding consent for post mortem?

   a) Consent is not required for the retention of tissues when the Coroner's authority ceases after a coronial post mortem
   b) Consent is not required for a Hospital post mortem
   c) Consent is not required for a post mortem authorised by the Coroner
   d) Relatives can withhold consent for a Coroner's post mortem

6) Which of the following factors does not affect estimation of the time of death?

   a) Cold ambient temperature
   b) Fever prior to death
   c) External heat sources

d) Alcohol intoxication

7) Which of the following options are included in the Codes of Practice issued by the Human Tissue Authority?

   a) Consent
   b) Research
   c) Post mortem examination
   d) All of the above

8) A 52 year old man with a long history of alcohol abuse is found dead at home after a concern for his welfare. A number of empty alcohol bottles are noted at the scene and the body smells of alcohol. Which of the following is *FALSE* regarding post mortem levels of blood alcohol?

   a) Post mortem decomposition may result in elevated levels of blood alcohol
   b) Prolonged immersion in water may result in lowered blood alcohol levels
   c) Post mortem decomposition may result in lowered blood alcohol levels
   d) Cardiac blood is the most reliable estimate of ante-mortem alcohol levels

9) Which of the following options is the commonest cause of death in children aged 0-4?

   a) Flu/pneumonia
   b) Epilepsy
   c) Congenital defects
   d) Meningitis

10) A 44 year old woman was admitted to hospital following an overdose of paracetamol (40 tablets) and died after 5 days. At post mortem, a sample of liver tissue was taken for histology. Which one of the following options would be the typical histological finding?

   a) Marked steatosis
   b) Centrilobular necrosis
   c) Periportal necrosis
   d) Bile duct damage.

11) In which of the following 'high risk' situations should the post mortem be carried out in a body bag?

   a) HIV infection
   b) Hepatitis B virus infection
   c) Hepatitis C virus infection
   d) Creutzfeldt- Jakob disease (CJD)

12) A 92 year old woman with known dementia is admitted to hospital with increasing confusion following a minor fall at home two weeks previously. A skull x-ray shows no fracture. She deteriorates and dies and, at post mortem, an intracranial haemorrhage is found. Choose from the following options the most likely type of haemorrhage.

   a) Extradural haemorrhage
   b) Subdural haemorrhage
   c) Subarachnoid haemorrhage

d) Intracerebral haemorrhage

13) In which one of the following circumstances should a death be reported to the Coroner?

a) Death occurred during an operation or before the person recovered from an anaesthetic
b) The person who died had not been seen by the doctor who signed the medical certificate within 7 days prior to death
c) The medical certificate suggests the death may have been due to an alcohol-related disease
d) Patient died within 24h of admission to hospital

14) A 75 year old male smoker dies in hospital of respiratory failure following a long history of chronic obstructive pulmonary disease. His family wish for a hospital (consent) post mortem to be carried out. Regarding gaining consent for the post mortem under the Human Tissue Act 2004, which of the following options shows the correct order of Qualifying Relationships (highest first)?

a) Parent/child, spouse/partner, brother/sister, grandparent/grandchild
b) Spouse/partner, parent/child, grandparent/grandchild, brother/sister
c) Spouse/partner, parent/child, brother/sister, grandparent/grandchild
d) Parent/child, spouse/ partner, grandparent/grandchild, brother/sister

15) A 12 year old boy with a known peanut allergy suffers a sudden collapse while at a friend's birthday party and cannot be revived. A post mortem examination is to be carried out. Which of the following investigations is relevant in a case of suspected acute anaphylaxis?

a) Histological blocks from the lungs and airways
b) Post mortem blood for serum mast cell tryptase
c) Histological blocks from vocal cords
d) All of the above

16) A 23 year old female body is recovered from the sea. At post mortem there are putrefaction changes. Which of the following statements is true regarding putrefaction in bodies recovered from water?

a) Putrefaction is more rapid in deep water than shallow water
b) Putrefaction is more rapid in fresh water as opposed to salt water
c) The rate of putrefaction in water is more rapid than in air
d) All of the above

17) A 55 year old man dies following a large haematemesis at home. At post mortem he is found to have a large gastrointestinal haemorrhage, cirrhosis of the liver, splenomegaly and ascites. No obvious gastric or duodenal ulceration is seen. Choose from the following options the most likely cause of death.

a) Ruptured aorto-duodenal fistula
b) Bleeding oesophageal varices
c) Oesophageal carcinoma
d) Angiodysplasia

18) A 39 year male with a known history of depression is found dead at home, suspended by a cord from an upstairs banister. In cases of death by suspension, which of the following mechanisms may elicit death?

a) Fracture or dislocation of upper cervical vertebrae
   b) Asphyxia due to compression and narrowing of the larynx/ trachea
   c) Reflex vagal inhibition
   d) All of the above

19) Which of the following macroscopic findings is typically the earliest sign of post mortem decomposition?

   a) Bloating
   b) 'Marbling' of the skin
   c) Skin slippage.
   d) Green discolouration of the abdomen.

20) Post mortem histology from an extensive left lung tumour in a 72 year old retired electrician and longstanding smoker who initially presented with breathlessness. Immunohistochemistry showed the tumour was CK7 (+), Calretinin (+), WT1 (+), TTF-1 (-), CK20 (-), Ber-EP4 (-). Choose from the following options the correct histological diagnosis.

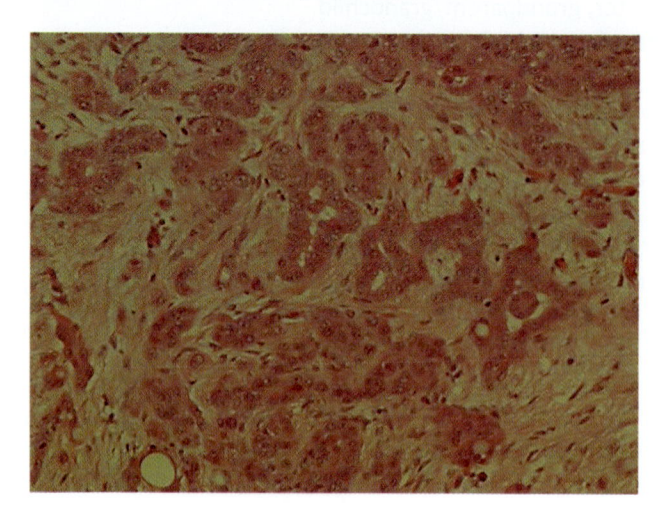

   a) Metastatic colorectal carcinoma
   b) Adenocarcinoma of lung
   c) Epithelioid mesothelioma
   d) Sarcomatoid mesothelioma

Scenario for questions 21-23: A 35 year old man collapses at home after dinner. Ambulance staff arrive and attempt resuscitation but sadly, the individual cannot be revived and is declared dead. The individual has been fit all of his life and has visited his GP only for minor illnesses. No significant medical history has been recorded with the GP.

21) Who do you think will certify the death (i.e. complete the Medical Certificate Certifying Death, MCCD)?

   a) GP
   b) Hospital doctor
   c) Police
   d) Registrar (registers births and deaths)
   e) Forensic Pathologist

f) None of the above

22) What will be the Coroner's approach to the case?

   a) Instruct GP to complete death certificate
   b) Instruct police to complete death certificate
   c) Instruct Pathologist to perform a post mortem
   d) Certify death himself/herself
   e) None of the above

23) In what other situations is the death referred to the Coroner, who then requests a post mortem examination?

   a) Occupational deaths
   b) Mesothelioma
   c) Industrial deaths
   d) Substance abuse
   e) Post operative deaths
   f) Road traffic accidents
   g) All of the above
   h) None of the above

Scenario for questions 24-27: Mr. DB, a 60 year old Asian man living alone was found dead by his son. The GP was unable to certify his death and the case was referred to the Coroner. A post mortem was carried out on instruction from the Coroner. There was no significant past medical history. There was history of alcohol consumption. There was no history of smoking. At post mortem, general examination revealed a markedly thin and emaciated individual with muscle wasting and very low BMI. There was mild pallor. No icterus, oedema or lymphadenopathy was seen. There was no evidence of trauma or injury.

24) Enumerate the potential causes of death to be considered prior to the internal examination.

   a) Pneumonia
   b) Alcoholic liver disease
   c) Ketoacidosis
   d) Alcohol toxicity
   e) All of the above

25) The only significant finding on internal examination was fatty liver. There was no evidence of cirrhosis or necrosis. The brain, lungs, heart and GI tract including the pancreas were normal. There was no significant atherosclerosis. The coronary arteries were normal. Based on these findings, what do you think *could* be the most likely cause of death?

   a) Hypothermia
   b) Ischaemic heart disease
   c) Pneumonia
   d) Alcohol toxicity
   e) None of the above

26) What should the pathologist do next to establish the cause of death?

a) Send brain for neuropathological examination
b) Send heart to cardiac pathologist
c) Send blood sample for genetic investigations
d) Send blood and urine for toxicological examination including ketone bodies

27) The toxicology results showed normal levels of blood glucose. No pharmaceutical or recreational drugs were found. Very high levels of ketones and ketone bodies were found in blood and urine. Alcohol levels were not significantly raised. What is the final cause of death in this case?

a) Alcoholic Ketoacidosis
b) Alcohol toxicit
c) Diabetic ketoacidosis
d) Drug toxicity
e) Alcoholic liver disease

Scenario for questions 28-31: A 30 year old lady presented to A&E with acute shortness of breath and sinking feeling in the epigastrium. General examination revealed hypotension, weak pulse and extreme pallor. A clinical diagnosis of circulatory collapse was made. Unfortunately, despite medical treatment, the lady died. The death could not be certified and a post mortem was requested by the Coroner.

28) With this information, before proceeding to post mortem, what do you think is the likely mode of death?

a) Cardiogenic shock
b) Sepsis
c) Coma
d) All of the above

29) At post mortem, extreme pallor was noted on general examination. Systemic examination revealed massive haemorrhage in the pericardial cavity. What is this condition?

a) Pericarditis
b) Aneurysm
c) Cor pulmonale
d) Haemopericardium

Further internal / systemic examination revealed:

- Thickened myocardium with multiple large yellow white solid nodules
- Transmural tear in one of these nodules associated with 500 ml of blood in the pericardial cavity
- Similar solid yellow white nodules were noted in the lungs and liver
- The right kidney was almost completely replaced by a similar looking solid yellowish mass

30) What is the immediate cause of death in this case?

a) Compression of heart due to ruptured metastatic tumour in myocardium
b) Renal failure due to renal cell carcinoma infiltrating kidney
c) Bleeding due to disseminated intravascular coagulopathy as result of cancer
d) Myocardial rupture due to myocardial infarction

31) What is the sequence of events leading to death? Choose the best option (as you would give in the death certificate) :
i) Disseminated renal cell carcinoma, ii) Haemopericardium and iii) Cardiac tamponade

   a)   iii, ii, i
   b)   i, ii, iii
   c)   ii,iii,i
   d)   ii, i, iii

Scenario for questions 32-37:   A 45 year old Afro-Caribbean man with no significant medical history collapsed and died suddenly at home.

32) What potential causes of sudden death will you keep in mind before proceeding with the post mortem examination?

   a) Stroke
   b) Pulmonary embolism
   c) Aortic dissection
   d) Ruptured aortic aneurysm
   e) Sudden death due to likely cardiac pathology
   f) All of the above

Post mortem examination in this case revealed: A massively enlarged heart with a weight of approximately 650g. The coronary arteries were unremarkable. The heart valves were normal. There was no significant large vessel atherosclerosis. The pulmonary arteries were clear. Slicing the heart showed asymmetrical hypertrophy of the left up to a thickness of 25mm. No pallor or softening of the myocardium was found.

33) With these findings, what is your preliminary diagnosis?

   a) Stroke
   b) Pulmonary embolism
   c) Aortic dissection
   d) Ruptured aortic aneurysm
   e) Sudden death due to likely cardiac pathology
   f) All of the above

34) Which of the following underlying cardiac pathologies best fits as the potential cause of death?

   a) Ischaemic heart disease
   b) Hypertensive heart disease
   c) Myxoma
   d) Amyloid
   e) Cardiomyopathy
   f) Consider mitral regurgitation

35) What will you do next?

   a) Take blood and urine for toxicology
   b) Send the dissected heart to a cardiac pathologist

c) Take photographs of the dissected heart
d) X-ray the heart

36) How many tissue blocks will you take from the heart?

a) At least 3
b) At least 4
c) At least 5
d) At least 6
e) At least 7
f) At least 8

37) What is usually the most likely event leading to death in this condition?

a) Arrhythmias
b) Cardiac tamponade
c) Myocardial infarction
d) Mitral regurgitation
e) Ventricular septal defect

Scenario for questions 38-42: An 82 year retired ship builder develops fever and cough with weight loss. Chest X-ray reveals massive bilateral pleural effusions. CT chest shows bilateral pleural thickening, more on the left side. The patient was very ill hence a biopsy could not be performed. A pleural tap was done and fluid was sent for cytological examination. Unfortunately, the patient's condition deteriorated and sadly he died. Post mortem revealed 800 ml bilateral cloudy pleural effusions. The left pleura was thickened, encasing the whole lung and replacing most of the parenchyma. The right pleura was thickened but to a much less degree.

38) What is the most likely diagnosis?

a) Lymphoma
b) Bronchogenic carcinoma
c) Mesothelioma
d) Tuberculosis
e) Chronic renal failure

39) What will you do next?

a) Take tissue for histological examination to ascertain cause of death
b) Take blood and urine samples for toxicological examination
c) Send lungs to expert for detailed examination
d) Send tissue to microbiology for culture

40) What samples will you choose for histological examination and what further investigations will you perform?

a) Pleura and lungs
b) Pleural, lungs, kidney
c) Pleural, lungs, kidney, heart
d) Pleural, lungs, kidney, heart, liver

e) Pleural, lungs, kidney, heart, liver, spleen

41) What do you think will be the final outcome of the autopsy investigation?

   a) No outcome
   b) Inquest
   c) Police investigation
   d) Refer for forensic examination

42) How long will the tissue blocks be retained and how will they be finally disposed?

   a) 1 month after post mortem
   b) Until the end of the post mortem
   c) Until the post mortem report is scrutinised by the lawyers representing the family
   d) Until the inquest is concluded and the death registered

Scenario for questions 43-45: A 30 year old sportsman had dinner and suddenly complained of chest pain and collapsed. Ambulance staff arrived following a 999 call and attempted resuscitation but in vain and he was pronounced dead. The only significant finding was critical stenosis of the left anterior descending artery and the left circumflex artery. The myocardium showed slight pallor.

43) Which of the following lumen diameters in coronary artery atherosclerosis amounts to critical stenosis of the artery and is thus significant to the cause of death?

   a) 2.0mm
   b) 0.75 mm
   c) 1.5mm
   d) 3.0mm

44) What will histology of the pale area of myocardium show?

   a) Oedema
   b) Contraction band necrosis
   c) Atherosclerosis
   d) Nutmeg appearance

45) What may you expect to find on histological examination of the coronary artery?

   a) Atherosclerotic plaque
   b) Thrombus
   c) Plaque haemorrhage
   d) All of the above

Scenario for questions 46-47: A 68 year old lady with chronic rheumatoid arthritis for 15 years, developed breathlessness and presented to A&E. Examination revealed bilateral chest crepitations, pedal oedema, ascites and pleural effusions. Despite medical treatment, her condition deteriorated and sadly she died. Post mortem examination confirmed ascites, pleural effusions and pulmonary oedema. The heart showed symmetrical hypertrophy. Slicing the myocardium showed a waxy cut surface. The spleen was enlarged with a smooth surface and deep red waxy-greasy cut surface. The liver was slightly enlarged with a smooth subcapsular surface and a greasy cut surface. The coronary arteries did not show significant atherosclerosis.

46) What is the immediate mode of death in this case?

   a) Pneumonia
   b) Congestive cardiac failure
   c) Ischaemic heart disease
   d) Tuberculosis
   e) Cardiomyopathy

47) Which of the following conditions could be the underlying cause of death in this case?

   a) Cardiac myxoma
   b) Amyloidosis
   c) Atherosclerosis
   d) Myocarditis
   e) Lymphoma

48) The police discover a male body floating in the river. The body is bloated and the skin is shedding. There are some bruises on the abdomen but no obvious signs of trauma are seen. What feature at post mortem will help to distinguish death due to drowning (i.e. the individual was alive when fell into the water) from death prior to entering the water?

   a) Serum amylase
   b) Diatom test
   c) Blood and urine ketones
   d) Blood culture

49) What other features indicate death due to drowning?

   a) Emphysema aquosum
   b) Paltauf's haemorrhages
   c) Champignon d'mousse
   d) Froth in tracheobronchial tree
   e) All of the above

50) What finding at post mortem will help to establish death due to smoke inhalation in a fire i.e. the person was alive at the time of fire?

   a) Dark soot in lower airways
   b) Pugilistic attitude
   c) Heat haematoma
   d) All of the above

Correct answers and explanations:

1) Correct answer: d) Violent activity before death will speed up the onset of rigor mortis.

Rigor mortis results from a depletion in ATP levels in the body post mortem. It typically commences in small muscles e.g. the eyelids, neck and jaw, followed by the larger muscles. Vigorous activity just prior to death results in the muscles being depleted of ATP and will increase the rapidity of onset of rigor mortis. Cold ambient temperature will delay the onset.

2) Correct answer: a) Urine.

Sweat and serum are considered to be acellular, while sperm cells fall within the remit of the Human Fertilisation and Embryology Act 1990, and are regulated by the Human Fertilisation and Embryology Authority (HFEA).

3) Correct answer: d) All of the above.

Vitreous humour is isolated and less susceptible to post mortem redistribution or microbiological degradation. Also, its composition is less affected after death than CSF or blood.

4) Correct answer: a) A blue/purple discolouration of the skin occurring where blood has settled after death.

Post mortem hypostasis or 'livor mortis' is caused by pooling of the blood as a result of gravity when the circulation has ceased, resulting in a blue/purple discolouration of the skin in dependent areas. The process starts immediately after death and may be visible after a few hours. In the early stages pressure on the skin may cause whitening but with the onset of decomposition leading to increased vascular permeability there is leakage of red blood cells into the tissues and staining becomes permanent.

5) Correct answer: c) Consent is not required for a post mortem authorised by the Coroner.

Relatives of the deceased are not able to prevent the Coroner requesting a post mortem examination. However, consent is required for a hospital post mortem. The storage and/or usage of tissues following a Coronial post mortem, after the Coroner's authority has ceased, will require consent.

6) Correct answer: d) Alcohol intoxication.

A cold ambient temperature will delay the onset of rigor mortis while fever prior to death and external heat sources all speed up the onset of rigor mortis, thus affecting the estimation of time of death.

7) Correct answer: d) All of the above.

The Human Tissue Authority (HTA) produces nine Codes of Practice to give professionals practical guidance on human tissue legislation.

8) Correct answer: d) Cardiac blood is the most reliable estimate of ante-mortem alcohol levels.

Blood should be taken from an isolated extremity vein, such as the femoral vein, as blood from the heart, pericardium and thoracic cavity are at an increased risk of contamination from alcohol from the gut. Microorganisms may both utilise alcohol as a substrate thus lowering blood alcohol levels but may also, conversely, cause post mortem production of alcohol.

9) Correct answer: c) Congenital defects.

Congenital malformations, deformations and chromosomal abnormalities are the leading cause of deaths in this age group (Office of National Statistics, UK, 2013) accounting for 13% of deaths in boys and 14% of girls.

10) Correct answer: b) Centrilobular necrosis.

In paracetamol overdose, an increased levels of paracetamol is converted via the cytochrome P450 pathway to a highly reactive intermediary metabolite, *N*-acetyl-*p*-benzoquinoneimine (NAPQI), which damages cell membranes and causes centrilobular hepatocyte necrosis.

11) Correct answer: d) Creutzfeldt- Jakob disease.

In suspected prion disease (transmissible spongiform encephalopathies), the post mortem should be carried out in a body bag, to avoid unnecessary contamination of the surroundings and instruments.

12) Correct answer: b) Subdural haemorrhage.

In the elderly, a subdural haemorrhage may occur with only minor trauma, often presenting with increasing confusion, especially in the elderly or in dementia where there is cerebral atrophy.

13) Correct answer: a) Death occurred during an operation or before the person recovered from an anaesthetic.

If the signing doctor had not seen the deceased within 7 days prior to death then the Coroner should be informed. Alcohol related diseases per se do not have to be reported to the Coroner, although suspected alcoholic poisoning would need to be referred.

14) Correct answer: c) Spouse/partner, parent/child, brother/sister, grandparent/grandchild.

If the deceased person had not indicated their consent (or refusal) to post mortem removal, storage or use of their body or tissue for scheduled purposes prior to death, then consent can be given by someone who was in a 'qualifying relationship' to the deceased immediately before their death. A Nominated Representative (a person nominated by the patient prior to death) may also give authorisation on behalf of the deceased person.

15) Correct answer: d) All of the above.

Histology from the upper respiratory tract may show oedema, as well as pulmonary oedema in the lungs and evidence of acute/chronic asthma. Sensitivity to drugs such as aspirin or NSAIDS may exacerbate asthma. Peripheral cadaveric spun blood for serum mast cell tryptase (MCT) may be useful for up to 3 days following death.

16) Correct answer: b) Putrefaction is more rapid in fresh water as opposed to salt water.

Deep water delays putrefaction, as does salt water compared to fresh water. The rate of putrefaction is more rapid in air compared to water.

17) Correct answer: b) Bleeding oesophageal varices.

The presence of cirrhosis with ascites and splenomegaly indicates the presence of portal hypertension. This eventually results in the formation of varices at porto-caval anastomoses such as oesophageal, gastric and rectal varices.

18) Correct answer: d) All of the above.

Venous congestion and compression of the larynx/trachea is common while fracture/dislocation of cervical vertebrae is usually associated with complete suspension and/or drop from a height. In addition, cerebral anoxia may contribute to death by compression of the neck arteries.

19) Correct answer: d) Green discolouration of the abdomen.

Decomposition occurs as a result of endogenous autolysis and putrefaction from intestinal micro-organisms. Abdominal discolouration usually starts in the right lower quadrant. Marbling occurs due to the lysis of red blood cells. Skin slippage occurs as the epidermis separates from the underlying dermis and bloating occurs as a result of gases produced by intestinal bacteria. Typically, discolouration of the abdomen will be visible after approximately 2-3 days. Decomposition changes begin shortly after death but the timeline is very variable, depending on a number of exogenous factors (ambient temperature, moisture, air, light exposure) and endogenous factors (infection, injuries, condition of the body).

20) Correct answer: c) Epithelioid mesothelioma.

Histology shows nested and tubulo-papillary epithelial groups with cytological atypia. Typically, the tumour forms a thick rind around the lung or may show multifocal studding of the lung or pleural surfaces. Clinical presentation is usually with breathlessness due to pleural effusion. An immunopanel including CK7, Calretinin, WT1, TTF-1, CK20 and Ber-EP4 will usually help differentiate between an epithelioid mesothelioma and a primary or metastatic adenocarcinoma. Sarcomatoid mesothelioma may be negative for calretinin and WT1 and the addition of other positive mesothelial markers such as podoplanin and D2-40 may be necessary.

21) Correct answer: f) None of the above

This is a sudden unexpected death in a young and fit individual with no known medical condition or illness which could explain the death. Since there is no known cause of death, none of stated individuals will be able to certify the death. The case will be referred to the jurisdictional Coroner.

22) Correct answer: c) Instruct Pathologist to perform a post mortem.

The Coroner will instruct the Pathologist to perform a post mortem to identify the cause of death.

23) Correct answer: g) All of the above.

All of the situations need to be referred to the coroner by law for investigation into the cause of death. The coroner will usually instruct the pathologist to conduct a post mortem to establish the cause of death in all of these situations.

24) Correct answer: e) All of the above

All of the causes must be kept in mind as potential causes of death in individuals with a history of excess alcohol consumption. Alcoholics are prone to malnourishment and hence pneumonia and ketoacidosis. Chronic alcohol abuse will also predispose to alcoholic liver disease and death. Acute alcohol toxicity is also a potential cause of death to be considered in this setting.

25) Correct answer: d) Alcohol toxicity.

The pathologist must exclude alcohol toxicity especially with the background history of excess alcohol consumption and also since all of the internal organs were essentially normal macroscopically except for fatty liver.

26) Correct answer: d) Send blood and urine for toxicological examination including ketone bodies.

Thus far, the pathologist has not been able to establish the cause of death with macroscopic internal examination alone and will need these ancillary tests to ascertain whether abnormal levels of alcohol/drugs/glucose/ketones could be the cause of death. Histology of the vital organs will also aid in arriving at the cause of death. Tissue should be taken from lung, liver, heart, kidneys, pancreas and brain. Alcohol-associated histological changes may be found. Despite the history of alcohol, this should be treated as a sudden adult death and investigated accordingly.

27) Correct answer: a) Alcoholic Ketoacidosis.

Alcoholic ketoacidosis most commonly occurs in adults with alcoholism, who typically have a history of binge drinking with little or no food intake. This leads to metabolic acidosis due to starvation and may be compounded by persistent vomiting leading to ketoacidosis. This must be thought of as a potential cause of death in thin emaciated individuals with low BMI i.e. malnourished individuals who drink excess alcohol. This precipitates ketoacidosis since without any source of glycogen, body fat is metabolized leading to metabolic acidosis. In such cases, the alcohol levels found in post mortem blood will not be markedly raised whilst ketones and ketone bodies will be much higher than the reference range.

28) Correct answer: a) Shock.

The mode of death is often confused with the cause of death. The mode of death is the direct pathophysiological state or mechanism by which death results, or which is present at the time of death. Examples include congestive cardiac failure, cardiac arrest or coma. It is different from the cause of death, which is the aetiology leading to and resulting in the mode of death.

In this case, all of the signs and symptoms are related to circulatory collapse i.e. cardiogenic shock. Shock is the final common pathway in a number of medical conditions. Depending on the underlying aetiology, types of shock include hypovolaemic shock, cardiogenic shock and septic shock. The aim of the post mortem is to establish the underlying aetiology or cause of shock in this case.

29) Correct answer: d) Haemopericardium.

Blood in the pericardial cavity will impede the cardiac functions and will eventually lead to cardiogenic shock.

30) Correct answer: a) Compression of heart due to ruptured metastatic tumour in myocardium

Since the mode of death is cardiogenic shock, the direct cause of death leading to this state is cardiac tamponade, which is the compression of the heart by blood in the pericardial cavity (haemopericardium).

This results from a rupture in the myocardium due to the presence of the nodules, which in the presence of a large renal mass with multiple nodules in the liver, lungs and heart indicate that the nodules are metastatic deposits from the kidney tumour. There is no evidence to suggest renal failure or myocardial infarction. Although carcinomas are associated with disseminated intravascular coagulopathy, the other findings indicate that myocardial rupture, rather than clotting abnormality, is the main cause of the haemopericardium.

31) Correct answer: a) iii, ii, i

This is the best option of sequence of events to give in the medical certificate of cause of death (MCCD). Immediate cause of death is cardiac tamponade due to haemopericardium, due to rupture of metastatic tumour in myocardium due to disseminated renal cell carcinoma.

32) Correct answer: f) All of the above

All the above conditions are important causes of sudden death to consider in a young individual who dies suddenly with no previous relevant medical history.

33) Correct answer: e) Sudden death due to likely cardiac pathology

The findings show cardiomegaly and hypertrophy of the ventricles, suggesting cardiac pathology. No evidence of a stroke, pulmonary embolism or aortic dissection/aneurysm.

34) Correct answer: e) Cardiomyopathy

Cardiomyopathy is the most likely cause of death in this case in the absence of atherosclerosis hence ischaemic heart disease and hypertensive heart disease. There was no evidence of valvular heart disease since heart valves were normal. No softening or pallor of the myocardium was seen thus myocarditis and myocardial infarction seems less likely but can be reliably excluded following histological examination of the heart. Additionally, the asymmetric enlargement of the left ventricle points more towards cardiomyopathy rather than towards ischaemia/hypertension/amyloid related cardiac enlargement which is usually symmetric.

35) Correct answer: c) Take photographs of the dissected heart

Taking photographs of the heart to demonstrate the asymmetrical enlargement is good practice to gather evidence to ascertain the cause of death.

Following this, take relevant blocks for histological examination of the myocardium as per the guidelines issued by the Royal College of pathologists. The histology of the myocardium will help to confirm a cardiomyopathy (dilated/hypertrophic/ARV cardiomyopathy) and to exclude other cardiac conditions like myocarditis, amyloid, infarction etc. Fresh samples of spleen or blood may also be taken and deep frozen for genetic investigations.

The intact heart (NOT dissected) can also be sent to a cardiac pathologist for examination following discussion with the Coroner who will then notify the deceased's family. If the cardiomyopathy is genetically linked, it has implications on the rest of the family who will then have to be investigated. Histological examination and tissue retention thus becomes crucial in such cases.

36) Correct answer: f) At least 8

When no cardiac lesion is identified macroscopically, then sampling for histology will be governed by the clinical judgement of the pathologist performing the autopsy. However, where possible as a minimum, it is recommended to take mapped blocks of anterior, lateral and posterior right and left ventricle and septum from a representative mid- ventricular transverse slice and right ventricular outflow tract where the AV node is located (i.e. 8 blocks at least).

Alternatively retention of the whole heart and sending it intact to a specialist centre for expert opinion should be considered

37) Correct answer: a) Arrhythmias

Cardiomyopathies i.e. dilated cardiomyopathy (DCM), hypertrophic cardiomyopathy (HCM), arrhythmogenic right ventricular cardiomyopathy (ARVC) and other degenerative cardiomyopathies are usually associated with progressive cardiac failure as well as a risk of cardiac dysrhythmias and death. The arrhythmias usually develop when there is significant enlargement of the left ventricle. Electrical instability develops in the hypertrophied myocardium making it prone to arrhythmias. Often, the conduction system may also be directly affected leading to arrhythmias.

In a case of sudden death where no cardiac pathology is found at post mortem, sudden arrhythmic death syndrome (SADS) is also a potential cause of death to be considered. The requisites of SADS as a cause of death are: negative general autopsy findings (macroscopic and histological), negative toxicology and a morphologically normal heart. In such cases, after discussions with the Coroner, either retain the heart in its entirety and send for expert cardiac opinion or examine histologically in detail in an attempt to identify any underlying cause.

38) Correct answer: c) Mesothelioma

The deceased's history of shipbuilding suggests asbestos exposure, as the material was commonly used in this trade in the past. Combined with the post mortem findings of marked pleural thickening, the suspicion of mesothelioma would be high.

39) Correct answer: a) Take tissue for histological examination to ascertain cause of death

Take tissue for confirmation of diagnosis and impression of mesothelioma as per the Royal College of Pathologists guidelines for investigation of death related to occupational diseases.

40) Correct answer: a) Pleura and lungs

Take tissue from pleural thickening/tumour and each lobe of both lungs. Perform immunohistochemical stains on pleural tumour with a panel of markers to confirm mesothelioma.

41) Correct answer: b) Inquest

The Coroner will hold an inquest into the death. The pathologist will present the findings of the post mortem and the Coroner will assess all the facts and give his verdict regarding the cause of death. If the death due to mesothelioma is attributed to occupation related asbestosis then the family of the deceased gets compensation from the employer's company.

42) Correct answer: d) Until the inquest is concluded and the death is registered.

Retention of tissue is under Coroner's amendment of rule 9, which determines period of retention. Pathologist retains relevant material which has bearing on cause/circumstances of death until the Coroner becomes functus officio i.e. cause of death is ascertained. In this case, until the inquest is held. The tissue blocks will be disposed by the pathology department as per the wishes/consent of the family. This could include the following options: 1) Retention for research, teaching, review, audit, quality control. 2) Respectful disposal.

43) Correct answer: b) 0.75 mm

0.75 mm i.e. <1mm is regarded as the critical lumen diameter of the coronary artery. In ischaemic heart disease related to atheroma, where there is >70% stenosis of the artery and lumen diameter is reduced to < 1mm, the coronary artery can be regarded as being of significance to the cause of death.

44) Correct answer: b) Contraction band necrosis

Contraction band necrosis is the hallmark of myocardial ischaemia/infarction. It is related to damage to the cardiac contractile proteins leading to loss of their alignment. In the early stages, there may be well be no findings at all.

45) Correct answer: d) All of the above

All of the events may be present in ischaemic heart disease. The atheroma in the coronary artery may gradually increase in size and encroach upon the lumen of the artery which may reach critical narrowing. The other possibility, in case of partial occlusion of the vessel with atheroma is, rupture of the atheromatous plaque with overlying thrombosis leading to complete occlusion of the vessel.

46) Correct answer: b) Congestive Cardiac failure

See answer to Question 28.

Congestive cardiac failure leads to pulmonary oedema and pleural effusions with ascites. The heart failure is related to underlying cardiac pathology as demonstrated by symmetrical enlargement of the heart associated with enlargement of other organs like the spleen and the liver. These findings suggest possible infiltration of the myocardium and systemic organs as a cause of the cardiac enlargement/hypertrophy. Histological examination of the myocardium and systemic organs should be undertaken to establish the aetiology of the suspected infiltrative disorder.

47) Correct answer: b) Amyloidosis

Systemic amyloidosis is often due to an underlying chronic inflammatory condition (rheumatoid arthritis in this case). Amyloid deposition in the heart leads to symmetric hypertrophy and a waxy greasy appearance of the heart. Lardaceous spleen is the appearance of a spleen with large lakes of amyloid within the parenchyma. Amyloid can be confirmed microscopically with Congo red stain and transthyretin on immunofluorescence.

48) Correct answer: b) Diatom test

If the individual was alive on entry into water, there will be aspiration of the water containing the regional diatoms into the lungs, which will penetrate the alveolar walls and enter into the bloodstream. These will then be transported to distant organs such as brain kidney, liver and bone marrow.

Diatoms are microscopic algae that live in water (2 microns- 1mm). Smaller diatoms can penetrate tissues. These can be demonstrated on microscopy in the different organ tissues. Other findings on microscopy of the lung in drowning include: alveolar dilatation, intra-alveolar haemorrhages and eosinophilic amorphous oedema fluid in alveoli.

49) Correct answer: e) All of the above

Champignon d'mousse is plume of white or red-tinged froth around the mouth and nose coming from the tracheobronchial tree. It is an admixture of alveolar fluid with surfactant and air in alveoli churned up by repeated attempts of the individual to breathe when coming to the surface of the water.

Emphysema aquosum is heavy waterlogged and hyperinflated lungs which appear pale and crepitant on examination. The indentations from the rib cage are noted on the pleural surface.

Paltauf's haemorrhages are localised non-petechial subpleural haemorrhages due to rupture of the alveolar walls.

50) Correct answer: d) All of the above

Dark soot all the way down into lower airways indicates suffocation and inhalation of smoke when the person was alive. The pugilistic attitude of the body indicates the person may have been alive in the fire as opposed to a dead body being burnt. Heat associated artefacts include heat haematoma especially in the skull which appears different from extradural haematoma related to trauma.

CHAPTER 20

CLINICAL MANAGEMENT ISSUES IN CELLULAR PATHOLOGY
Dr Valerie Thomas & Dr Martin Young

Introduction:
The importance of developing leadership and skills in management is becoming ever more important in specialty training. Demonstration of these competencies has been a part of the FRCPath Curricula from 2010 as required by the GMC (UK). These situation-based judgement scenarios have been written to guide your thinking about management issues that you might encounter as a senior trainee or a newly appointed consultant. They also may become part of the appointment process.

Note that Questions 1-8 are scenarios for a senior trainee, Questions 9-20 are scenarios for an established consultant.

1) You (as a trainee) are the leave co-ordinator for a group of trainees. One trainee insists on having Christmas and New Year off for annual leave through a late request but this leave time is no longer available available because other trainees have already been granted leave according to the departmental leave policy. policy. Leave was not granted by the Trust but the trainee does not attend for work. Please choose the most appropriate response:

   a) Report the absence to the lead educational supervisor and redistribute their duties to the trainees who are at work.
   b) Phone the trainee and tell him/her to come into work.
   c) Report the trainee to the clinical director and medical director.
   d) Do nothing and assume the trainee will arrive later.
   e) Take the trainee's work and do it yourself along with your own work.

2) You have failed the Part II FRCPath on the 3rd sitting. Your educational supervisor does not have the time to meet you. Please choose the most appropriate response:

   a) Arrange to take study leave and attend cytology and histopathology revision courses.
   b) Unilaterally take yourself off the reporting rota for private study leave.
   c) Meet with the Director of Medical Education/clinical tutor to ask for time off.
   d) Meet with your training programme director to discuss your approach for preparation for the next exam sitting.
   e) Contact the exam department and request a remark.

3) One of your trainee colleague is taking home office stationery and you know that they are running a separate business venture from their home. Please choose the most appropriate response:

   a) Confront the trainee and ask them what they are doing with the material.
   b) Tell the trainees to keep an eye on the trainee in question.
   c) Phone the NHS fraud line.
   d) Inform your educational supervisor.
   e) Contact the police.

4) You need to discuss some cases with your clinical supervisor. On entering his office, you see pornographic images on his private laptop computer. Please choose the most appropriate response:

   a) Excuse yourself from disturbing a senior colleague and leave the room.
   b) Ignore what you see and start double headed reporting.
   c) Discuss with your peers to see if they have had similar experiences.
   d) Phone the Sun newspaper.
   e) Discuss the situation with your educational supervisor.

5) You have just heard that a close friend has undergone a wide local excision of a breast lump in your hospital and you are worried about the findings. She has not discussed the surgery with you. Please choose choose the most appropriate response:

   a) Check the result for a "just in case" conversation.
   b) Confront your friend and ask why she has not asked you to help.
   c) Look up the result and tell your friend's boyfriend.
   d) Do nothing.
   e) Ask your educational supervisor to look up the result.

6) Your educational supervisor has a busy private practice and has asked you if you can help with grossing the private cases. This work is undertaken in a private laboratory away from the NHS department and is only only open from 9am to 5pm. Please choose the most appropriate response:

   a) Agree and fit the work in around your other activity.
   b) Decline the invitation.
   c) Decline the invitation and discuss the request with the clinical director.
   d) Report the request to the Director of Medical Education at your Trust.
   e) Discuss the request with Human Resources.

7) A clinical supervisor has encouraged you to apply for a training fellowship in molecular pathology in a different department in your Trust. Your application has been successful. You still have to take and pass the the Part I FRCPath examination. Please choose the most appropriate response:

   a) Arrange to take the post without any further discussion.
   b) Defer taking the Part I exam.
   c) Contact the Royal College of Pathologists.
   d) Contact your educational supervisor.
   e) Contact the Deanery to arrange an OOP (Out of Programme period).

8) You are working as a full time trainee. You have an elderly parent who lives alone and has recently suffered a stroke. You are the primary carer. You have applied to take the Part II FRCPath examination and and feel that you do not have the time to deliver your service commitment and at the same time prepare for for the exam. Please choose the most appropriate response:

   a) Resign from your post and take the exam and look after your parent as best you can.
   b) Carry on working and take the examination and look after your parent as best you can.
   c) Discuss your problems with your educational supervisor.
   d) Discuss your problems with your clinical head of department.
   e) Discuss your problems with your training programme director.

9) As a new consultant, you are undertaking an audit into the reporting of cervical loop cone biopsies to see if the reports comply with NHSCSP and RCPath recommendations. You find that one surgeon has a higher proportion of positive margins. Please choose the most appropriate response:

    a) Work up the audit and present the findings at the local audit meeting in the pathology dept.
    b) Discuss your findings with your clinical supervisor.
    c) Inform the medical director.
    d) Inform the hospital based programme coordinator.
    e) Invite the gynaecology clinical director to the audit meeting.

10) You are the educational supervisor for a trainee who has had brain surgery for a meningioma affecting the frontal lobe. He has returned to work having been signed off by the neurosurgeon in charge of the case case but he is erratic in attendance at work, his concentration is poor and he is leaving cases unfinished in time for MDT meetings. Please choose the most appropriate response:

    a) Send the trainee home and tell him ask for the GP to send in a sick note.
    b) Meet with trainee and give them the Trust Sickness and Leave policy and tell it they must adhere to it.
    c) Meet with the trainee to discuss their progress and agree an action plan with occupational health input.
    d) You contact the neurosurgeon in charge of the case and ask them to see the trainee to sort out appropriate behaviour.
    e) Refer the trainee to the post-graduate dean as there are revalidation issues.

11) One of your colleagues is becoming increasingly erratic in their behaviour. You suspect that they are coming in drunk to work. Which one of the following best describes how you would address this issue?

    a) Go to the medical director to report your concerns.
    b) Ignore it as it is not your responsibility.
    c) Raise your concerns with her line manager/head of department.
    d) Report the individual to the GMC.
    e) Search for any supportive evidence in her office.

12) Whilst preparing for the lung multidisciplinary team meeting, on reviewing the slides the day before the meeting, you find tumour cells in a bronchial biopsy reported as negative. Which one of the following best outlines how you would proceed?

    a) Assume you are mistaken in your opinion as the case was reported by the lead lung pathologist.
    b) Issue a supplementary report changing the diagnosis.
    c) Discuss with the reporting pathologist and issue a joint report.
    d) Tell the MDM that you disagree with the original report.
    e) Report your colleague to the medical director.

13) You are duty consultant for frozen sections. You are asked to examine a node on which the request form states? TB? lymphoma. Which one of the following best describes what would you do?

    a) Put the specimen into formalin and ring the surgeon to inform them that there will be no frozen section.

b) Using appropriate personal protection equipment, slice the node and select a representative part for culture, place into a labelled empty pot. Place the rest in formalin. Ring the surgeon to inform them that there will be no frozen section.
c) Using appropriate personal protection equipment, slice the node and select a representative part for culture. Place part of the residual tissue in formalin for histology and send the rest fresh for flow cytometry. Ring the surgeon to inform them that there will be no frozen section.
d) Using appropriate personal protection equipment, slice the node and select an appropriate area for frozen section. Report the frozen section to the surgeon.
e) Using appropriate personal protection equipment, slice the node and take a representative part for culture, place into a labelled empty pot. Select an area for frozen section. Report the frozen section to the surgeon.

14) You are reporting a small pile of breast core biopsies. Good clinical data has been given on each biopsy. As you work through the cases you become aware that there is some lack of correlation between the clinical and imaging data provided and the pathology. On checking, you find there is mismatch in five of the ten cases you have looked at. Which one of the following options outlines how you should proceed?

a) Issue reports on all the cases and take them for discussion at the multi-disciplinary meeting.
b) After checking the labeling and blocks, hold back all the cases and take them for discussion at the multi-disciplinary meeting.
c) After checking the labeling and blocks, hold back the five mismatched cases and take them for discussion at the multi-disciplinary meeting.
d) After checking the labeling and blocks, contact the lead radiologist and arrange to go through all the cases prior to the multi-disciplinary meeting.
e) After checking the labeling and blocks, issue reports on all the cases and take them for discussion at the multi-disciplinary meeting.

15) Whilst you are cutting up a rather fibrous uterus, your knife slips and you cut your right thumb quite deeply. Which one of the following best describes what you would do?

a) Wash the cut hand under the tap and put a plaster on it.
b) Wash the cut hand with antiseptic and put a plaster on it.
c) Wash the cut hand with antiseptic and put a plaster on it. Report the incident in the accident book. Report the incident to the local first aider.
d) Wash the cut hand with antiseptic and put a plaster on it. Report the incident to the local first aider.
e) Wash the cut hand with antiseptic and put a plaster on it. Report the incident in the accident book. Report the incident to the consultant in occupational health.

16) You are in the cut-up room working through a series of cases and two other consultants are also cutting-up: two biomedical support workers are assisting. One of the consultants drops a large specimen pot containing about 2l of 10% buffered formalin, this soaks his trousers and spreads into a large puddle across the cut-up room. Which one of the following is advised?

a) Shut all the windows and doors. Mop up the spillage. Clean spillage area thoroughly with plenty of water.
b) Shut the doors and open the windows. Mop up the spillage. Clean spillage area thoroughly with plenty of water.
c) Mop up the spillage. Clean spillage area thoroughly with plenty of water.

d) Absorb with inert, absorbent material. Sweep up. Transfer to suitable, labelled containers for disposal. Clean spillage area thoroughly with plenty of water.
e) Absorb with inert absorbent material. Sweep up. Transfer to suitable, labelled containers for disposal as clinical waste. Clean spillage area thoroughly with plenty of water.

17) Whilst selecting a piece of tissue for a frozen section of a tumour margin, the fire alarm starts ringing continuously. It is a Thursday afternoon which is regular testing time for the fire alarm system. Which one of of the following options is advised?

a) Leave the department immediately and go to the designated fire assembly area.
b) Ring theatres immediately to tell the surgeon that there will be no frozen section. Leave the department and go to the designated fire assembly area.
c) Ignore the alarm as it is the usual test time.
d) Wait 10 minutes to see if the alarm stops meanwhile carry out the frozen section and then leave the department.
e) Leave the department immediately. Ring theatres to tell the surgeon that a frozen section is not possible. Go to the designated fire assembly area.

18) A hospital post-mortem has been requested on a patient who is HIV positive where the cause of death is not apparent to the clinical team caring for them. The patient does not have an AIDS identifying illness. Which of the following best summarises how would you proceed?

a) Carry out a full post-mortem taking blocks of all tissues for analysis having established that full consent has been signed.
b) Discuss with the clinicians, then carry out a full-post mortem taking blocks of all tissues for analysis, having established that full consent has been signed.
c) Refer the case to the Coroner.
d) Discuss with the clinicians then carry out a full-post mortem, in a high risk PM suite with full personal protection equipment, taking blocks of all tissues for analysis, having established that full consent has been signed.
e) Discuss with the clinicians then carry out a post-mortem, in a high risk PM suite with full personal protection equipment, taking blocks of all tissues for analysis targeting areas identified by the clinicians, having established that consent has been signed.

19) You are asked to sign a part 5 (confirmatory medical certificate) cremation form on a patient who died whilst undergoing a diagnostic bronchoscopy. Which one of the following options reflects best how you would proceed?

a) Having established the identity of the deceased, perform an external examination prior to signing the cremation form 5.
b) Having established the identity of the deceased, examine the notes and perform an external examination prior to signing the cremation form 5.
c) Having established the identity of the deceased, sign the Medical Certificate of Cause of Death and then sign the cremation form 5.
d) Decline to sign the cremation form 5 as you were not involved in the management of the patient's care.
e) Having established the identity of the deceased, perform an external examination and sign the cremation form 5 having read the Medical Certificate of Cause of Death which indicates that the case has been discussed with the Coroner.

20) You are educational supervisor to three trainees. You receive an email from the Deanery telling you that one of the trainees has complained that you are sexually harassing them. In reality, you have not been harassing them; you have never had a similar complaint made against you or received any intimation from any other trainees that your behaviour is in anyway inappropriate. Which one of the following would you do initially?

    a) Go immediately to the trainee to discuss the allegation.
    b) Immediately ring your medical protection organisation and email the Deanery back to let them know you have given up being educational supervisor with immediate effect.
    c) Immediately arrange to meet with the clinical director responsible for training to discuss the best course of action.
    d) Immediately ask all three trainees to a meeting to discuss the allegation.
    e) Immediately email the Deanery back to let them know you have given up being educational supervisor with immediate effect.

Correct answers and explanations:

1) Correct answer: a) Report the absence to the lead educational supervisor and redistribute their duties to the trainees who are at work.

Absence from work is a serious matter and constitutes a poor attitude to one's professional responsibilities to work and there may be patient safety issues. The immediate task in hand is to ensure that there is no delay to signing out the cases. Further discussions with the trainee should be left to senior members of the department who may need to explore wider issues that may include personal problems that have played a part in this seemingly unprofessional behaviour. Referral to a Professional Support Unit for confidential advice may be appropriate.

2) Correct answer: d) Meet with your training programme director to discuss your approach for preparation for the next exam sitting.

Failing an examination is always hugely disappointing for a trainee and there may be a number of contributory factors. Although the NHS is looking always to deliver a high quality and efficient service it is also committed to supporting excellence in training. All training departments should aspire to this principle including your own. The fact that your educational supervisor does not have the time to meet would imply that the department is under pressure and you may not be receiving as much support that you need. Discussion of the situation with some-one from outside the department can be helpful and allow you to agree a plan which will consider your best interests. The other options described above may be part of your agreed training plan, but additional ideas can be considered too that might include coaching and mentoring that can be delivered from the resources of the Deanery.

3) Correct answer: d) Inform your educational supervisor.

The NHS is subject to considerable loss from fraudulent activity that ranges from trivial loss to serious financial abuse. Theft of office stationery may amount to a small loss to the Trust but such behaviour is wrong. The person best placed to deal with the issue is the Clinical Head of Department having been informed by your lead educational supervisor. It is likely that the trainee would receive a warning from Management and a note would be kept on file. The trainee should be asked also to write a reflective note as a work based assessment to be included in the LEPT e-portfolio. Any further transgressions would be likely to result in questions around revalidation of the trainee and the responsible officer (Post-Graduate Dean) should be informed either by the Trust Directory of Medical Education or Training Programme Director.

4) Correct answer: e) Discuss the situation with your educational supervisor.

Viewing pornographic images during work time is unprofessional behaviour. As a trainee you have been placed in a difficult position through no fault of your own. You should not discuss this unfortunate episode with your peer colleagues as this would not be in line with GMC guidelines around good working relations with colleagues. It would be appropriate to report the incident to a senior member of the department and the best person to inform would be your educational supervisor who should then inform the Clinical Head of service. Further investigation should be followed through by Human Resources.

5) Correct answer: d) Do nothing.

There are strict rules around protecting the confidentiality of personal identifiable data and this applies to pathology results. However well-meaning you wish to be and frustrated you are in your desire to help your friend she has not turned to you for advice, support or guidance. You have to take a step back and to look up the result as this would be a breach of trust and confidentiality.

6) Correct answer: c) Decline the invitation and discuss the request with the clinical director.

There is a strict rule around undertaking privately remunerated work during NHS contracted time and in general it is not allowed. Your educational supervisor should not have asked you to undertake this activity and you have been placed in a difficult position. As you are unable to discuss the problem with your educational supervisor the person who is best placed to take the matter forward would be the clinical director. It is likely that matter would be referred on to Human Resources.

7) Correct answer: d) Contact your educational supervisor.

You should already have discussed this application with your educational supervisor who has overall responsibility for your training along with your Training Programme Director. When you take time out of training (which is the case for this training fellowship) it needs to be agreed prospectively by the Deanery (HEE) and the RCPath as an out of programme (OOP). The person who instigates this process is your educational supervisor.

8) Correct answer: e) Discuss your problems with your training programme director.

You find yourself in a very difficult position through no fault of your own. The person who is best placed to provide guidance and advice is you Training Programme Director as he/she has specialist knowledge of support that is available from the Deanery/HEE. It is likely that the professional support unit would be able to provide objective coaching/mentoring from a generic perspective. Deferring taking the exam and arranging a career break (OOPC) may be a helpful suggestion that can be arranged by the Training Programme Director.

9) Correct answer: e) Invite the gynaecology clinical director to the audit meeting.

Clinical audit is a powerful tool in validation of good practice and identifying areas where there may be patient safety issues. In this case you have found that there appears to be difference in outcome from one of gynecologists who has a higher rate of positive margins compared with his colleagues. The finding needs to investigated in an objective fashion and the best starting point is at the audit meeting. It would be helpful to have a senior clinician present too (Clinical Director or a representative) who can interpret the results with a clinical perspective: it is possible that the clinician concerned has been referred particularly challenging cases and the raised margin rate may not reflect poor practice. If the meeting however cannot exclude patient safety issues the clinical director would need to escalate the findings to the Medical Director and institute an investigation through the Clinical Governance reporting route.

10) Correct answer: c) Meet with the trainee to discuss their progress and agree an action plan with occupational health input.

Management of a trainee with Health Issues can be complex. It is important to follow the Trust Health and Safety policy and this will call upon the services of Occupational Health (OH). OH will have to assess fitness to practice and decide what are reasonable expectations of the Trainee from an objective point of view, but taking account of the needs of the service and employment legislation. From the Department perspective it is appropriate to agree targets to oversee the return to work and these should be reviewed at a follow up meeting.

11) Correct answer: c) Raise your concerns with her line manager/head of department.

You should not ignore your suspicions as this puts patients at risk and it is your responsibility to something to ensure that your suspicions are investigated. However, your suspicions are just suspicions.

The appropriate person to discuss this with would be the head of department in the first instance. There may be another reason for the colleague's behaviour. Searching someone's room on the basis of casual suspicion is untenable. There are no grounds to go to the GMC.

12) Correct answer: c) Discuss with the reporting pathologist and issue a joint report.

You should not under-value your own opinion but ask the reporting consultant to look at the case with you to assure you what the diagnosis is. You should not simply issue a separate report changing the diagnosis which will create confusion in the clinical team. Telling the MDM you disagree with the original report is acceptable but it would be better to express uncertainty and a solution in the form a review, rather than baldly stating your disagreement. Although histopathology is held up by some as the 'gold standard', pathologists know that pathology is another opinion-based part of medicine. Mis-diagnoses should be dealt with by openness admitting the mistake and correcting in a timely fashion. See 'Duty of Candour' guidelines Care Quality Commission in England.

13) Correct answer: b) Using appropriate personal protection equipment, slice the node and select a representative part for culture, place into a labeled empty pot. Place the rest in formalin. Ring the surgeon to inform them that there will be no frozen section.

You are being asked to put laboratory staff at risk of infection with tuberculosis by performing a frozen section on potentially infected tissue. The largest risk is the aerosol of frozen fragments during sectioning being inhaled by the person cutting the sections. Cutting slices in a laminar flow hood in a category 3 room, wearing gloves, protective spectacles, plastic apron and laboratory coat is sufficient PPE for slicing a fresh node. Whilst culture of fresh tissue remains standard microbiology practice (rather than PCR) providing fresh tissue to microbiology improves diagnostic yield on which patient treatment is reliant.

Flow cytometry relies on analysis of cell surface markers on fresh cells. Sending potentially infected tissue for flow cytometry could expose haematology laboratory staff to risk of infection.

You could also ring the surgeon to ask for more clinical information – the most junior member of the team often fills out the request form whilst the more senior may have a better understanding of what the excision has been requested for. Not undertaking flow cytometry on tissue where lymphoma is expected will potentially impair the diagnosis and certainly provide less information than routine immunohistochemistry on fixed tissue.

14) Correct answer: d) After checking the labeling and blocks, contact the lead radiologist and arrange to go through all the cases prior to the MDM.

When results that usually correlate well with clinical data or are explicable by the nature of the lesional tissue suddenly do not match up, you should suspect a problem. It may be that a new clinician is taking biopsies and is not accurately interpreting imaging or clinical appearances but it is also possible that there has been a systematic labeling issue perhaps someone has pre-labeled the pots and the biopsies have errantly been placed in the wrong specimen pot for example.

The MDM forum in breast screening is really a final check that pathology correlates (or doesn't correlate) so that the next step in the patient pathway can be taken (return to normal recall, further investigation or refer for treatment). These decisions are explained within a day or two of the MDM to the patient who will already have an arranged follow-up appointment. If 10 results are held back this would impact on patients and clinic appointment systems. So, you would want to take cases and I would take all cases not just the obvious problems to a separate early meeting with the lead radiologist to clarify. I would not authorize any of the cases without the discussion.

15) Correct answer: c) Wash the cut hand with antiseptic and put a plaster on it. Report the incident in the accident book. Report the incident to the local first aider.

Any incident such as this should be recorded in the accident book or equivalent held within the laboratory. Laboratories in the UK are required to have identified members of staff who have instruction in first aid and they will be able to provide first aid assistance and equipment.

Occupational physicians usually have roles in health promotion, financial/business management, and in improving access to work for individuals with health problems. They are able to advise in relation to toxicology, health and safety legislation, and employment law/ethics. Whilst the overall incidence of needle-stick and other injuries is part of their role, reporting is usually via systematic collection of incidents rather than individuals notifying the consultant themselves.

16) Correct answer: d) Absorb with inert, absorbent material. Sweep up. Transfer to suitable, labelled containers for disposal. Clean spillage area thoroughly with plenty of water.

Although opening windows may help ventilate the area, the large surface area presented by the puddle will lead to rapid evaporation and increase the formalin vapour in the room potentially causing respiratory and eye problems. Mopping is not effective in removing formalin: various inert absorbent materials are available and there should be a formalin spillage kit in each accredited laboratory area where formalin is used. Containers should be clearly labeled as containing absorbed formalin (not clinical waste).

In this scenario, the drenched consultant should remove trousers into a plastic bag for disposal (or into the fume hood until dry then wash) and have a shower to remove irritant formalin from the skin. 'Control of Substances Harmful to Health' (CoSHH) protocols are available in all accredited laboratories for safe handling of formalin and other harmful substances with instruction on dealing with spillages etc.

17) Correct answer: e) Leave the department immediately. Ring theatres to tell the surgeon that a frozen section is not possible. Go to the designated fire assembly area.

By staying in the department carrying on working, you put not only yourself but any laboratory staff assisting in danger. You also potentially put both local laboratory nominated fire officers and the fire brigade at risk. Once you have left the area with continuous alarm, you should phone theatres and speak to the surgeon. Other options are ill advised.

18) Correct answer: c) Refer the case to the Coroner.

If a cause of death is not identifiable, the case should be referred to the coroner.

If the Coroner requests a Coronial PM then the Coroner's pathologist will carry out a post-mortem, in a high risk PM suite with full personal protection equipment. Any blocks taken will target areas of gross abnormality potentially supplemented by additional samples from lung, brain and other organs where HIV associated abnormalities and infections are common.
No consent is required for a coronial post-mortem in the UK.

19) Correct answer: e) Having established the identity of the deceased, perform an external examination and sign the cremation form 5 having read the Medical Certificate of Cause of Death which indicates that the case has been discussed with the Coroner.

This patient has died of complications during a diagnostic procedure and therefore the case should be discussed with the coroner. There may be other evidence of disease e.g. a tumour mass on imaging or clinical evidence of metastatic disease which informs the Coroner of a natural cause of death and a post-mortem examination may not be necessary. The patient should have been informed of possible complications during the consent process.

In the UK, there are two cremation forms (form 4 and form 5) each of which are required to be completed by a different medical practitioner. This is a legal requirement without which cremation cannot proceed.

Form 4 can be completed by a 'registered medical practitioner with a licence to practise with the General Medical Council. This includes those who hold a provisional or temporary registration'... In order to complete the form 4, , you should have attended the deceased during their last illness.

Form 5 is the 'confirmatory medical certificate' which must be completed by 'a
fully registered medical practitioner of at least 5 years' standing'. It is also required that
'the registered medical practitioner who completes the confirmatory medical certificate should not be a relative of the deceased, or a relative or partner or colleague in the same practice or clinical team of the medical practitioner who has given the medical certificate, form Cremation 4'.

'Where a junior hospital medical practitioner has completed the certificate in form Cremation 4, the medical practitioner who signs the form Cremation 5 should not have been in charge of the case, nor directly concerned in the patient's treatment'.

Reference: The Cremation (England and Wales) Regulations 2008. Guidance to medical practitioners completing forms Cremation 4 and 5.
This guidance was published by the Ministry of Justice in 2008 and reviewed in 2012.

20) Correct answer: c) Immediately arrange to meet with the clinical director responsible for training to discuss the best course of action.

You should not go straight to the trainee or arrange a meeting with all trainees both of which could be considered further examples of harassment. Giving up being an educational supervisor immediately is a knee jerk reaction which may not be necessary and could damage the educational progress of at least the other two trainees under your care. It may be advisable to contact your medical protection organization at some point. However, the starting point would be to meet with the Clinical Director responsible for Training to discuss the best course of action as this will likely have previously occurred and sound advice will be available from the Clinical Director and Human Resources department.